Various Adsorbents for Water Purification Processes

Various Adsorbents for Water Purification Processes

Editors

Monika Wawrzkiewicz
Anna Wołowicz

MDPI • Basel • Beijing • Wuhan • Barcelona • Belgrade • Manchester • Tokyo • Cluj • Tianjin

Editors

Monika Wawrzkiewicz	Anna Wołowicz
Department of Inorganic Chemistry	Department of Inorganic Chemistry
Institute of Chemical Sciences	Institute of Chemical Sciences
Maria Curie-Sklodowska University in Lublin	Maria Curie-Sklodowska University in Lublin
Lublin	Lublin
Poland	Poland

Editorial Office
MDPI
St. Alban-Anlage 66
4052 Basel, Switzerland

This is a reprint of articles from the Special Issue published online in the open access journal *Processes* (ISSN 2227-9717) (available at: www.mdpi.com/journal/processes/special_issues/adsorbents_water).

For citation purposes, cite each article independently as indicated on the article page online and as indicated below:

LastName, A.A.; LastName, B.B.; LastName, C.C. Article Title. *Journal Name* **Year**, *Volume Number*, Page Range.

ISBN 978-3-0365-5628-4 (Hbk)
ISBN 978-3-0365-5627-7 (PDF)

Contents

About the Editors

Monika Wawrzkiewicz

Monika Wawrzkiewicz has been working as a professor in the Department of Chemistry, Faculty of Chemistry, Maria Curie-Sklodowska University in Lublin (MCSU) since 2016. She obtained her Ph.D. in chemistry (2005) and then habilitation (2015) at MCSU. She is a lecturer in inorganic chemistry for students of the Faculty of Chemistry at MCSU. Her research interests are sorption and separation of inorganic and organic compounds with particular emphasis on dyes and heavy metal ions on various sorbents, synthesis, and characterization of hybrid adsorbents and composites, as well as environmental protection. She is the author of more than 64 international peer-reviewed papers and book chapters. She is a member of the Polish Chemical Society and the European Chemical Society.

Anna Wołowicz

Anna Wolowicz is an assistant professor at the Department of Chemistry, Faculty of Chemistry, Maria Curie-Sklodowska University in Lublin (MCSU). She obtained her M.Sc. in Chemistry (2006) and Ph.D. in Chemistry (2010) from MCSU. Since then, she has continued her postdoctoral research on the evaluation of the sorption properties of various materials used to remove inorganic and organic pollutants from water and wastewater. Her research interests are in the sorption and separation of noble metals and heavy metals with a focus on ion exchange chromatography and environmental protection. She is the author of more than 38 international peer-reviewed papers and chapters in books. She is a member of the Polish Chemical Society and the European Chemical Society.

Preface to "Various Adsorbents for Water Purification Processes"

Water resources in the world are becoming increasingly scarce, and the demand for water is constantly increasing due to the growing population, rapid industrial development, or the resulting climate changes. On the other hand, the quality of available water is threatened not only by natural processes taking place, e.g., weathering of rocks, ocean evaporation, volcanic eruptions, etc., but primarily by anthropogenic human activity, which contributes significantly to the deterioration of water quality. Direct discharge of wastewater from municipal and industrial treatment plants without going through any treatment processes, or only after preliminary treatment, leads to many environmental problems, including deterioration of aerobic conditions for aquatic organisms, eutrophication of water bodies, and increase in pathogens and toxic compounds causing serious diseases [1]. Soil and water are most often contaminated with heavy metals, polycyclic aromatic hydrocarbons, chlorophenols, petroleum compounds, organic compounds, pesticides, pharmaceuticals, hormones, etc. Pb(II), Cd(II), Cr(III), Co(II), As(V), Zn(II), Hg(II), and Ni(II) are the most common heavy metals in wastewater. Like organic compounds, they can penetrate the human food chain. The non-biodegradability and bioaccumulative properties of these pollutants cause serious health problems in the human body, such as bone defects, elevated blood pressure, chronic asthma, and coughing [1–3]. Therefore, the treatment of wastewater before it is discharged into water bodies has become a necessity, and the development of effective treatment methods or the synthesis of new effective adsorbents capable of selective sorption of toxic substances is now widely and intensively studied [1]. Over the past few decades, physical, chemical, biological, as well as electrochemical and mixed methods have been used to remove organic and inorganic contaminants. Each of these has its advantages as well as disadvantages, so further research is needed to overcome them, and the topic of wastewater treatment is still very much in the news.

This Special Issue contains 12 articles [1–12] on wastewater treatment topics and is available online at: *Processes* | Special Issue: Various Adsorbents for Water Purification Processes (mdpi.com) (accessed on 26 May 2022). Among the papers, the vast majority are devoted to the removal of heavy metal ions, i.e., Ni(II) [2, 6], Pb(II) [3–5], Cu(II) [3, 8], Co(II) [7], As(V) [8], and Cd(II) [9] from synthetic and real solutions, and two papers on the removal of anabolic hormones, i.e., trenbolone [11] and β-estradiol [12] from water and wastewaters. Here, a comprehensive review concerning heavy metal ions removal by nano-adsorbents [1], as well as cost estimation of activated carbon production from waste nutshells by physical activation [10] could be found.

It was shown that the efficiency of the removal process of the above mentioned contaminants on synthetic ion exchangers [2], natural and synthetic aluminosilicates [3], zeolites [3], magnetic multiwall carbon nanotubes [4], leaves [5], date seeds powder [6], K_2HPO_4-pretreated Duckweed Lemna gibba [7], chitosan [8], fungi [9], imprinted polymer [10], and magnetic nano-Akaganeite [12] depends on many factors, including the chosen sorbent and its properties, as well as experimental conditions and the type and concentration of pollutants, but the proposed solutions are effective and interesting from a cognitive point of view. We invite you to take a detailed look at this work.

We are hoped that the papers collected in this Special Issue will inspire you for further studies on development of wastewater treatment techniques and may provide a basis for developing other creative approaches.

We congratulate and thank all the authors who made valuable contributions to this Special Issue. We also sincerely thank all the expert members and reviewers who spent their valuable time and make an effort in reviewed all the papers. The Guest Editors thank Amelia Qie for her dedicated and assistance in the development of this Special Issue. Thank you and best wishes for continued success.

References:

1. Kumar, R.; Protima, R.; Erwan, R. Nanoadsorbants for the Removal of Heavy Metals from Contaminated Water: Current Scenario and Future Directions, Processes, 2021, 9(8), 1379. https://doi.org/10.3390/pr9081379.

2. Wołowicz, A.; Wawrzkiewicz, M. Screening of Ion Exchange Resins for Hazardous Ni(II) Removal from Aqueous Solutions: Kinetic and Equilibrium Batch Adsorption Method. Processes, 2021, 9(2), 285. https://doi.org/10.3390/ pr9020285.

3. Kozera-Sucharda B.; Gworek, B.; Kondzielski, I.; Chojnicki, J. The Comparison of the Efficacy of Natural and Synthetic Aluminosilicates, Including Zeolites, in Concurrent Elimination of Lead and Copper from Multi-Component Aqueous Solutions. Processes, 2021, 9(5), 812. https://doi.org/10.3390/ pr9050812.

4. Hanbali, G.; Jodeh, S.; Hamed, O.; Bol, R.; Khalaf, B.; Qdemat, A.; Samhan, S. Dagdag, O., Magnetic Multiwall Carbon Nanotube Decorated with Novel Functionalities: Synthesis and Application as Adsorbents for Lead Removal from Aqueous Medium, Processes, 2020, 8(8), 986; https://doi.org/10.3390/pr8080986.

5. Elkhaleefa, A.; Ali, I.H.; Brima, E.I.; Shigidi, I.; Elhag, A..B.; Karama, B. Evaluation of the Adsorption Efficiency on the Removal of Lead(II) Ions from Aqueous Solutions Using Azadirachta indica Leaves as an Adsorbent. Processes, 2021, 9(3), 559. https://doi.org/10.3390/pr9030559.

6. Elkhaleefa, A.; Ali, I.H.; Brima, E.I.; Elhag, A.B.; Karama, B., Efficient Removal of Ni(II) from Aqueous Solution by Date Seeds Powder Biosorbent: Adsorption Kinetics, Isotherm and Thermodynamics, Processes, 2020, 8(8), 1001; https://doi.org/10.3390/pr8081001.

7. Reyes-Ledezma,J.L.; Cristiani-Urbina, E.; Morales-Barrera, L. Biosorption of Co^{2+} Ions from Aqueous Solution by K_2HPO_4-Pretreated Duckweed Lemna gibba, Processes, 2020, 8(12), 1532. https://doi.org/10.3390/pr8121532.

8. An, B. Cu(II) and As(V) Adsorption Kinetic Characteristic of the Multifunctional Amino Groups in Chitosan, Processes, 2020, 8(9), 1194. https://doi.org/10.3390/pr8091194.

9. Kumar, R.; Sharma, P.; Umar, A.; Kumar, R.; Singh, N.; Joshi, P. K.; Alharthi, F.A.; Alghamdi, A.A.; Al-Zaqri, N. In Vitro Bioadsorption of Cd^{2+} Ions: Adsorption Isotherms, Mechanism, and an Insight to Mycoremediation, Processes, 2020, 8(9), 1085; https://doi.org/10.3390/pr8091085.

10. León, M.; Silva, J.; Carrasco, S.; Barrientos, N., Design, Cost Estimation and Sensitivity Analysis for a Production Process of Activated Carbon from Waste Nutshells by Physical Activation, Processes, 2020, 8(8), 945; https://doi.org/10.3390/pr8080945.

11. Mpupa, A.; Dinc, M.; Mizaikoff, B.; Nomngongo, P.N. Exploration of a Molecularly Imprinted Polymer (MIPs) as an Adsorbent for the Enrichment of Trenbolone in Water. Processes, 2021, 9(2), 186. https://doi.org/10.3390/ pr9020186.

12. Mpupa, A.; Mizaikoff, B.; Nomngongo, P.N., Enhanced Adsorptive Removal of β-Estradiol from Aqueous and Wastewater Samples by Magnetic Nano-Akaganeite: Adsorption Isotherms, Kinetics, and Mechanism, Processes, 2020, 8(9), 1197. https://doi.org/10.3390/pr8091197.

Author Contributions: Writing—original draft preparation, A.W., M.W; writing—review and editing, A.W., M.W. All authors have read and agreed to the published version of the manuscript.

Monika Wawrzkiewicz and Anna Wołowicz
Editors

processes

Review

Nanoadsorbants for the Removal of Heavy Metals from Contaminated Water: Current Scenario and Future Directions

Rohit Kumar, Protima Rauwel and Erwan Rauwel *

Institute of Technology, Estonian University of Life Sciences, Kreutzwaldi 56/1, 51014 Tartu, Estonia; rohit.kumar@student.emu.ee (R.K.); protima.rauwel@emu.ee (P.R.)
* Correspondence: erwan.rauwel@emu.ee

Abstract: Heavy metal pollution of aquatic media has grown significantly over the past few decades. Therefore, a number of physical, chemical, biological, and electrochemical technologies are being employed to tackle this problem. However, they possess various inescapable shortcomings curbing their utilization at a commercial scale. In this regard, nanotechnology has provided efficient and cost-effective solutions for the extraction of heavy metals from water. This review will provide a detailed overview on the efficiency and applicability of various adsorbents, i.e., carbon nanotubes, graphene, silica, zero-valent iron, and magnetic nanoparticles for scavenging metallic ions. These nanoparticles exhibit potential to be used in extracting a variety of toxic metals. Recently, nanomaterial-assisted bio-electrochemical removal of heavy metals has also emerged. To that end, various nanoparticle-based electrodes are being developed, offering more efficient, cost-effective, ecofriendly, and sustainable options. In addition, the promising perspectives of nanomaterials in environmental applications are also discussed in this paper and potential directions for future works are suggested.

Keywords: nanomaterials; heavy metals; remediation; bioelectrochemical systems; wastewater; adsorption; nanocomposites

Citation: Kumar, R.; Rauwel, P.; Rauwel, E. Nanoadsorbants for the Removal of Heavy Metals from Contaminated Water: Current Scenario and Future Directions. *Processes* **2021**, *9*, 1379. https://doi.org/10.3390/pr9081379

Academic Editors: Monika Wawrzkiewicz and Anna Wołowicz

Received: 18 June 2021
Accepted: 30 July 2021
Published: 6 August 2021

1. Introduction

Even though 70% of the Earth's surface is composed of water, freshwater resources are rapidly dwindling. The latter accounts for approximately 1% of total water bodies. In addition, the contamination of these aqueous resources with a broad range of pollutants due to rapid industrialization and lifestyle changes has further reduced the availability of clean fresh water sources. In particular, mining, volcanic eruptions, farming, and excessive dumping of hazardous chemicals has resulted in a significant invasion of organic compounds, pharmaceuticals, customer care products, pathogens, and heavy metals into water reservoirs [1]. Among the persistent pollutants, the existence of heavy metals in water is considered as a prime global concern due to their considerable role in environmental degradation [2,3].

In general, heavy metals (HM) can be defined as a group of transition metals, metalloids, actinides, and lanthanides with a density greater than 4000 kg/m^3, and can be further categorized into the groups of essential and non-essential metals [4]. Under both categories, they are heavily involved in a vast number of industries and crafts and are an essential part of various biological processes/reactions [5]. However, long-term, and sometimes short-term, exposure to them, even in trace amounts, can lead to serious health implications [6]. The most commonly-encountered heavy metals in water include Pb, Cd, Cr, Co, As, Zn, Hg, and Ni. They are considered as particularly problematic due to their non-biodegradability and bioaccumulation behavior when ingested; thus, they are listed in the Environmental Protection Agency's list of priority pollutants [7,8]. More specifically, the intake of Cd, Pb, Hg, Cr, and As beyond the standard limits can give rise to serious health implications such as bone defects, increased blood pressure, lung cancer, nervous system damage, neurological depositions, gastrointestinal disorders, and many more significant diseases [9–11]. Therefore, considering the complex chemistry and carcinogenicity of HM, there is an urgent need to develop appropriate methods for their extraction from water sources. Several physical, chemical, biological, and electrochemical methods, along with their combinations, which have been studied and researched for HM removal over the past few decades are briefly summarized in Table 1.

Table 1. Conventional approaches for the remediation of heavy metals and their associated advantages and disadvantages.

Technology	Advantages	Disadvantages	References
Physical 1. Membrane separation 2. Adsorption/Physiosorption 3. Filtration 4. Sedimentation	1. Very effective in treatment of a variety of metals. 2. Easily applicable 3. Economically acceptable	1. Production of hazardous by-products. 2. Energy intensive	[12,13]
Chemical 1. Adsorption/Chemisorption 2. Ion-exchange 3. Chemical precipitation 4. Flocculation/Coagulation	1. Easy to apply and very effective. 2. Applicable to a broad range of inorganic/organic pollutants.	1. Applicable at small scale. 2. Formation of more toxic chemical by-products.	[14–16]
Biological (Microbes assisted remediation)	1. Environmental-friendly and cost effective. 2. Applicable at large scale. 3. Less disruptive.	1. Well-defined growth conditions required for microbes. 2. Slow process. 3. Continuous monitoring required.	[17–19]
Electrochemical	1. Very effective and efficient in treating a vast variety of pollutants including heavy metals. 2. Production of energy.	1. Energy and cost intensive process. 2. Applicable at small scale. 3. Chemically intensive process.	[20–22]

The conventional technologies summarized in Table 1 are effective but present some unavoidable drawbacks, such as high costs, energy intensiveness, tediousness, low efficiency, metal specificity, and unsustainability, thereby rendering them ineffective in meeting environmental standards and, in turn, too difficult to implement at the industrial scale [23]. Therefore, considering the negative impacts of HM on human health and the environment, one deems it necessary to introduce a cost-effective, environmental-friendly and efficient processes for the removal of HM from the contaminated water. Adsorption is a mass transfer process, where the adsorbate molecules are attracted to the surface of an adsorbent, resulting in either a physical or chemical interaction. It is one of the most favored processes in the water treatment industry, especially due to the regenerative capacity of the adsorbents [24]. Therefore, among the technologies reviewed, adsorption is considered the most efficient, safe, and technically feasible process due to its facile operation and higher efficiency [25]. In fact, the adsorption capacity varies with the type of adsorbent. Generally, activated carbon based adsorbents are widely used for the removal of heavy metals but due to their clogging, inability to recover them from the treated water, waste generation, and biofouling, they are ineffective for large scale applications [26,27]. Therefore, the search for new and effective adsorbent materials has always been an active field of research [28]. Nanomaterials, due to their nanoscale dimensions (ranging from 1–100 nm), show some unique physical, chemical, and biological properties. These properties result in the modification of their structure and specific surfaces [29]. Different kinds of nano-adsorbents and nanocomposites have been extensively researched and are used for the treatment of organic dyes, inorganic compounds, heavy metals, and other micropollutants (customer-care products, biocides and hormone active substances) from water/wastewater [30]. This manuscript will firstly provide a review of the different types of nanomaterial-based adsorbents used for heavy metal extraction. The main focus of this review is to provide a basic insight of the conventional nanomaterials used (CNT-GO, silica-based, ZVI) and show how magnetic NP exhibit a significant advantage over these nanomaterials. However, the main disadvantages of magnetic nanomaterials are also considered and discussed. The possible combination of these nanomaterials with existing technology is then discussed and the last part of the review particularly discusses the industrially active adsorbents and how the electro-chemical technologies combined with nanomaterials can provide a sustainable, cost effective, and ecofriendly approach in the future. The second part of the review will the concentrate on the nanomaterial assisted bioelectrochemical remediation of metallic ions with an emphasis on the current developments and next generation nanoadsorbants. Several studies suggest that there are improvements in the metal reduction efficiency when nanomaterials are introduced in bioelectrochemical systems (BES).

2. Nanomaterials Applied for the Removal of Metallic Ions from Water

Nanomaterials (NM), due to properties such as high specific surface, porosity, surface functionalities, and ion binding capabilities, have been widely researched over the past two decades in water and wastewater treatment applications. In fact, they also show a high potential in the removal of metallic ions even in trace amounts [31,32]. Nanomaterials are classified into different categories, i.e., carbon based, silica based, metal and metal oxide nanoparticles, including zero-valent iron (ZVI), iron-oxide based magnetic nanomaterials, and nanocomposites, as shown in Figure 1. There are several factors, such as concentration of adsorbent, contact time, and flow rate, that control the adsorption process of metallic ions. However, the concentration of nanoparticles (adsorbent) plays a crucial role in the removal process. Some studies suggest an increase in the removal efficiency when the adsorbent dosage increases, whereas some studies report a decrease in the removal when the adsorbent concentration increases due to a possible agglomeration process. For example, Lei et al. reported a decrease in the ad-sorption capacity of Cd^{2+} ions when the amount of Dopamine Modified Magnetic Nano-Adsorbent was increased from 10–50 mg [33]. The probable reason mentioned was the agglomeration of the adsorbent at higher concentrations which inhibits the adsorption process due to surface adsorption decrease. In fact, in another study an increase in the removal efficiency of Cd^{2+} ions (83% to 89%) was observed when the concentration of gas industry-based adsorbent was increased from 0.25–1.25 g/100 ML [34]. In gas phase, it was found that no agglomeration takes place, so the adsorption only depends on the amount of adsorbent added. Based on these studies, it can be concluded that the concentration of adsorbent affects the removal process, but it depends on the nature and chemical properties of the ad-sorbent itself, suggesting that either an increase or decrease in the removal will occur. However, higher and lower volumes of NP with equal concentration will lead to the same results.

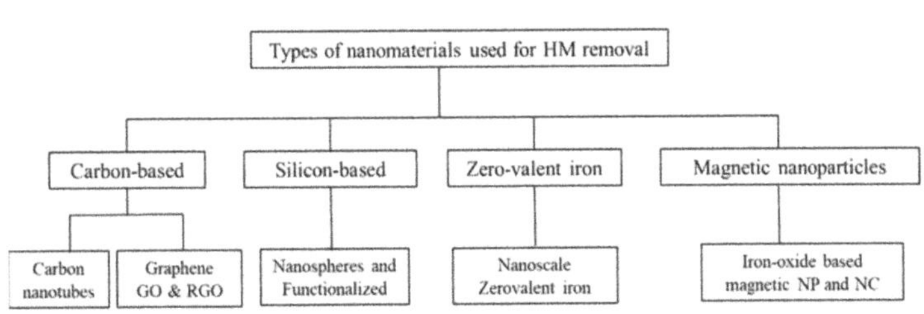

Figure 1. Nanomaterials for heavy metal remediation in aqueous media.

2.1. Carbon Based Nanomaterials

Carbon-based NM, such as carbon nanotubes, Fullerenes, activated carbon, graphene, and graphene oxide, have been widely used in energy storage, sensors, electronics, water purification, drug delivery, and disease diagnosis [35]. In addition, their unique properties also allow the removal of both organic and inorganic pollutants, making them a promising alternative for treating wastewater. They are therefore considered as one of the most promising adsorbents for metallic pollutants [36,37].

2.1.1. Carbon Nanotubes

Carbon nanotube (CNT) adsorbents are widely employed for metallic pollutant extraction [38]. They are broadly classified into single walled carbon nanotubes (SWCNT) and multi-walled carbon nanotubes (MWCNT), and both are extensively tested for the removal of heavy metals [39–41]. In the case of CNT, the adsorption process is usually controlled by four possible active sites: (i) the hollow interior of individual CNT designated as internal sites; (ii) the interstitial channels between CNT in the stacks; (iii) the grooves between adjacent CNT; and (iv) the external surface of individual CNT [42,43] (Figure 2). The sites of interstitial channels and grooves are responsible for initializing the process of adsorption, which is then followed by the adsorption of the contaminants on the external walls and the accumulation of molecules on internal axial sites. From a kinetic point of view, internal sites are more inclined to acquire the equilibrium state than the external sites under the same conditions. Therefore, increasing the binding sites on the CNT surface can undoubtedly enhance the saturation capability and kinetic rate. The applicability of CNT in wastewater treatment is dependent on several factors including the cost to complexity of CNT functionalization, the necessity for solid and liquid segregation, the type of wastewater, and the recycling cost and efficiency.

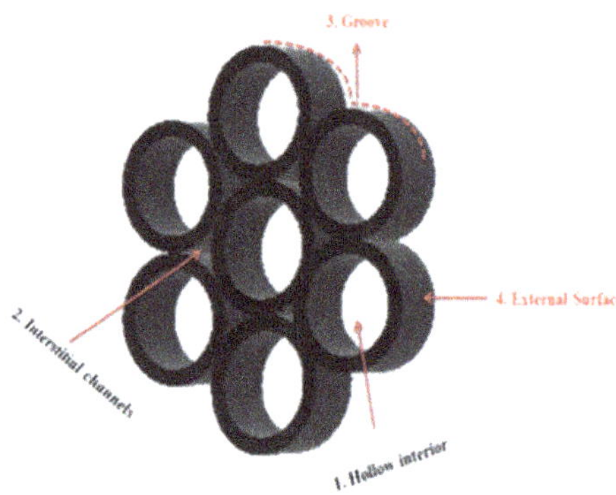

Figure 2. Possible active adsorption sites on the carbon nanotubes (CNT) surface.

Many studies have been performed on CNT-based composites for the remediation of trace elements in water such as heavy metal ions (See Table 2).

Table 2. Summary of the types of CNT used for the removal of different metallic ions.

Type of CNT	Target Metal/s	Initial Concentration (mg/L)	%Removal Efficiency	References
SWCNT	Hg^{2+}	(1–2000) mg/L	4.16%	[44]
MSWCNT-CoS	Hg^{2+}	(1–2000) mg/L	166.6%	[39]
MWCNT-SH	Hg^{2+}	10 mg/L	15.15%	[45]
SWCNTs-polysulfone nanocomposite-based membrane	Pb^{2+}, As^{3+}	1 mg/L	94.2%, 87.6%	[46]
MWCNT-COOH functionalized nanotube	Pb^{2+}	(10–100) mg/L	99.1%	[47]
Acidified MWCNT	Pb^{2+}, Cu^{2+}, Ni^{2+}	100 mg/L	93%, 78%, 83%	[48]
MWCNT	Mn^{7+}	(50–800) mg/L	71.5%	[49]
MWCNTs	Cd^{2+}	100 mg/L	10.7% (pH = 2) 94.2% (pH = 7) 100% (pH = 10)	[48]
Oxidized MWCNT	Cu^{2+}	100 mg/L	78%	[48]
MWCNT	Fe^{2+}	200 mg/L	52%	[49]
Oxidized MWCNT	Ni^{2+}	100 mg/L	83%	[48]

The adsorption affinity of CNT towards HM ions varies as a function of the ionic radius of metallic ions, electronegativity of metals, and solubility of metal hydroxides. These factors play an important role in understanding the interaction between metal ions and adsorbents. The ionic radii, in increasing order, are Pb^{2+} (120 pm) > Sr^{2+} >Ca^{2+} (100 pm) > Cd^{2+} (99 pm) > Mn^{2+} (80 pm) > Cu^{2+} (77 pm) > Zn^{2+} (74 pm) and Co^{2+} (72 pm) > Ni^{2+} (69 pm) [50]. This suggests that higher ionic radii generate higher steric overcrowding, which in turn, lowers the maximum adsorption capacity. For example, Brasquet et al. investigated the adsorption affinities of Pb^{2+} and Cu^{2+} towards activated carbon cloth and established a higher adsorption capacity for Cu^{2+}. The larger ionic radius of Pb^{2+} induces a quicker saturation of the adsorption sites, probably due to overcrowding of Pb^{2+} ions on the adsorbent surface [51]. On the other hand, the surface adsorption available for the Cu^{2+} ions are larger, due to its smaller ionic radius, resulting in its higher maximum adsorption capacity. The metal ions that exhibit higher electronegativity will have a stronger adsorption capacity towards the negatively charged sites on the CNT surface. The adsorption capacities of Pb^{2+} and Cu^{2+} were investigated in regard to MWCNT and since Cu^{2+} has a lower electronegativity (1.99) than Pb^{2+} (2.33), it therefore leads to weaker interactions towards the CNT surface and maximum adsorption levels of Cu^{2+} decreased by approximately 12% in comparison with Pb^{2+} [52]. Various functionalized and pure CNT have been extensively investigated, and authors report an efficient extraction of heavy metals from contaminated water (discussed in Table 2). They present high adsorption efficiencies in comparison with activated carbon, which is widely used in various water treatment industries. Moreover, their regeneration and reuse provide a special benefit over other conventional adsorbents such as activated carbon, clays, and biosorbents [53]. Besides their notable potential for HM extraction, there are some unavoidable shortcomings associated with them. One of the main problems of using CNT is associated with their agglomeration or bundling resulting in a lower specific surface for adsorption. Moreover, when it comes to their applicability at an industrial scale, their attachment to filters should be strong enough to prevent their release into treated waters [54].

2.1.2. Graphene Based Adsorbents

Graphene is another carbon-based material with high adsorption capabilities in removing HM from wastewater [55–57]. It is a 2D material in which carbon atoms are arranged in a hexagonal, honeycomb lattice. Graphene can be classified into two major forms, graphene oxide (GO), the oxidized form, and reduced graphene oxide (RGO); the latter shows potential towards remediation of various environmental pollutants [58]. In fact, the availability of several functional groups, i.e., hydroxyl, carboxyl, and epoxide on the surfaces of GO and RGO serve as the active sites for the removal of metallic ions from water [59,60]. The functional groups, e.g., –CH(O)CH–, –OH, and –COOH have high negative charge densities along with hydrophilic characteristics, which promote their interactions with positively charged metal ions in order to facilitate the extraction mechanisms [61]. Graphene and GO/RGO-based materials have been extensively studied and are reported to have excellent properties for heavy metal ion extraction from water/wastewater (Table 3) [61]. For example, few-layered graphene oxide nanosheets were investigated for the adsorption of Cd^{2+} and Co^{2+} from the contaminated water in batch mode [62]. This study highlighted a strong influence of pH and the presence of humic acid in adsorption of Cd^{2+} and Co^{2+}. However, the maximum adsorption capacities of Cd^{2+} and Co^{2+} were found to be 106.3 mg/g and 68.2 mg/g respectively. Wang et al. studied different factors including pH, the dosage of the adsorbent, contact time, temperature, and competing ions on the adsorption performance of GO for the removal of zinc ions and observed an adsorption capacity of 246 mg/g of Zn^{2+} [63]. The removal of Pb^{2+} ions from water was investigated using a nanocomposite of Fe_3O_4 and reduced graphene oxide (Fe_3O_4@RGO). This study showed an adsorption capacity of 373.14 mg/g with an initial Pb^{2+} concentration of 97.68 mg/L [64]. Figure 3 shows different thermodynamic parameter values represented by the Van't Hoff curve at room temperature, confirming that the adsorption is a spontaneous exothermic process with a relatively low entropy. The absolute value of ΔH was found to be higher than 20 kJ.mol^{-1}, implying a strong chemical interaction between adsorbent and adsorbate. The experimental data fitted Temkin's model which assumes the adsorption heat of all the molecules in the layer decreases linearly with coverage due to chemical interactions between adsorbent and adsorbate, and that the adsorption is characterized by a uniform distribution of the binding energies, up to a maximum binding energy. Based on this assumption, they concluded that monolayer chemisorption is the main adsorption mechanism. In fact, in addition to the latter, liquid film diffusion was also found to have interplay with this mechanism. Porous graphene was designed and studied for the adsorption of As^{3+} from water [65]. This synthesized adsorbent showed an adsorption capacity >90% and retained its water treatment properties even after regeneration and recycling. However, porous graphene applied to the cleaning of real wastewater that contains various competing elements poses a notable disadvantage in the case of large-scale applications. The authors concluded that monolayer chemisorption takes place during the extraction process. Functionalized reduced graphene oxide with 4-sulfophenylazo (RGOs) was studied for the removal of heavy metal ions from aqueous solution. The nanomaterial designed showed a maximum adsorption capacity of 689 mg/g, 59 mg/g, 66 mg/g, 267 mg/g, and 191 mg/g for the Pb^{2+}, Cu^{2+}, Ni^{2+}, Cd^{2+}, and Cr^{3+} respectively [66]. The adsorption of RGOs for Pb^{2+}, Cd^{2+} and Cr^{3+} was ascribed to coordination reaction of N atoms with heavy metal ions. Awad et al. found a significant increase in the adsorption capacity of Hg^{2+} onto the modified GO surface with COOH groups (24 mg/g to 122 mg/g), which suggests stronger interactions of Hg^{2+} ions towards negatively charged COOH groups [67].

Table 3. Graphene derived nanoparticles for the removal of heavy metals.

Adsorbate	Target Metal	Initial Concentration	Removal %	References
Tea polyphenols—rGO-ZnO	Pb^{2+}	20 mg/L	98.9%	[68]
Porous Graphene	As^{3+}	130 mg/L	>90%	[65]
rGO-Fe_3O_4	Pb^{2+}	20 mg/L	37.314%	[64]
rGO-Sulfophenylazo	Cu^{2+}, Ni^{2+}	40 mg/L	5.9 %, 6.6 %	[66]
GO embedded calcium alginate (GOCA) beads	Pb^{2+}, Cd^{2+}, Hg^{2+}	50 mg/L	60.2%, 18.1%, 37.4%	[69]
rGO-Sulfophenylazo	Cd^{2+}	40 mg/L	26.7%	[66]
GO-alpha cyclodextrin-polypyrrole	Cr^{6+}	100–700 mg/L	66.67%	[70]
rGO-Sulfophenylazo	Cr^{3+}	40 mg/L	19.1%	[66]
-COOH functionalized GO	Hg^{2+}	400 mg/L	12.2%	[67]
Chitosan/GO composite nanofibrous adsorbent	Cr^{6+}	10–1000 mg/L	31.04%	[71]

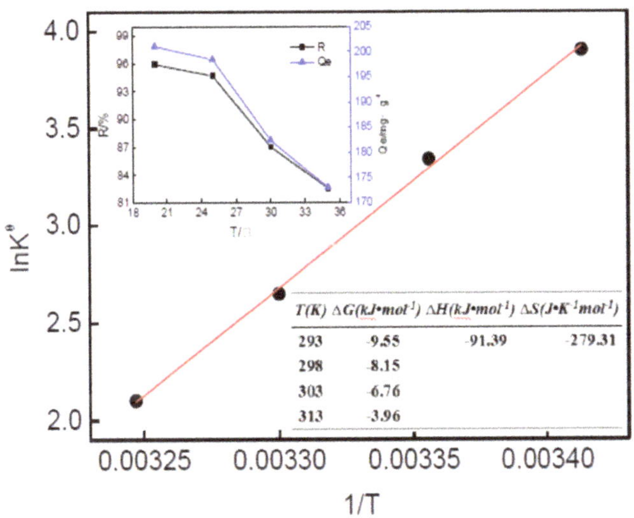

Figure 3. Van't Hoff fit curve (inset at the upper left corner: dependence of adsorption on temperature, at the lower right corner: values of adsorption thermodynamic functions). Reproduced under creative commons agreement from [64].

2.2. *Silica Based Nanomaterials*

Another important category of nanomaterials are Silica-based nanomaterials, which are widely used for removing HM ions owing to their non-toxicity and excellent surface characteristics [72,73]. The different possible interactions between mesoporous silica and metallic ions are shown in Figure 4. The surface of the mesoporous silica can also be functionalized with groups, such as amine (–NH_2) and thiol (–SH), which enhance interactions with heavy metal ions and their possible extraction from water. Table 4 summarizes previous studies reporting the functionalization of mesoporous silica applied to heavy metal ion removal.

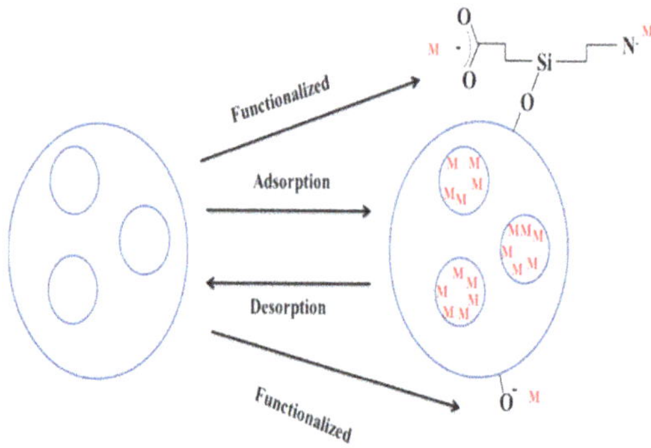

Figure 4. Adsorption sites of a mesoporous silica nanoparticles and functionalization with different groups.

The efficiency of amino-functionalized silica gel, silica hollow nanospheres, and amino-functionalized hollow silica nanospheres was investigated for the removal of Cd^{2+}, Ni^{2+}, and Pb^{2+} in batch mode. The adsorption capacity was found to decrease in the order of $Pb^{2+} > Cd^{2+} > Ni^{2+}$ for all of the adsorbents, which was attributed by the authors to the higher electronegativity of Pb^{2+} leading to stronger interactions with the negatively charged adsorbent surface. The maximum adsorption capacities (q_m) for uncoated silica hollow nanospheres as an adsorbent were found to be 8.375 mg/g (Ni^{2+}), 25.924 mg/g (Cd^{2+}), and 31.291 mg/g (Pb^{2+}). This adsorption increased to 26.858 mg/g, 54.351 mg/g, and 96.786 mg/g for amino-functionalized silica gel [74]. In another study, Nanopolyaniline and crosslinked nanopolyaniline based nanocomposites were studied for the removal of Cu^{2+}, Cd^{2+}, Hg^{2+}, and Pb^{2+} using a batch technique [72]. The adsorption capacity values for Pb^{2+}, Cu^{2+}, Hg^{2+}, and Cd^{2+} were found to be 341.4 mg/g, 289.8 mg/g, 162.9 mg/g, and 146.7 mg/g respectively. Selective Hg^{2+} removal by thiol-functionalized, porous, organic polymer-based nano-trap has an adsorption capacity of >1000 mg/g. Particularly, the material showed high stability in water under a wide pH range, which was attributed to its stable C–C bond; it also remained stable at high temperatures of up to 270 °C. The removal of 10 ppm of Cu^{2+} and cationic thiazine dyes using 3-aminopropyl and phenyl groups-based silica nanospheres was investigated in static mode. The absorption of Cu^{2+} ions reached 70% in 1 h, and thereafter increased to 80% after 2.5 h [75].

Table 4. Silica based materials for the adsorption of heavy metals.

Adsorbent	Target Metal	Initial Concentration	Removal%	References
Thiol and Amino functionalized SBA-15 Silica	Hg^{2+}	10.1 mg/L	29.2%	[76]
Amino functionalized mesoporous silica	Cr^{6+}	40 mg/L	8.205%	[77]
Functionalized silica with -SH	Hg^{2+}	-	50.5%	[78]
Amino functionalized silica gel in Tea Polyphenol extracts	Pb^{2+}	5–1200 mg/L (Pb^{2+}) 5–800 mg/L (Cu^{2+})	98.1%	[79]
Amino-functionalized and pure silica nano hollow sphere (NH_2-SNHS, SHNS) and silica gel(NH_2-SG)	Ni^{2+}	100 mg/L	0.84% (SHNS), 2.59% (NH_2-SG), and 3.13% (NH_2-SNHS) mg/g	[74]

Table 4. *Cont.*

Adsorbent	Target Metal	Initial Concentration	Removal%	References
Amino-functionalized and pure silica nano hollow sphere (NH_2-SNHS, SHNS) and silica gel(NH_2-SG)	Pb^{2+}	100 mg/L	26.85%(SHNS), 54.35%(NH_2-SG), and 96.78%(NH_2-SNHS)	[74]
Amino functionalized silica gel in Tea Polyphenol extracts	Cd^{2+}	5–800 mg/L	99.78%	[74]
Organically functionalized silica gel	Cu^{2+}	63 mg/L	1.99%	[80]
Amino-functionalized and pure silica nano hollow sphere (NH_2-SNHS, SHNS) and silica gel(NH_2-SG)	Cd^{2+}	100 mg/L	2.6% (SHNS), 3.2% (NH_2-SG), and 4.1% (NH_2-SNHS)	[74]
Ionic liquid-functionalized silica	Pb^{2+}	50–200 mg/L	20.23% and	[81]
Amino functionalized silica gel in Tea Polyphenol extracts	Cu^{2+}	5–800 mg/L	99.59%	[79]
Ionic liquid-functionalized silica	Cd^{2+}	50–200 mg/L	15%	[76]

This highlights that heavy metal ions have different affinities for organic groups and their extraction can be tuned depending on the targeted metal ions. One shortcoming of the silica-based materials is the poor stability of the –Si–O–Si– bond in basic conditions, which may cause leaching of the surface-grafted functional groups.

2.3. *Zero-Valent Iron Nanoparticles*

Zero-valent metal nanoparticles have demonstrated their ability in remediating a variety of pollutants in wastewaters [82]. For instance, Ag NP are more particularly used for disinfecting wastewater owing to their antimicrobial/antifungal properties [83]. Nanoscale zero-valent iron (nZVI) can be defined as a composite consisting of Fe(0) and ferric oxide coating [84]. The nanosize of nZVI results in a higher surface to volume ratio or higher specific surface, which leads to an increased removal of pollutants [85]. Therefore, they have received considerable attention in the scientific community as a novel adsorbent to remediate a variety of heavy metals, including Hg^{2+}, Cr^{6+}, Cu^{2+}, Ni^{2+}, and Cd^{2+} [86–88]. In fact, nZVI is an excellent electron donor (see Equation (1)) with a reduction potential greater than −0.447 V, highly capable of reducing pollutants; the feasibility of the reduction depends on the redox potential of particular metal ions. For example, metal ions such as Zn or Cd have lower or almost similar reduction potential (i.e., −0.76 V and −0.40 V vs. Standard Hydrogen Electrode (SHE), respectively) than that of Fe, which interrupts the surface redox phenomenon of nZVI [89]. Subsequently, the removal of a particular pollutant (i.e., heavy metal ions) can occur through various processes such as precipitation, complexation, adsorption, oxidation, and reduction, which are briefly schematized in Figure 5.

$$\mathrm{Fe(II)} + 2e \rightarrow \mathrm{Fe(0)};\ E(\mathrm{red}) = -0.447\mathrm{V\ vs.\ (SHE)} \quad (1)$$

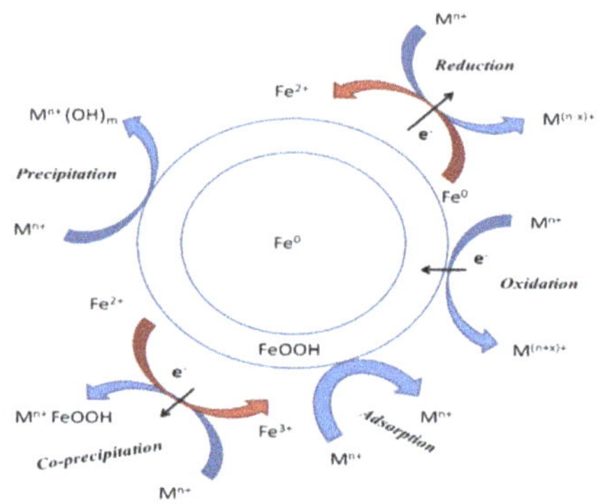

Figure 5. Schematic of a ZVI depicting several interaction mechanisms with heavy metals.

In spite of the noteworthy effectiveness of nZVI in the removal of heavy metal ions, its shortcomings cannot be neglected. For example, it was reported that nZVI oxidizes in air and in aqueous solutions, which results in the slowdown of the reduction processes of heavy metal ions [90]. Moreover, some scientific investigations have also suggested a tendency to agglomerate, which consequently decreases the reaction surface area and mobility [84]. Therefore, to improve its performance, several modification strategies, such as surface chemical modification or doping nZVI with other metals (Pd, Cu, Ni, and Pt) were investigated [91]. These modified nZVI showed an improvement in the removal of metallic pollutants. For example, in order to evaluate removal capacity of Cr^{6+}, Huang et al. studied a surface modified nZVI material formed by combining nZVI with sodium dodecyl sulfate (SDS). They measure an adsorption capacity of 253.68 mg/g with 300 mg/L initial Cr^{6+} concentration in batch adsorption experiments. This study demonstrated an improved adsorption capacity and a decreased aggregation of the modified nanomaterial [92]. In another study, nZVI and Au-doped nZVI nanoparticles were investigated for the removal of both Cd^{2+} and nitrates from water in batch mode [93]. They highlighted that increased pH and negative charge contributed to a significant increase in the Cd^{2+} removal capacity (from 40 mg/g to 188 mg/g) if nitrates are present in the water. The nZVI deposited with 1 wt.% Au reduced the nitrate quantity to less than 3% of the initial value, while maintaining a high Cd^{2+} removal capacity.

2.4. Magnetic Nanoparticles

Even though the nanomaterials described above showed an enormous capacity in the extraction of heavy metals, they carry certain limitations with regards to their cost-effectiveness, reusability, separation from aqueous solutions, and complex synthesis routes, which impede their utilization at a commercial scale. During the last two decades, micro and nano-scaled magnetic particles have attracted attention as adsorbents for eliminating the biological molecules, organic pollutants, and heavy metal ions from water and wastewater [94]. The major advantage with magnetic nanomaterials lies in their easy recovery after exhaustion from the treated solution by applying an external magnetic field, as shown in one of the studies carried out using magnetic mesoporous silica nanospheres for the extraction of Pb^{2+}, Hg^{2+}, and Pd^{2+} (Figure 6) [95].

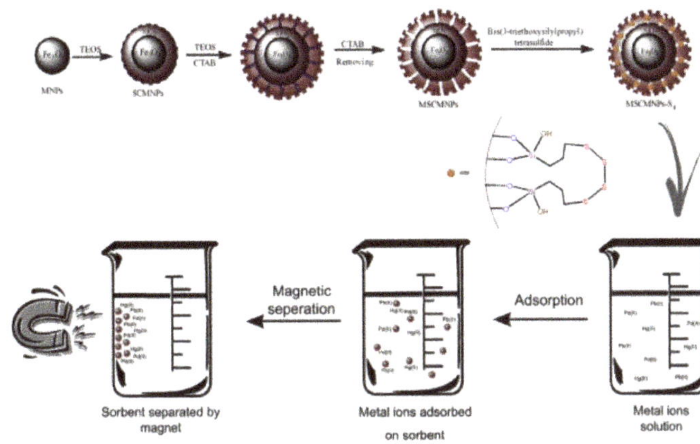

Figure 6. Schematic showing the mechanism of the magnetic nanoparticles in the removal of heavy metals. Reproduced with permission from [95].

2.4.1. Magnetic Iron-Oxide Nanoparticles

Among the magnetic materials, magnetic iron-oxide NP, i.e., magnetite (Fe_3O_4) and maghemite (γ-Fe_2O_3) present substantial dominance over conventionally used metal and alloy based magnetized nanomaterials in terms of facile synthesis, corrosion, and abrasion resistance. Maghemite and magnetite NP are extensively studied for wastewater treatment and heavy metal ion removal in particular [96]. Some studies investigating magnetic iron oxide NP for heavy metal ion removal are summarized in Table 5. Maghemite NP possess a large surface area, which contributes to their high adsorption capacity and are moreover environmental-friendly [97].

Akhbarizadeh et al. studied the removal of Cu^{2+}, Ni^{2+}, Mn^{2+}, Cd^{2+}, and Cr^{6+} using maghemite NP with initial concentration of 50 mg/L. They observed removal efficiencies of 88.2% for Cu, 84.4% for Cr, 18.3% for Mn, 15.7% for Ni, and 8.4% for Cd [98]. This study also indicated that removal of these five different metal ions by maghemite reached an equilibrium after a short period (10 min). Moreover, the absorption showed a good fit for Langmuir isotherm verifying the monolayer coverage of metallic ions on maghemite surface. In fact, authors highlighted a pH dependence during the extraction, when other effective parameters were kept constant (see Figure 7). The removal efficiency of copper increased from 8.4% to 50% when the pH was increased from 3 to 5. This study shows that selective removal of particular metal ions is feasible with these magnetic nanomaterials.

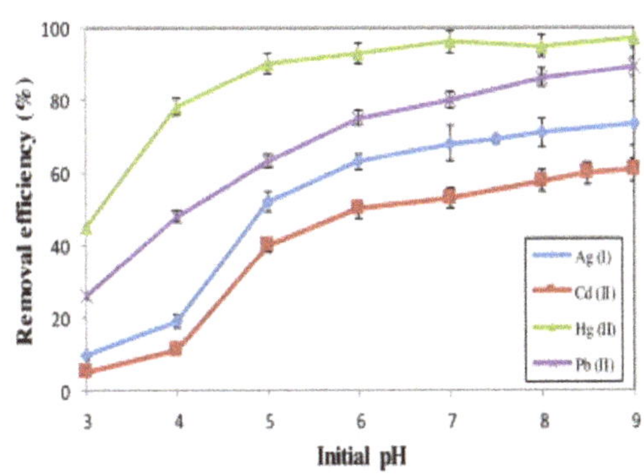

Figure 7. Removal efficiency with varying pH (polymer encapsulated maghemite). Reproduced with permission from [98].

In recent years, several reports on polymer-modified maghemite nanomaterials have been published. For example, Madrakian et al. prepared a novel mercaptoethylamino monomer-modified maghemite nanomaterial (MAMNPs) [99]. The maximum removal capacities of Ag^+, Hg^{2+}, Pb^{2+}, and Cd^{2+} were 260.55 mg/g, 237.60 mg/g, 118.51 mg/g, and 91.55 mg/g respectively and the adsorption fitted the Sips isotherm model, which is the combination of Langmuir and Freundlich models. The Sips isotherm model is used for predicting the adsorption in heterogeneous systems and circumventing the limitation of the rising adsorbate concentration associated with the Freundlich isotherm [100]. The removal efficiency for cadmium increased from 8.7% to 50% when the pH was increased from 3 to 5 (Figure 7). This illustrates the efficiency of polymer-coated maghemite NP over pure maghemite NP. These modified maghemite NP showed good removal capabilities and selectivity towards various heavy metal ions. The removal of Cu^{2+} and other metallic ions using mesoporous magnetite NP was also investigated and the study showed a removal efficiency of 90% for a solution containing 50 mg/L of Cu^{2+} ions [101]. However, the adsorption followed the pseudo-second-order kinetic model deciphering a chemisorptive adsorption of metal ions on the magnetite NP. The experimental data followed the Langmuir isotherm, indicating the monolayer adsorption mechanism. Interestingly, the magnetic NP could be used for five consecutive cycles, demonstrating their high removal capacity. Shipley et al. studied the removal of Cu^{2+} using hematite NP and obtained a removal efficiency of 89% for an initial metal concentration ranging from 0.016 mg/L [102], which is lower in comparison with magnetic NP used by Fato et al., for Cu^{2+} removal. These results conclude the effectiveness and efficiency of magnetic NP over other adsorbents.

Superparamagnetic iron oxide NP (Fe_3O_4) were investigated for the treatment of synthetic water contaminated with Mn^{2+}, Zn^{2+}, Cu^{2+}, and Pb^{2+} ions [103]. In their study, the authors estimated an adsorption capacity of 11.5 mg/g 12.4 mg/g, 14.5 mg/g, and 16.4 mg/g for Mn^{2+}, Zn^{2+}, Cu^{2+}, and Pb^{2+} ions respectively. The superparamagnetic ascorbic acid-coated Fe_3O_4 NP were synthesized and studied for their potential to remove arsenic ions from wastewater [104]. They found a maximum adsorption capacity of 46.06 mg/g and 16.56 mg/g for As^{3+} and As^{5+} respectively, as followed by Langmuir isotherm. In another study, Fe_2O_3 NP encapsulated in cellulose matrix were investigated for the removal of arsenic from aqueous solution [105]. The experimental results showed an excellent adsorption capacity towards As^{3+} and As^{5+} (23.16 mg/g and 32.11 mg/g, respectively) and the experimental data followed both Langmuir and Freundlich isotherms. These studies show that cellulose encapsulated Fe_2O_3 NP exhibit a higher As^{5+} adsorption capacity with multilayered adsorption compared to superparamagnetic ascorbic acid-coated Fe_3O_4 NP. This increased adsorption might be due to the stronger interactions of As^{5+} with OH groups by forming a sodium arsenate complex. However, in environmental conditions arsenic is present in the form of hydrides. The higher adsorption for As^{3+} with ascorbic acid coated magnetic NP is probably due to its higher affinity to form methyl arsenic acid. Through these studies, we can conclude that polymer coated Fe_3O_4 NP can increase the adsorption affinity for some specific metallic ions and can be used for selective removal of heavy metals.

Table 5. Iron-oxide magnetic NP utilized for the removal of metallic ions from contaminated water.

Adsorbent	Target Metal/s	Initial Concentration	Removal%	References
Hematite-Magnetite hybrid	Pb^{2+}	2 mg/L	97.67%,	[106]
Hematite-Magnetite hybrid	Cd^{2+}	2 mg/L	99.84%	[106]
Maghemite NP	As^{5+}	1–11 mg/L	50%	[97]
Hematite-Magnetite hybrid	Cr^{3+}	2 mg/L	99.50%	[106]
Maghemite NP	Cr^{6+}	5–200 mg/L	1.92%	[107]
Biogenic nano-magnetite	Cr^{6+}	16.69 mg/L	3.2%	[108]
Magnetite NP	Pb^{2+}	10–600 mg/L	3730%	[103]

Table 5. *Cont.*

Adsorbent	Target Metal/s	Initial Concentration	Removal%	References
Magnetite NP	Mn^{2+}	10–600 mg/L	7700%	[103]
Carboxyl functionalized magnetite NP	Cu^{2+}	10 mg/L	0.983%	[109]
Carboxyl functionalized magnetite NP	Cd^{2+}	10 mg/L	1.03%	[109]
Magnetite NP	Zn^{2+}	10–600 mg/L	1046%	[103]

$CuFe_2O_4$ magnetic nanoparticles were prepared from industrial sludge and used as an adsorbent for the removal of Cd^{2+} ions. Experimental results show that at very low pH ~ 2.0, they observed negligible adsorption, whereas almost 99.9% adsorption occurred at pH 6.0 [110]. They observed fast removal of cadmium ions during the initial 10 min and thereafter a decrease in removal was observed until equilibrium was attained 20 min later. The Cd^{2+}-$CuFe_2O_4$ interaction was described by the pseudo-second-order mechanism. The adsorption process followed the Langmuir adsorption isotherm with the monolayer adsorption capacity 13.87 mg/g at 298 K, which increased slightly to 17.54 mg/g at 318 K, indicating endothermic interactions. The removal of Ni^{2+} was investigated using Fe_3O_4 magnetic nanoparticle impregnated tea waste [111]. The removal efficiency decreased from 99 to 87% with the increase of initial concentration of Ni^{2+} in solution from 50 to 100 mg/L. In addition, the adsorption of Ni^{2+} increased with the increasing temperature from 303 to 323 K, which again highlights the endothermic nature of the adsorption process. However, in this study, the experimental data fit with both Langmuir and Freundlich models revealing a maximum adsorption capacity of 38.3 mg/g.

The efficient mixing of nanoparticles in contaminated water is seen as one of the major challenges for magnetic extraction of pollutants. It seems complicated to use magnetic nanomaterials for heavy metals removal at industrial scales. In fact, the efficient mixing of magnetic nanoparticles in solution requires appropriate micromixers. With that aim, Karvelas et al. investigated a Y-shaped micromixer to remove heavy metals from the contaminated water [112]. They observed that a higher mixing in the micromixer leads to slower adsorption, whereas the adsorption capacity increases with lower mixing of the NP with heavy metals. In their study, the faster adsorption leads to lower adsorption capacity.

2.4.2. Magnetic Nanocomposites

Magnetic nanocomposites are receiving an increased attention because of their easy recovery after exhaustion from water with the help of a magnet [113]. In general, magnetic nanocomposites are mostly based on magnetic iron oxide nanomaterials. In fact, the fabrication of these magnetic nanocomposites can be achieved through three approaches: (1) modifying the NP surface with functional groups such as $-NH_2$ and –SH; (2) coating the iron/iron oxide nanoparticles with other materials, such as humic acid, polyethylenimine, polyrhodanine, MnO_2 and polypyrrole, in order to make a core-shell structure; and (3) decorating porous materials such as graphene oxide and CNT with iron/iron oxide nanoparticles [114,115].

Polymer-functionalized Fe_2O_3 nanomaterials have been investigated for the removal of divalent Cu, Ni, and Co metal NP over a pH range of 3.5 to 10 [116]. The experimental results showed the highest removal efficiency for Cu(II) with the adsorption capacity of 6.98 mg/g at pH of 5.3. Ferric and ferrous chlorides are usually used for the synthesis of magnetite/zeolite nanocomposites. The synthesized nanocomposite possessed the composite ratio of 3:1 (zeolite: iron oxide) with saturation magnetization of 19.03 emu/g [117].

These NC were tested for the removal of Cs^+ and Sr^{2+}. They provided a maximum adsorption capacity of 207.4 mg/g and 83.68 mg/g for Cs^+ and Sr^{2+} respectively. Fe@MgO nanocomposites were investigated for the removal of Pb^{2+} ions and methyl orange from contaminated water. These nanocomposites are a combination of strong magnetic nZVI nanoparticles and MgO, which exhibit a good adsorption capacity of 1476.4 mg/g and

6947.9 mg/g for Pb^{2+} ions and methyl orange respectively [118]. The experimental data followed the Langmuir linear regression model representing monolayer chemisorption of Pb^{2+} ions, as shown in Figure 8. Similarly, Guo et al. studied Pb^{2+} ion removal using rGO-Fe_3O_4 composite and reported an adsorption capacity of 373.14 mg/g [64], which is lower compared to other cited works. The different examples of magnetic nanocomposite-based adsorption studies are summarized in Table 6. The efficiency of superparamagnetic Iron Oxide Nanoparticles (SPIONs) and Chitosan-Coated SPIONs was tested for the removal of hexavalent chromium ions (with initial concentration of 1 mg/L) [119]. The experimental results showed a removal of 80.44% and 99.7% of Cr^{6+} ions for uncoated and Chitosan-coated SPIONs, respectively.

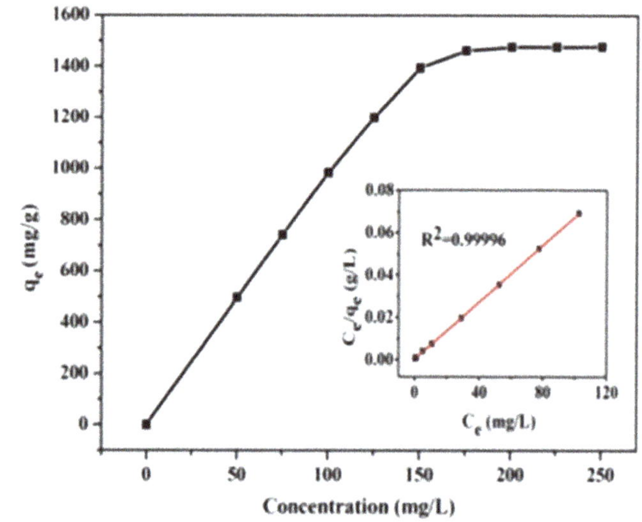

Figure 8. Langmuir fit for lead removal using Fe@MgO nanocomposites. Reproduced with permission from [118].

However, the complexity of using magnetic nanoparticles for water purification makes them complicated to apply at a larger scale and the overall cost-benefit ratio of magnetic nanoparticles is very high, which is a significant barrier for utilization at industrial scale. For this reason, research combining nanoparticles with membranes has spurred attention at it is seen as a more promising solution for larger scale water remediation [120]. For example, hydrogels have a 3D network structure that is capable of retaining water in their nanoporous structure. During the recent years, various composite membrane based nanoadsorbants were studied for the removal of heavy metals. Zhang et al. reported the removal of Cs ions using Clay–Hexacyanoferrate composite hydrogels [121]. The experimental results showed an adsorption capacity of 173 mg/g, even in 0.2 mg/L Cs contaminated water. Many nanocomposites have been studied for the adsorption of $^{137}Cs^{+}$ at larger scales [122]. Various other hydrogel such as polyacrylic acid hydrogel, 2-acrylamido-2-methyl-1-propane sulfonic acid-based magnetic responsive hydrogels, magnetic-vinyl pyridine-based hydrogel are investigated for the removal of other metallic ions like Cu^{2+}, Ni^{2+}, Cd^{2+}, Co^{2+}, U^{6+} etc. [123]. These hydrogels can be recycled and reused, but the scientific findings reported a decrease in the removal efficiency. Therefore, an in-depth research is required on these hydrogel-based nanoadsorbants that can sustain their removal capacities.

Table 6. Summary of the removal studies on heavy metals using magnetic nanocomposites.

Adsorbent	Target Metal	Initial Concentration or Concentration Range	Removal Efficiency (%)	References
Silica coated magnetic nanocomposites	Pb^{2+}	5–120 mg/L	1.49%	[124]
Silica based hybrid organic inorganic magnetic nanocomposites(MNPs@SiO_2-TSD-TEOS)	Pb^{2+}	100 mg/L	41.7%	[125]
Silica based hybrid organic inorganic magnetic nanocomposites(MNPs@SiO_2-TSD-TEOS)	Ni^{2+}	100 mg/L	35.7%	[125]
o-Vanillin functionalized mesoporous silica–coated magnetite nanoparticles (Fe_3O_4@MCM-41)	Pb^{2+}	120 mg/L	15.57%	[126]
Silica coated iron oxide magnetic nanocomposites(Fe_3O_4@SiO2)	Pb^{2+}	10 mg/L	97%	[127]
Silica coated iron oxide magnetic nanocomposites(Fe_3O_4@SiO2)	Hg^{2+}	10 mg/L	94.12%	[127]
Polythiophene modified chitosan/magnetite nanocomposites	Hg^{2+}	0.02–100 mg/L	5.28%	[128]
Bismuthiol-II-immobilized magnetic nanoparticles	Cr^{3+}	-	>90%	[129]
Bismuthiol-II-immobilized magnetic nanoparticles	Cu^{2+}	-	>90%	[129]
Thiol-lignocellulose sodium bentonite (TLSB) nanocomposites	Cd^{2+}	(0.20–1.70) × 10^3 mg/L	45.832%	[130]
Thiol-lignocellulose sodium bentonite (TLSB) nanocomposites	Zn^{2+}	(0.20–1.70) × 10^3 mg/L	35.729%	[130]
Magnetic Chitosan Nanocomposites	Cd^{2+}	10 mg/L	92.1%	[131]
Water-soluble magnetic graphene nanocomposites	Cd^{2+}	NA	>85%	[114]
Functionalized magnetic Fe_3O_4	Cu^{2+}	10 mg/L	96%	[132]
Functionalized magnetic Fe_3O_4	Hg^{2+}	10 mg/L	96%	[132]
Fe@MgO nanocomposite	Pb^{2+}	100 mg/L	147.64%	[118]
Magnetic MWCNT	Cr^{6+}	1–25 mg/L	1.14%	[133]

2.4.3. Reusability of Magnetic Nanoparticles

Various studies have demonstrated that magnetic NP can be applied to the removal of metallic ions from contaminated water. However, for a cost-effective adsorbent, it is essential to study their reusability or recyclability. For example, Rivera et al. studied Cr^{6+} ion removal using 2 g/L of the magnetic magnetite NP. They measured an adsorption capacity of up to 12.4 mg/g. The regeneration cycles of the magnetic nanomaterials showed a slight decrease (about 4%) in the removal efficiency, concluding that the nanomaterial can be reused for up to 4 cycles continuously [134]. Tao et al., showed an effective removal of heavy metal ions (Hg^{2+} and Pb^{2+}) using thiol-functionalized magnetic mesoporous microspheres [135]. After water treatment, the NP were separated from the water using a magnet and washed with a suitable eluent. The recycled NP were tested against similar heavy metals and were able to remove the metal ions with a similar efficiency. Recyclability tests were also conducted by Chen et al. on functionalized magnetic Fe_3O_4 nanoparticles. In order to study the possibility of reusing Fe_3O_4 NP, they were subjected to several loading and elution operations. The elution operations were carried out by shaking Fe_3O_4 NP, which contained maximum amounts of metal ions in 100.00 mL of 0.50 M HCl. Interestingly, they showed no significant decrease in the removal efficiencies even after seven adsorption-desorption cycles, with a decrease from 100% to 93% [132]. Fato et al. investigated the reusability of MNP using nitric acid as eluent and reported that even after five successive cycles, the MNP can be reused with no significant decrease in the removal efficiency [101]. In addition, these magnetic NP can be easily separated from the treated water, which turned out to be their major advantage over other technologies. However, the behavior of engineered nanoparticles and their regeneration and desorption capacities in real conditions need to be evaluated before commercialization.

3. Nanomaterial Modified Bioelectrochemical Systems for Enhanced Power Production and the Remediation of Heavy Metals

During the past few years, a technique based on the integration of electrochemical and biological processes, usually referred to as bioelectrochemical systems (BESs), has emerged for wastewater treatment. A typical BES setup consists of anodic and cathodic chambers separated by an ion-selective membrane with microorganisms as catalysts for the redox reactions. The organic matter is oxidized in the anode chamber owing to microbial metabolic activity responsible for producing electrons. These electrons then circulate through an external circuit towards the cathode where the reduction of target species i.e., nitrates, H^+, CO_2 or HM ions occurs [136,137]. In particular, BES can be operated in two different modes: microbial fuel cell (MFC) and microbial electrolysis cell (MEC), depending upon the targeted process. The electrons produced on substrate oxidation at the anode, accelerate the reduction of targeted heavy metal ions [138]. Even though BES poses a great advantage in scavenging metallic ions, their performance is highly dependent on the electrode material, redox potential, and substrate oxidation. These influence the reduction efficiency of HM at the cathode and the output power of BES. During the past few years, various electrode materials have been investigated for enhancing the overall performance, especially for the reduction of HM in BES. Due to their unique properties, nanomaterials have attracted significant attention in the scientific world. Therefore, researchers are investigating nanostructured bioelectrochemical systems (BESs) that utilize microorganisms along with nanomaterials to mediate, facilitate, or catalyze the redox reactions at one or both the electrodes, which in turn enhances the power output of the BES.

3.1. Electrode Materials for Anode in BESs

3.1.1. Conventional Anode Materials

Generally, carbon based materials, i.e., graphite plates [139], graphite rods [140], graphite felt [141], carbon brush [142], and carbon paper [143], are widely used as anode materials due to their good conductivity, biofilm formation, and non-corrosive behavior [144]. For example, carbon brush was used as an anode in the MFC and their performance was evaluated under different temperature conditions ranging from 300–750. The maximum power output increased form 1160 mW/m^2 to 1561 mW/m^2 with variation in temperature [142]. An enhancement in the electrical output was observed when plasma-modified carbon paper was used as an anode in an air-cathode cylindrical shaped MFC setup [143]. The maximum power output obtained was 107 mW/m^2. However, these materials exhibit some limitations, such as lack of durability and higher cost. Carbon felt with 95.5% of carbon content is considered as a good candidate for anode but simultaneously offers a large resistance. Graphite rods manifest certain advantages in terms of good electrical conductivity, chemical stability, and lower costs for MFC applications. However, the availability of large surface areas for biofilm growth and redox reactions is a major drawback for them. Hence, anodes with graphite fiber brush were designed that showed enhanced MFC power production [145]. However, several metal-based materials have also been investigated for MFC anodes with advantages in terms of conductivity, robustness, and cost-effectiveness. It has also been reported that the doping of graphite anodes with Mn^{4+}, Fe_3O_4, and Ni^{2+} increases power generation of MFCs [146]. Generally, MFCs generate lower output voltage in comparison with the overall cell potential because of several energy losses, i.e., activation, bacterial metabolism, mass transfer, and ohmic, which effect the performance of an MFC. In this regard, strategies, such as introducing mediators and increasing electrode surface area are employed in order to overcome these losses. However, one of the strategies includes the invasion of nanomaterials. Nevertheless, there are some reports on nanomaterial-based anodes that have demonstrated an increment in the power output of the MFCs.

3.1.2. Nanomaterial Modified Anode Materials

The anode in MFC can be upgraded by using nanomaterial coatings, which provide a high specific surface, an improved electron transfer and promote electroactive biofilm. These enhance the power output of the MFC, as shown in Figure 9 [147]. Iron oxide can facilitate the extracellular electron transfer through two pathways; within the biofilm in the form of an electrical conduit or at the interface by accumulation on the cell surface [148]. Most of the metal oxides are not conductive but nevertheless promote and enhance biofilm formation. Thus, in this regard, carbon conductive materials or fillers e.g., CNT, Graphene, etc., can enhance the electrical conductivity of the metal oxides and thereby increase the extracellular electron transfer from the organism to the anode. Zhao et al. studied the possible increase of bio-current generation using ionic liquid functionalized graphene nanosheets as an anode [149]. They showed that current density increases from 0.40 mA/m^2 to 2.8 mA/m^2 within 30 h if the carbon paper anode is replaced with graphene nanosheets. The continuous testing on bare pencil graphite electrode (PGE), α-MnO_2/PGE, PANI/PGE, and α-MnO_2/PANI/PGE were evaluated in glucose-fed-*Escherichia coli*-based MFC.

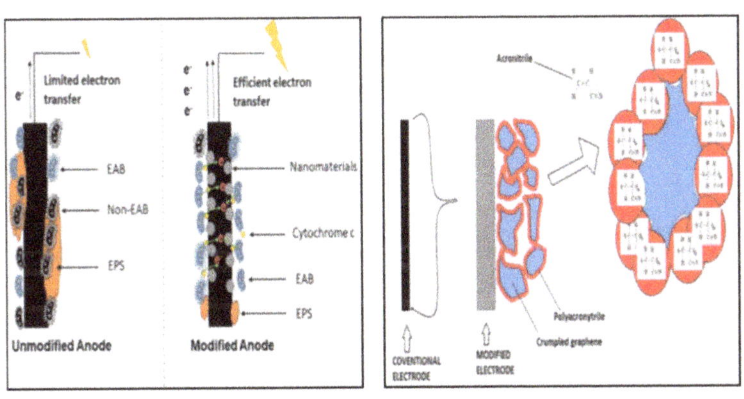

(a) (b)

Figure 9. (**a**) General picture of nanomaterial modified anode and (**b**) graphene modified anode. Reproduced from [147] under creative commons attribution license.

The experimental results showed that the α-MnO_2/PANI/PGE-coated anode showed highest current and power density, which was 6 times higher than for uncoated PGE. Polyaniline networks grown on graphene nanoribbon-coated carbon paper as anode was reported to potentially increase the performance of MFC. When the carbon paper anode was coated with graphene nanoribons the current density increased from 0.52 mA/^{2} to 1.8 mA/m^2 [150]. The current density is dependent on the type of material used, as shown in Figure 10. Reduced graphene was also coated with Polyaniline (PANI) in carbon cloth [151]. Biosynthesized α-MnO_2-based polyaniline binary composite were investigated as a biocatalyst in MFC [152].

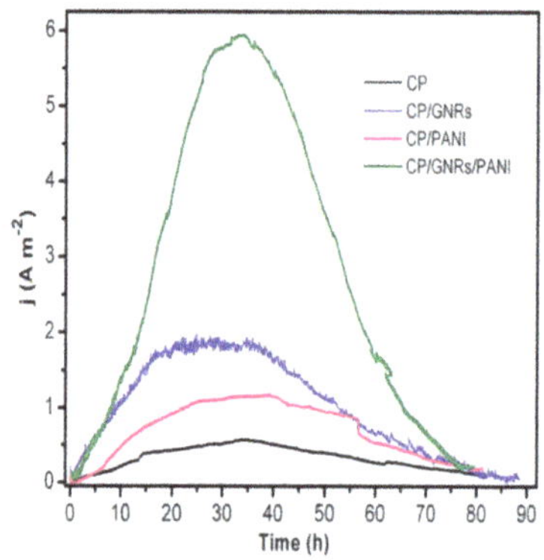

Figure 10. Current generation by *S. oneidensis* in ECs equipped with different anodes [150].

Similarly, the anode modified with reduced graphene oxide/tin oxide nanocomposite showed a significant increase of 4.8 times in the power density of MFC compared to its bare counterpart, as shown in the cyclic voltammetry curves reproduced with permission from [153] (see Figure 11a).

3.2. *Electrode Materials for Cathode in BESs*

3.2.1. Conventional Cathode Materials

Cathode performance is considered to be the main obstacle in the development of MFCs. Thus, the fabrication of cathode materials having high power generation and columbic efficiency is the most crucial and challenging aspect for the development of successful MFC technology, more particularly for the treatment of metal-contaminated water. Graphite-based electrodes have been more extensively investigated for the removal of various heavy metal ions in water because of their low cost, high electrical conductivity, and biocompatibility. Several studies have reported findings that support electrochemical reduction of heavy metal ions at the biotic and abiotic cathode, summarized in Table 7 [154]. For instance, the removal of Cu^{2+} ions was studied via electrodeposition process at the cathode of MFC. A removal efficiency of >96% of Cu was estimated [155]. Moreover, the XRD analysis confirmed the reduction of Cu^{2+} to Cu^{+} and Cu^{0} on the electrode surface. The reduction was successfully achieved within 24 h of operation at the reduction potential of 0.347 V vs. Ag/AgCl. The reduced product was analyzed using ICP-MS, which indicated the deposition of elemental mercury over the cathode surface. Huang et al. studied the reduction of Cr ($Cr^{6+/3+}$) having initial concentration of 39.2 mg/L at the bio-cathode of MFC with a power production of 2.4 ± 0.1 W/m^3 and at a current density of 6.9 A/m^3 [156]. They concluded that initial Cr^{6+} concentration and solution conductivity have a clear effect on the biocathode MFC performance. In another study, carbon fiber cathode was used for the removal of chromium ions [157]. As a result, they observed 75.4% reduction of chromium ions with the maximum power density and current density of 970 mW/m^2 and 2462 mA/m^2. During the reduction process, the pH of the catholyte increased from 2 to

3.76, most likely due to the formation of $Cr(OH)_3$ responsible for the basic environment. Based on the studies discussed, the bioelectrochemical remediation brings an alternative approach towards a cost-effective and ecofriendly degradation of heavy metals from the water. However, the reduction efficiency can be improved by introducing nanoadsorbents as the cathode.

Table 7. Carbon based materials as electrodes for bioelectrochemical reduction of heavy metals.

Target Metal	BES Configuration	Anode	Cathode	Power Output	Reference
Cr^{6+}	Double chambered MFC	Graphite plate	Graphite plate	150 mW/m^2	[158]
Hg^{2+}	Double chambered MFC	Grpahite rod	Graphite rod	32.6 ± 0.5 W/m^2	[140]
Hg^{2+}	Double chambered MFC	Graphite felt	Carbon paper	433.1 mW/m^2	[159]
Cu^{2+}	Double chambered MFC	Carbon felt	Carbon plate	5.5 W/m^2	[160]
Ag^{+}	Double chambered MFC	Carbon brush	Carbon felt	4.25 mW/m^2	[161]
Se	Single chambered MFC	Carbon cloth	Carbon cloth	13–1500 mW/m^2 depending on initial concentration	[162]
Co	Double chambered MFC	Graphite felt	Graphite felt	258 mW/m^2	[163]
Ag^{+}	Double chambered MFC	Carbon felt	Carbon felt	0.109 mW/m^2	[164]

3.2.2. Nanomaterials Modified Cathode Materials

There are a very limited number of studies which have reported the use of nanomaterials as cathodes for the reduction of heavy metals. For the first time, Alumina-nickel nanoparticles-dispersed carbon nanofiber electrode were investigated for the removal of chromium ion in MFC. The experimental results showed a high reduction rate of 2.13 g/m^3-h with achievable maximum power density and current density of 1540 mW/m^2 and 4560 mA/m^2 (see Figure 11) [165]. Shi et al. investigated graphite electrode modified with natural pyrrhotite for the reduction of Cr^{6+} in the MFC [166]. A removal of 99.59% was achieved within 10.5 h of operation with a maximum power density of 45.4 mW/m^2. However, the uncoated graphite cathode generated a maximum power density of 35.5 mW/m^2. This study demonstrated the efficiency of a naturally-occurring mineral in the effective removal of Cr ions. However, it also produced a lower output power and maximum removal was achieved at pH = 3, which suggest the use of additional chemicals to facilitate the reduction. Therefore, other cathode materials are explored to obtain a higher output power under ambient environmental conditions. Recently, graphene oxide-based catalysts were used at the cathode of MFC to enhance the recovery of Cu ions [167]. This study showed that Cu^{2+} could be efficiently removed in the rGO-MFC within 8 h of operation at the external resistance of 1000 Ω. The removal efficiency reached 98% with the rGO-MFC for an initial Cu^{2+} concentration of 4, 8, and 12 mg/L and a maximum power output of 1.38 W/m^2. Nanocomposite-based electrocatalysts have also been investigated for the extraction of some heavy metals. For example, carbon cloth modified with α-Fe_2O_3/polyaniline nanocomposites were investigated for enhanced bioelectricity production and hexavalent chromium removal at the cathode of MFC [168]. The MFC showed a complete Cr^{6+} reduction at 50 mg/L with a reduction rate of 1.39 $g/m^3/h$ within 36 h of operation. However, they reported a maximum power density of 1502.78 mW/m^2, which was 1.753 times higher than carbon cloth cathode. Besides, α-Fe_2O_3/polyaniline modified electrode showed a reduction in the over-potential and favored electrocatalytic reduction of Cr with an increase in the current density as shown in the cyclic voltammetry curves with different electrode materials (Figure 11b). In an another study, graphite felt modified with iron sulfide wrapped with reduced graphene oxide (FeS@rGO) nanocomposites are investigated for the removal of Cr^{6+} having initial concentration of 15 mg/L in MFC [169].

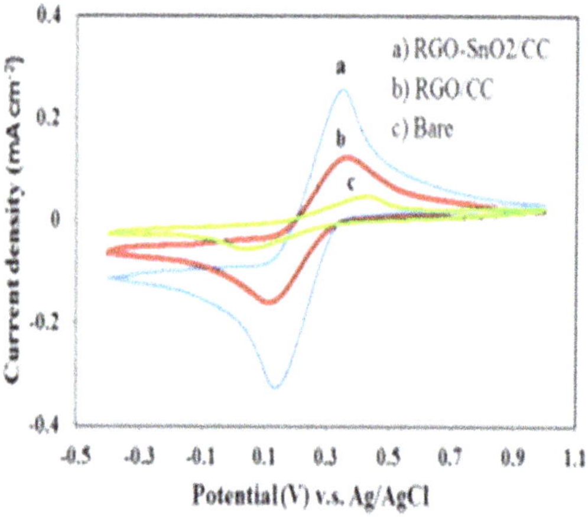

(a)

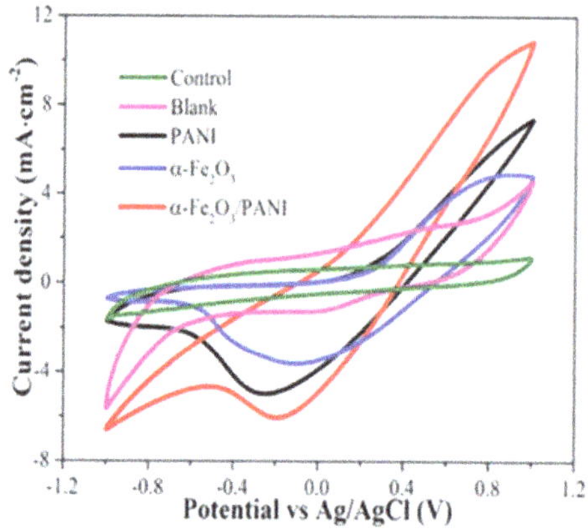

(b)

Figure 11. (**a**) Cyclic voltammograms of the bare CC, RGO/CC, and RGO–SnO_2/CC electrodes in 10 mM $K_3[Fe(CN)_6]$ containing 0.1 M KCl (scan rate: 0.1 V/s). Reproduced with permission from [153] (**b**) Cyclic voltammetry curves with a potential range from −1.0 V to 1.0 V of various cathodes at the scanning rate of 1 mV/s in catholyte. Reproduced with permission from [165].

These nanocomposites showed 100% Cr^{6+} removal efficiency with the reduction rate of 1.43 mg/L/h. Overall, the improved electrochemical performance of MFC-FeS@rGO was expected due to the high conductivity, low internal resistance, and better reaction kinetics of FeS@rGO nanocomposites, as shown in Figure 12. A novel study representing the integrated setup of ZVI and MFC was investigated for the removal of As^{3+} with an initial concentration of 0.3 mg/L [170].

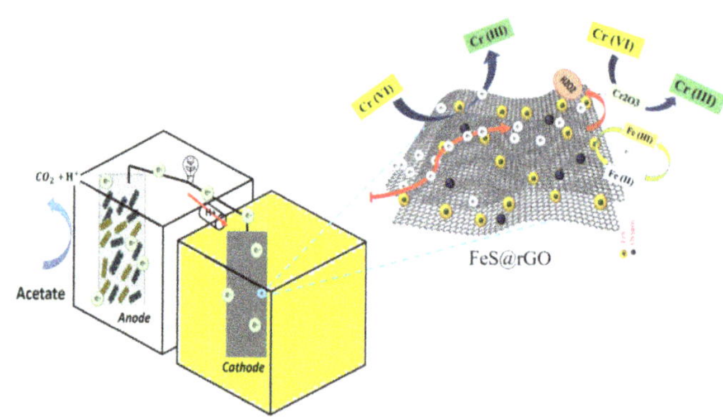

Figure 12. Possible mechanism of Cr(VI) reduction at FeS@rGO decorated cathode in MFC. Reproduced with permission from [169].

This study showed a more significant decrease in the concentrations of both the As^{3+} and total As using MFC–ZVI hybrid process than solely using ZVI. After 2 h of operation, the total As concentration remaining in the solution was only 9.8 μg/L and the As^{3+} concentration was below the detection limit in the case of the MFC–ZVI hybrid process. On the other hand, the residual total As concentration and As^{3+} concentration were 180 μg/L and 106 μg/L, respectively in the case of ZVI process. This corresponds to an improvement of 40% in the removal efficiency of total As. These results highlight a notable increase in the efficiency of As removal with the MFC-ZVI based hybrid. Such a combined technology was studied only for the lower concentrations of artificially prepared As contaminated water. However, in real situations, the presence of other contaminants can interfere with the selective removal of As and in turn reduce the applicability of this technology for environmental samples.

4. Conclusions and Future Perspectives

4.1. Roadmap of the Nanomaterial Based Adsorbents for the Extraction of Heavy Metals

Due to the prevalent water crisis, nanoadsorbants are widely being investigated for the treatment of metal-contaminated waters owing to their exceptional properties, as described in this review. A variety of nanomaterials, including carbon-based nanotubes, graphene, silica-derived nanomaterials, ZVI, magnetic nanoparticles, and nanocomposites have been extensively discussed and critically analyzed. These nanomaterials exhibit a great potential as adsorbents for heavy metal ion removal from wastewater. However, there are still some notable shortcomings associated with these NM that need to be addressed in order to make them suitable for wastewater treatment at the industrial scale. For instance, the natural agglomeration of CNT results in a lower specific surface for adsorption, which reduces their adsorption capacities. Moreover, when it comes to their applicability at an industrial scale, their attachment to the filters should be strong enough to prevent their release in the treated water, as their toxicity remains debatable. Graphene-based nanomaterials face several limitations at the industrial scale. The large-scale production of GO nanoparticles requires a suitable reaction media. It is a chemically intensive process, which generates toxic by-products and can lead to serious environmental implications. CNT and graphene are also subject to clogging, similarly to activated carbon. These downsides of graphene-based adsorbents limit their large-scale implementation. Moreover, in the case of water treatment, it is difficult to separate them from the aqueous solution swiftly and efficiently due to their size. For this reason, the concept of nanocomposites seems to be a promising solution to solve the shortcomings of freestanding nanoscale adsorbents. Amongst these nanocomposites, magnetic nanocomposites possess some advantages in terms of their easy separation from treated waters using an external magnetic field, which shows potential in industrial applications.

4.2. Challenges and Opportunities for Large Scale Implementation of BESs for Heavy Metals Removal

Over the past few years, bioelectrochemical systems have emerged as a promising technology for heavy metal ion removal from contaminated water. However, their utilization at an industrial scale is still quite challenging due to their slow extraction rate and insufficient knowledge about the electron transfer mechanisms. The major challenge for implementing MFCs at larger scales is their inability to treat all type of metallic ions. The MFC setup can be used to reduce some of the heavy metals, i.e., Cu, Co, Hg, Cr, and Ag. It should be considered that only those metals which have a positive reduction potential can be spontaneously removed without any external applied voltage. For instance, Ni^{2+} has a reduction potential of -0.25 V and thus, cannot be reduced in MFC mode; therefore, it was investigated in MEC mode. The experimental results showed 33% and 9% removal efficiency of Ni^{2+} ions in MEC and MFC modes, respectively [171]. Interestingly, to overcome such issues, an innovative reactor configuration consisting of Cr-MFC and Cd-MFC in series was investigated for the removal of Cd ions with an achievable maximum output of 22.5 W/m^2 [172]. The MFCs in series enhance the power production and reduction of Cd ions. Another challenge that has been holding back the application of MFCs for the removal/recovery of heavy metals is the deterioration of performances and removal efficiencies based on the reactor configuration [173]. Both single-chambered MFC (SCMFC) and double-chambered MFC (DCMFC) are the most used designs for the removal of heavy metals. SCMFC has some notable advantages in terms of easy operation, cost effectiveness, and direct utilization of the substrate; but the lower coloumbic efficiency contributes to the major drawback of this design [174,175]. In fact, the presence of membranes can maintain the ion balance and reduce the oxygen diffusion, which improves the coloumbic efficiency. However, the pH imbalance due to this membrane leads to potential losses and thus results in the deterioration of performances and removal efficiencies. Finally, besides the pH imbalance, the most important downside of MFCs is possible membrane fouling. This phenomenon occurs in the long-term operation of MFCs because of the unavoidable growth of biofilm on and outside the membrane which directly affects the overall power and current outputs of MFCs. Additionally, the separation membrane is expensive and contributes to about 38% of the total capital cost of MFCs [176], which is an obstacle, especially for scaling-up and future practical applications of MFCs. Owing to the interconnection between different components including anode, cathode, biofilm formation and power generation of MFC, the scaling-up of MFCs should not be limited to the augmentation of the anode volume. Additionally, appropriate amplification of electrodes, membranes and catalyst coatings should be analyzed.

4.3. Future Outlook for Nanomaterial Assisted BESs

Based on the studies reviewed, bioelectrochemical systems appear to be a promising technology for the removal of heavy metal ions. However, nanomaterial assisted bioelectrochemical-based hybrid technology shows an improved efficiency in terms of enhancing output power and accelerating the metal reduction reaction at the cathode. The nanomaterial-modified cathode manifested great potential for HM remediation, but the regeneration of the electrodes is a potential hindrance for the upscaling of this hybrid technology. Therefore, more extensive research on suitable recyclable electrode materials is required. Magnetic nanoadsorbants, owing to their facile recovery from treated water, are currently studied more particularly. They demonstrate potential for the selective removal of single component systems as well as multicomponent systems. However, despite promising pollutant removal capacities, the overall cost-benefit ratio of magnetic nanoparticles is very high, which is a significant barrier to utilization as main mean. To improve the cost effectiveness, it will be necessary to combine these magnetic nanoparticles with existing technologies. Based on the critical review on the various nanomaterials and nanomaterials based bioelectrochemical removal of HM, magnetic nanoparticles based cathodes can be further investigated for the treatment of metal-contaminated water. This can be achieved in two ways: (1) directly adding the MNP into the catholyte solution, which are easy to separate using an external magnetic field; and (2) coating the cathode with MNP, which may enhance the reduction efficiency on one hand but on the other hand, impede the recycling of MNP. However, this combined technology can be advantageous in terms of efficiency, cost-effectiveness, sustainability, and environmental friendliness. An additional advantage is the possibility to harvest energy during the removal process.

Author Contributions: R.K. has conceptualized and written the manuscript. E.R. and P.R. have extensively reviewed the manuscript and have contributed significantly to it. All authors have read and agreed to the published version of the manuscript.

Funding: This research has been supported by the European Regional Development Fund project grant number TK134 "Emerging orders in quantum and nanomaterials" EQUiTANT, EMÜ Astra project EMBio "Value-chain based bio-economy", and Eesti Maaülikool (EMÜ) Bridge Funding grant number P200030TIBT.

Institutional Review Board Statement: Not applicable.

Informed Consent Statement: Not applicable.

Data Availability Statement: Not applicable.

Conflicts of Interest: The authors declare no conflict of interest.

References

1. Rogowska, J.; Cieszynska-Semenowicz, M.; Ratajczyk, W.; Wolska, L. Micropollutants in treated wastewater. *Ambio* **2020**, *49*, 487–503. [CrossRef] [PubMed]

2. Ali, H.; Khan, E.; Ilahi, I. Environmental Chemistry and Ecotoxicology of Hazardous Heavy Metals: Environmental Persistence, Toxicity, and Bioaccumulation. *J. Chem.* **2019**, *2019*, 6730305. [CrossRef]
3. Gautam, P.K.; Gautam, R.; Banerjee, S.; Chattopadhyaya, M.; Pandey, J. Heavy metals in the environment: Fate, transport, toxicity and remediation technologies. In *Heavy Metals*; Nava Science Publishers: Plzen, Czech Republic, 2016; pp. 101–130.
4. Vardhan, K.H.; Kumar, P.S.; Panda, R.C. A review on heavy metal pollution, toxicity and remedial measures: Current trends and future perspectives. *J. Mol. Liq.* **2019**, *290*, 111197. [CrossRef]
5. Sharma, R.; Agrawal, M. Biological effects of heavy metals: An overview. *J. Environ. Biol. Acad. Environ. Biol. India* **2005**, *26*, 301–313.
6. Tchounwou, P.B.; Yedjou, C.G.; Patlolla, A.K.; Sutton, D.J. Heavy metal toxicity and the environment. *Exp. Suppl.* **2012**, *101*, 133–164. [CrossRef] [PubMed]
7. Dominguez-Benetton, X.; Varia, J.C.; Pozo, G.; Modin, O.; Ter Heijne, A.; Fransaer, J.; Rabaey, K. Metal recovery by microbial electro-metallurgy. *Prog. Mater. Sci.* **2018**, *94*, 435–461. [CrossRef]
8. Lesmana, S.O.; Febriana, N.; Soetaredjo, F.E.; Sunarso, J.; Ismadji, S. Studies on potential applications of biomass for the separation of heavy metals from water and wastewater. *Biochem. Eng. J.* **2009**, *44*, 19–41. [CrossRef]
9. Morais, S.; Costa, F.; Pereira, M. Heavy Metals and Human Health. *Environ. Health Emerg. Issues Pract.* **2012**, *10*, 227–245.
10. Tokar, E.J.; Benbrahim-Tallaa, L.; Waalkes, M.P. Metal ions in human cancer development. *Met. Ions Life Sci* **2011**, *8*, 375–401. [PubMed]
11. Jomova, K.; Valko, M. Advances in metal-induced oxidative stress and human disease. *Toxicology* **2011**, *283*, 65–87. [CrossRef]
12. Cruz-Olivares, J.; Martínez-Barrera, G.; Pérez-Alonso, C.; Barrera-Díaz, C.E.; Chaparro-Mercado, M.d.C.; Ureña-Núñez, F. Adsorption of Lead Ions from Aqueous Solutions Using Gamma Irradiated Minerals. *J. Chem.* **2016**, *2016*, 8782469. [CrossRef]
13. Khulbe, K.C.; Matsuura, T. Removal of heavy metals and pollutants by membrane adsorption techniques. *Appl. Water Sci.* **2018**, *8*, 19. [CrossRef]
14. Lee, I.H.; Kuan, Y.-C.; Chern, J.-M. Equilibrium and kinetics of heavy metal ion exchange. *J. Chin. Inst. Chem. Eng.* **2007**, *38*, 71–84. [CrossRef]
15. Burakov, A.; Galunin, E.; Burakova, I.; Memetova, A.; Agarwal, S.; Tkachev, A.; Gupta, V. Adsorption of heavy metals on conventional and nanostructured materials for wastewater treatment purposes: A review. *Ecotoxicol. Environ. Saf.* **2017**, *148*, 702–712. [CrossRef] [PubMed]
16. Abdullah, N.; Yusof, N.; Lau, W.J.; Jaafar, J.; Ismail, A. Recent trends of heavy metal removal from water/wastewater by membrane technologies. *J. Ind. Eng. Chem.* **2019**, *76*. [CrossRef]
17. Kumar, V.; Shahi, S.; Singh, S. Bioremediation: An eco-sustainable approach for restoration of contaminated sites. In *Microbial Bioprospecting for Sustainable Development*; Springer: Berlin/Heidelberg, Germany, 2018; pp. 115–136.
18. Donati, E.R.; Sani, R.K.; Goh, K.M.; Chan, K.-G. Editorial: Recent Advances in Bioremediation/Biodegradation by Extreme Microorganisms. *Front. Microbiol.* **2019**, *10*. [CrossRef]
19. Hlihor, R.M.; Gavrilescu, M.; Tavares, T.; Favier, L.; Olivieri, G. Bioremediation: An Overview on Current Practices, Advances, and New Perspectives in Environmental Pollution Treatment. *Biomed. Res. Int.* **2017**, *2017*, 6327610. [CrossRef] [PubMed]
20. Jin, W.; Zhang, Y. Sustainable Electrochemical Extraction of Metal Resources from Waste Streams: From Removal to Recovery. *ACS Sustain. Chem. Eng.* **2020**, *8*, 4693–4707. [CrossRef]
21. Jin, W.; Hu, M.; Hu, J. Selective and Efficient Electrochemical Recovery of Dilute Copper and Tellurium from Acidic Chloride Solutions. *ACS Sustain. Chem. Eng.* **2018**, *6*, 13378–13384. [CrossRef]
22. Karbasi, M.; Alamdari, E.K.; Dehkordi, E.A. Electrochemical performance of PbCo composite anode during Zinc electrowinning. *Hydrometallurgy* **2019**, *183*, 51–59. [CrossRef]
23. Tahoon, M.A.; Siddeeg, S.M.; Salem Alsaiari, N.; Mnif, W.; Ben Rebah, F. Effective Heavy Metals Removal from Water Using Nanomaterials: A Review. *Processes* **2020**, *8*, 645. [CrossRef]
24. Babel, S.; Kurniawan, T.A. Low-cost adsorbents for heavy metals uptake from contaminated water: A review. *J. Hazard. Mater.* **2003**, *97*, 219–243. [CrossRef]
25. Yan, H.; Li, H.; Tao, X.; Li, K.; Yang, H.; Li, A.; Xiao, S.; Cheng, R. Rapid Removal and Separation of Iron(II) and Manganese(II) from Micropolluted Water Using Magnetic Graphene Oxide. *ACS Appl. Mater. Interfaces* **2014**, *6*, 9871–9880. [CrossRef] [PubMed]
26. Marsh, H.; Rodríguez-Reinoso, F. Chapter 8—Applicability of Activated Carbon. In *Activated Carbon*; Marsh, H., Rodríguez-Reinoso, F., Eds.; Elsevier Science Ltd.: Oxford, UK, 2006; pp. 383–453.
27. Saleem, J.; Shahid, U.B.; Hijab, M.; Mackey, H.; McKay, G. Production and applications of activated carbons as adsorbents from olive stones. *Biomass Convers. Biorefin.* **2019**, *9*, 775–802. [CrossRef]
28. Sarma, G.K.; Sen Gupta, S.; Bhattacharyya, K.G. Nanomaterials as versatile adsorbents for heavy metal ions in water: A review. *Environ. Sci. Pollut. Res.* **2019**, *26*, 6245–6278. [CrossRef] [PubMed]
29. Theron, J.; Walker, J.A.; Cloete, T.E. Nanotechnology and Water Treatment: Applications and Emerging Opportunities. *Crit. Rev. Microbiol.* **2008**, *34*, 43–69. [CrossRef] [PubMed]
30. Yaqoob, A.A.; Parveen, T.; Umar, K.; Mohamad Ibrahim, M.N. Role of Nanomaterials in the Treatment of Wastewater: A Review. *Water* **2020**, *12*, 495. [CrossRef]

31. Parvin, F.; Rikta, S.Y.; Tareq, S.M. 8—Application of nanomaterials for the removal of heavy metal from wastewater. In *Nanotechnology in Water and Wastewater Treatment*; Ahsan, A., Ismail, A.F., Eds.; Elsevier: Amsterdam, The Netherlands, 2019; pp. 137–157.
32. Yang, J.; Hou, B.; Wang, J.; Tian, B.; Bi, J.; Wang, N.; Li, X.; Huang, X. Nanomaterials for the Removal of Heavy Metals from Wastewater. *Nanomaterials* **2019**, *9*, 424. [CrossRef] [PubMed]
33. Lei, T.; Li, S.-J.; Jiang, F.; Ren, Z.-X.; Wang, L.-L.; Yang, X.-J.; Tang, L.-H.; Wang, S.-X. Adsorption of Cadmium Ions from an Aqueous Solution on a Highly Stable Dopamine-Modified Magnetic Nano-Adsorbent. *Nanoscale Res. Lett* **2019**, *14*, 352. [CrossRef] [PubMed]
34. Kumar, P.; Kumar, P. Removal of cadmium (Cd-II) from aqueous solution using gas industry-based adsorbent. *SN Appl. Sci.* **2019**, *1*, 365. [CrossRef]
35. Pyrzyńska, K.; Bystrzejewski, M. Comparative study of heavy metal ions sorption onto activated carbon, carbon nanotubes, and carbon-encapsulated magnetic nanoparticles. *Colloids Surf. A Physicochem. Eng. Asp.* **2010**, *362*, 102–109. [CrossRef]
36. El-sayed, M.E.A. Nanoadsorbents for water and wastewater remediation. *Sci. Total Environ.* **2020**, *739*, 139903. [CrossRef] [PubMed]
37. Smith, S.C.; Rodrigues, D.F. Carbon-based nanomaterials for removal of chemical and biological contaminants from water: A review of mechanisms and applications. *Carbon* **2015**, *91*, 122–143. [CrossRef]
38. Menezes, B.R.C.d.; Rodrigues, K.F.; Fonseca, B.C.d.S.; Ribas, R.G.; Montanheiro, T.L.d.A.; Thim, G.P. Recent advances in the use of carbon nanotubes as smart biomaterials. *J. Mater. Chem. B* **2019**, *7*, 1343–1360. [CrossRef] [PubMed]
39. Mujawar, M.; Sahu, J.; Abdullah, E.; Natesan, J. Removal of Heavy Metals from Wastewater Using Carbon Nanotubes. *Sep. Purif. Rev.* **2014**, *43*, 311–338. [CrossRef]
40. Bassyouni, M.; Mansi, A.E.; Elgabry, A.; Ibrahim, B.A.; Kassem, O.A.; Alhebeshy, R. Utilization of carbon nanotubes in removal of heavy metals from wastewater: A review of the CNTs' potential and current challenges. *Appl. Phys. A* **2019**, *126*, 38. [CrossRef]
41. Yu, G.; Lu, Y.; Guo, J.; Patel, M.; Bafana, A.; Wang, X.; Qiu, B.; Jeffryes, C.; Wei, S.; Guo, Z.; et al. Carbon nanotubes, graphene, and their derivatives for heavy metal removal. *Adv. Compos. Hybrid. Mater.* **2018**, *1*, 56–78. [CrossRef]
42. Arora, B.; Attri, P. Carbon Nanotubes (CNTs): A Potential Nanomaterial for Water Purification. *J. Compos. Sci.* **2020**, *4*, 135. [CrossRef]
43. Zhang, X.; Cui, H.; Gui, Y.; Tang, J. Mechanism and Application of Carbon Nanotube Sensors in SF(6) Decomposed Production Detection: A Review. *Nanoscale Res. Lett* **2017**, *12*, 177. [CrossRef]
44. Alijani, H.; Shariatinia, Z. Synthesis of high growth rate SWCNTs and their magnetite cobalt sulfide nanohybrid as super-adsorbent for mercury removal. *Chem. Eng. Res. Des.* **2018**, *129*, 132–149. [CrossRef]
45. Gupta, A.; Vidyarthi, S.R.; Sankararamakrishnan, N. Enhanced sorption of mercury from compact fluorescent bulbs and contaminated water streams using functionalized multiwalled carbon nanotubes. *J. Hazard. Mater.* **2014**, *274*, 132–144. [CrossRef] [PubMed]
46. Gupta, S.; Bhatiya, D.; Murthy, C.N. Metal Removal Studies by Composite Membrane of Polysulfone and Functionalized Single-Walled Carbon Nanotubes. *Sep. Sci. Technol.* **2015**, *50*, 421–429. [CrossRef]
47. Robati, D. Pseudo-second-order kinetic equations for modeling adsorption systems for removal of lead ions using multi-walled carbon nanotube. *J. Nanostruct. Chem.* **2013**, *3*, 55. [CrossRef]
48. Farghali, A.A.; Abdel Tawab, H.A.; Abdel Moaty, S.A.; Khaled, R. Functionalization of acidified multi-walled carbon nanotubes for removal of heavy metals in aqueous solutions. *J. Nanostruct. Chem.* **2017**, *7*, 101–111. [CrossRef]
49. Elsehly, E.M.I.; Chechenin, N.G.; Bukunov, K.A.; Makunin, A.V.; Priselkova, A.B.; Vorobyeva, E.A.; Motaweh, H.A. Removal of iron and manganese from aqueous solutions using carbon nanotube filters. *Water Supply* **2015**, *16*, 347–353. [CrossRef]
50. Mallakpour, S.; Khadem, E. 8—Carbon nanotubes for heavy metals removal. In *Composite Nanoadsorbents*; Kyzas, G.Z., Mitropoulos, A.C., Eds.; Elsevier: Amsterdam, The Netherlands, 2019; pp. 181–210.
51. Faur-Brasquet, C.; Kadirvelu, K.; Le Cloirec, P. Removal of metal ions from aqueous solution by adsorption onto activated carbon cloths: Adsorption competition with organic matter. *Carbon* **2002**, *40*, 2387–2392. [CrossRef]
52. Hur, J.; Shin, J.; Yoo, J.; Seo, Y.-S. Competitive Adsorption of Metals onto Magnetic Graphene Oxide: Comparison with Other Carbonaceous Adsorbents. *Sci. World J.* **2015**, *2015*, 836287. [CrossRef] [PubMed]
53. Borji, H.; Ayoub, G.M.; Bilbeisi, R.; Nassar, N.; Malaeb, L. How Effective Are Nanomaterials for the Removal of Heavy Metals from Water and Wastewater? *Water Air Soil Pollut.* **2020**, *231*, 330. [CrossRef]
54. Rodríguez, C.; Briano, S.; Leiva, E. Increased Adsorption of Heavy Metal Ions in Multi-Walled Carbon Nanotubes with Improved Dispersion Stability. *Molecules* **2020**, *25*, 3106. [CrossRef]
55. Amin, M.T.; Alazba, A.A.; Manzoor, U. A Review of Removal of Pollutants from Water/Wastewater Using Different Types of Nanomaterials. *Adv. Mater. Sci. Eng.* **2014**, *2014*, 825910. [CrossRef]
56. Chowdhury, S.; Balasubramanian, R. Recent advances in the use of graphene-family nanoadsorbents for removal of toxic pollutants from wastewater. *Adv. Colloid Interface Sci.* **2014**, *204*, 35–56. [CrossRef] [PubMed]
57. Ali, I.; Basheer, A.A.; Mbianda, X.Y.; Burakov, A.; Galunin, E.; Burakova, I.; Mkrtchyan, E.; Tkachev, A.; Grachev, V. Graphene based adsorbents for remediation of noxious pollutants from wastewater. *Environ. Int.* **2019**, *127*, 160–180. [CrossRef]

58. Woo, Y.C.; Kim, S.-H.; Shon, H.K.; Tijing, L.D. Introduction: Membrane Desalination Today, Past, and Future. In *Current Trends and Future Developments on (Bio-) Membranes*; Basile, A., Curcio, E., Inamuddin, Eds.; Elsevier: Amsterdam, The Netherlands, 2019; pp. xxv–xlvi.
59. Smith, A.T.; LaChance, A.M.; Zeng, S.; Liu, B.; Sun, L. Synthesis, properties, and applications of graphene oxide/reduced graphene oxide and their nanocomposites. *Nano Mater. Sci.* **2019**, *1*, 31–47. [CrossRef]
60. Azizighannad, S.; Mitra, S. Stepwise Reduction of Graphene Oxide (GO) and Its Effects on Chemical and Colloidal Properties. *Sci. Rep.* **2018**, *8*, 10083. [CrossRef] [PubMed]
61. Xu, L.; Wang, J. The application of graphene-based materials for the removal of heavy metals and radionuclides from water and wastewater. *Crit. Rev. Environ. Sci. Technol.* **2017**, *47*, 1042–1105. [CrossRef]
62. Zhao, G.; Li, J.; Ren, X.; Chen, C.; Wang, X. Few-Layered Graphene Oxide Nanosheets As Superior Sorbents for Heavy Metal Ion Pollution Management. *Environ. Sci. Technol.* **2011**, *45*, 10454–10462. [CrossRef] [PubMed]
63. Wang, H.; Yuan, X.; Wu, Y.; Huang, H.; Zeng, G.; Liu, Y.; Wang, X.; Lin, N.; Qi, Y. Adsorption characteristics and behaviors of graphene oxide for Zn(II) removal from aqueous solution. *Appl. Surf. Sci.* **2013**, *279*, 432–440. [CrossRef]
64. Guo, T.; Bulin, C.; Li, B.; Zhao, Z.; Yu, H.; Sun, H.; Ge, X.; Xing, R.; Zhang, B. Efficient removal of aqueous Pb(II) using partially reduced graphene oxide-Fe3O4. *Adsorpt. Sci. Technol.* **2017**, *36*, 1031–1048. [CrossRef]
65. Tabish, T.A.; Memon, F.A.; Gomez, D.E.; Horsell, D.W.; Zhang, S. A facile synthesis of porous graphene for efficient water and wastewater treatment. *Sci. Rep.* **2018**, *8*, 1817. [CrossRef]
66. Zhang, C.-Z.; Chen, B.; Bai, Y.; Xie, J. A new functionalized reduced graphene oxide adsorbent for removing heavy metal ions in water via coordination and ion exchange. *Sep. Sci. Technol.* **2018**, *53*, 2896–2905. [CrossRef]
67. Awad, F.S.; AbouZied, K.M.; Abou El-Maaty, W.M.; El-Wakil, A.M.; Samy El-Shall, M. Effective removal of mercury(II) from aqueous solutions by chemically modified graphene oxide nanosheets. *Arab. J. Chem.* **2020**, *13*, 2659–2670. [CrossRef]
68. Zheng, S.; Hao, L.; Zhang, L.; Wang, K.; Zheng, W.; Wang, X.; Zhou, X.; Li, W.; Zhang, L. Tea Polyphenols Functionalized and Reduced Graphene Oxide-ZnO Composites for Selective Pb(2+) Removal and Enhanced Antibacterial Activity. *J. Biomed. Nanotechnol.* **2018**, *14*, 1263–1276. [CrossRef]
69. Arshad, F.; Selvaraj, M.; Zain, J.; Banat, F.; Haija, M.A. Polyethylenimine modified graphene oxide hydrogel composite as an efficient adsorbent for heavy metal ions. *Sep. Purif. Technol.* **2019**, *209*, 870–880. [CrossRef]
70. Chauke, V.P.; Maity, A.; Chetty, A. High-performance towards removal of toxic hexavalent chromium from aqueous solution using graphene oxide-alpha cyclodextrin-polypyrrole nanocomposites. *J. Mol. Liq.* **2015**, *211*, 71–77. [CrossRef]
71. Hadi Najafabadi, H.; Irani, M.; Roshanfekr Rad, L.; Heydari Haratameh, A.; Haririan, I. Removal of Cu2+, Pb2+ and Cr6+ from aqueous solutions using a chitosan/graphene oxide composite nanofibrous adsorbent. *RSC Adv.* **2015**, *5*, 16532–16539. [CrossRef]
72. Mahmoud, M.E.; Fekry, N.A.; El-Latif, M.M.A. Nanocomposites of nanosilica immobilized-nanopolyaniline and crosslinked nanopolyaniline for removal of heavy metals. *Chem. Eng. J.* **2016**, *304*, 679–691. [CrossRef]
73. Nguyen, T.T.; Ma, H.T.; Avti, P.; Bashir, M.J.K.; Ng, C.A.; Wong, L.Y.; Jun, H.K.; Ngo, Q.M.; Tran, N.Q. Adsorptive Removal of Iron Using SiO_2 Nanoparticles Extracted from Rice Husk Ash. *J. Anal. Methods Chem.* **2019**, *2019*, 6210240. [CrossRef]
74. Najafi, M.; Yousefi, Y.; Rafati, A.A. Synthesis, characterization and adsorption studies of several heavy metal ions on amino-functionalized silica nano hollow sphere and silica gel. *Sep. Purif. Technol.* **2012**, *85*, 193–205. [CrossRef]
75. Kotsyuda, S.S.; Tomina, V.V.; Zub, Y.L.; Furtat, I.M.; Lebed, A.P.; Vaclavikova, M.; Melnyk, I.V. Bifunctional silica nanospheres with 3-aminopropyl and phenyl groups. Synthesis approach and prospects of their applications. *Appl. Surf. Sci.* **2017**, *420*, 782–791. [CrossRef]
76. Liu, A.M.; Hidajat, K.; Kawi, S.; Zhao, D.Y. A new class of hybrid mesoporous materials with functionalized organic monolayers for selective adsorption of heavy metal ions. *Chem. Commun.* **2000**, 1145–1146. [CrossRef]
77. Li, X.; Han, C.; Zhu, W.; Ma, W.; Luo, Y.; Zhou, Y.; Yu, J.; Wei, K. Cr(VI) Removal from Aqueous by Adsorption on Amine-Functionalized Mesoporous Silica Prepared from Silica Fume. *J. Chem.* **2014**, *2014*, 765856. [CrossRef]
78. Li, R.; Zhang, L.; Wang, P. Rational design of nanomaterials for water treatment. *Nanoscale* **2015**, *7*, 17167–17194. [CrossRef]
79. Huang, X.; Wang, L.; Chen, J.; Jiang, C.; Wu, S.; Wang, H. Effective removal of heavy metals with amino-functionalized silica gel in tea polyphenol extracts. *J. Food Meas. Charact.* **2020**, *14*, 2134–2144. [CrossRef]
80. Wang, H.; Kang, J.; Liu, H.; Qu, J. Preparation of organically functionalized silica gel as adsorbent for copper ion adsorption. *J. Environ. Sci.* **2009**, *21*, 1473–1479. [CrossRef]
81. Wieszczycka, K.; Filipowiak, K.; Wojciechowska, I.; Buchwald, T.; Siwińska-Ciesielczyk, K.; Strzemiecka, B.; Jesionowski, T.; Voelkel, A. Novel highly efficient ionic liquid-functionalized silica for toxic metals removal. *Sep. Purif. Technol.* **2021**, *265*, 118483. [CrossRef]
82. Crane, R.A.; Scott, T.B. Nanoscale zero-valent iron: Future prospects for an emerging water treatment technology. *J. Hazard. Mater.* **2012**, *211–212*, 112–125. [CrossRef] [PubMed]
83. Srinivasan, N.R.; Shankar, P.A.; Bandyopadhyaya, R. Plasma treated activated carbon impregnated with silver nanoparticles for improved antibacterial effect in water disinfection. *Carbon* **2013**, *57*, 1–10. [CrossRef]
84. O'Carroll, D.; Sleep, B.; Krol, M.; Boparai, H.; Kocur, C. Nanoscale zero valent iron and bimetallic particles for contaminated site remediation. *Adv. Water Resour.* **2013**, *51*, 104–122. [CrossRef]
85. Phenrat, T.; Lowry, G.V.; Babakhani, P. Nanoscale zerovalent iron (NZVI) for environmental decontamination: A brief history of 20 years of research and field-scale application. In *Nanoscale Zerovalent Iron Particles for Environmental Restoration: From Fundamental*

Science to Field Scale Engineering Applications; Phenrat, T., Lowry, G.V., Eds.; Springer International Publishing: Cham, Switzerland, 2019; pp. 1–43.
86. Liu, T.; Wang, Z.-L.; Sun, Y. Manipulating the morphology of nanoscale zero-valent iron on pumice for removal of heavy metals from wastewater. *Chem. Eng. J.* **2015**, *263*, 55–61. [CrossRef]
87. Zhang, Z.; Hao, Z.W.; Liu, W.L.; Xu, X.H. Synchronous treatment of heavy metal ions and nitrate by zero-valent iron. *Huan Jing Ke Xue* **2009**, *30*, 775–779.
88. Seyedi, S.M.; Rabiee, H.; Shahabadi, S.M.S.; Borghei, S.M. Synthesis of Zero-Valent Iron Nanoparticles Via Electrical Wire Explosion for Efficient Removal of Heavy Metals. *CLEAN Soil Air Water* **2017**, *45*, 1600139. [CrossRef]
89. Guo, J.J.; Fan, M.D. Progress of Removing Heavy Metals by Zero Valent Iron Nanoparticles from Wastewater. *Adv. Mater. Res.* **2013**, *726–731*, 2563–2566. [CrossRef]
90. Sarathy, V.; Tratnyek, P.G.; Nurmi, J.T.; Baer, D.R.; Amonette, J.E.; Chun, C.L.; Penn, R.L.; Reardon, E.J. Aging of Iron Nanoparticles in Aqueous Solution: Effects on Structure and Reactivity. *J. Phys. Chem. C* **2008**, *112*, 2286–2293. [CrossRef]
91. Fu, F.; Dionysiou, D.D.; Liu, H. The use of zero-valent iron for groundwater remediation and wastewater treatment: A review. *J. Hazard. Mater.* **2014**, *267*, 194–205. [CrossRef]
92. Huang, D.-L.; Chen, G.-M.; Zeng, G.-M.; Xu, P.; Yan, M.; Lai, C.; Zhang, C.; Li, N.-J.; Cheng, M.; He, X.-X.; et al. Synthesis and Application of Modified Zero-Valent Iron Nanoparticles for Removal of Hexavalent Chromium from Wastewater. *Water Air Soil Pollut.* **2015**, *226*, 375. [CrossRef]
93. Su, Y.; Adeleye, A.S.; Huang, Y.; Sun, X.; Dai, C.; Zhou, X.; Zhang, Y.; Keller, A.A. Simultaneous removal of cadmium and nitrate in aqueous media by nanoscale zerovalent iron (nZVI) and Au doped nZVI particles. *Water Res.* **2014**, *63*, 102–111. [CrossRef]
94. Tamjidi, S.; Esmaeili, H.; Kamyab Moghadas, B. Application of magnetic adsorbents for removal of heavy metals from wastewater: A review study. *Mater. Res. Express* **2019**, *6*, 102004. [CrossRef]
95. Vojoudi, H.; Badiei, A.; Bahar, S.; Mohammadi Ziarani, G.; Faridbod, F.; Ganjali, M.R. A new nano-sorbent for fast and efficient removal of heavy metals from aqueous solutions based on modification of magnetic mesoporous silica nanospheres. *J. Magn. Magn. Mater.* **2017**, *441*, 193–203. [CrossRef]
96. Etale, A.; Tutu, H.; Drake, D.C. The effect of silica and maghemite nanoparticles on remediation of Cu(II)-, Mn(II)- and U(VI)-contaminated water by *Acutodesmus* sp. *J. Appl. Phycol.* **2016**, *28*, 251–260. [CrossRef]
97. Tuutijärvi, T.; Lu, J.; Sillanpää, M.; Chen, G. As(V) adsorption on maghemite nanoparticles. *J. Hazard. Mater.* **2009**, *166*, 1415–1420. [CrossRef] [PubMed]
98. Akhbarizadeh, R.; Shayestefar, M.R.; Darezereshki, E. Competitive Removal of Metals from Wastewater by Maghemite Nanoparticles: A Comparison Between Simulated Wastewater and AMD. *Mine Water Environ.* **2014**, *33*, 89–96. [CrossRef]
99. Madrakian, T.; Afkhami, A.; Zadpour, B.; Ahmadi, M. New synthetic mercaptoethylamino homopolymer-modified maghemite nanoparticles for effective removal of some heavy metal ions from aqueous solution. *J. Ind. Eng. Chem.* **2015**, *21*, 1160–1166. [CrossRef]
100. Tzabar, N.; ter Brake, H.J.M. Adsorption isotherms and Sips models of nitrogen, methane, ethane, and propane on commercial activated carbons and polyvinylidene chloride. *Adsorption* **2016**, *22*, 901–914. [CrossRef]
101. Fato, F.P.; Li, D.-W.; Zhao, L.-J.; Qiu, K.; Long, Y.-T. Simultaneous Removal of Multiple Heavy Metal Ions from River Water Using Ultrafine Mesoporous Magnetite Nanoparticles. *ACS Omega* **2019**, *4*, 7543–7549. [CrossRef]
102. Shipley, H.J.; Engates, K.E.; Grover, V.A. Removal of Pb(II), Cd(II), Cu(II), and Zn(II) by hematite nanoparticles: Effect of sorbent concentration, pH, temperature, and exhaustion. *Environ. Sci. Pollut. Res.* **2013**, *20*, 1727–1736. [CrossRef]
103. Giraldo, L.; Erto, A.; Moreno-Piraján, J. Magnetite nanoparticles for removal of heavy metals from aqueous solutions: Synthesis and characterization. *Adsorption* **2013**, *19*. [CrossRef]
104. Feng, L.; Cao, M.; Ma, X.; Zhu, Y.; Hu, C. Superparamagnetic high-surface-area Fe3O4 nanoparticles as adsorbents for arsenic removal. *J. Hazard. Mater.* **2012**, *217–218*, 439–446. [CrossRef]
105. Yu, X.; Tong, S.; Ge, M.; Zuo, J.; Cao, C.; Song, W. One-step synthesis of magnetic composites of cellulose@iron oxide nanoparticles for arsenic removal. *J. Mater. Chem. A* **2013**, *1*, 959–965. [CrossRef]
106. Ahmed, M.A.; Ali, S.M.; El-Dek, S.I.; Galal, A. Magnetite–hematite nanoparticles prepared by green methods for heavy metal ions removal from water. *Mater. Sci. Eng. B* **2013**, *178*, 744–751. [CrossRef]
107. Hu, J.; Chen, G.; Lo, I.M. Removal and recovery of Cr(VI) from wastewater by maghemite nanoparticles. *Water Res.* **2005**, *39*, 4528–4536. [CrossRef] [PubMed]
108. Watts, M.P.; Coker, V.S.; Parry, S.A.; Pattrick, R.A.; Thomas, R.A.; Kalin, R.; Lloyd, J.R. Biogenic nano-magnetite and nano-zero valent iron treatment of alkaline Cr(VI) leachate and chromite ore processing residue. *Appl. Geochem.* **2015**, *54*, 27–42. [CrossRef]
109. Shi, J.; Li, H.; Lu, H.; Zhao, X. Use of Carboxyl Functional Magnetite Nanoparticles as Potential Sorbents for the Removal of Heavy Metal Ions from Aqueous Solution. *J. Chem. Eng. Data* **2015**, *60*, 2035–2041. [CrossRef]
110. Tu, Y.-J.; You, C.-F.; Chang, C.-K. Kinetics and thermodynamics of adsorption for Cd on green manufactured nano-particles. *J. Hazard. Mater.* **2012**, *235–236*, 116–122. [CrossRef] [PubMed]
111. Panneerselvam, P.; Morad, N.; Tan, K.A. Magnetic nanoparticle (Fe3O4) impregnated onto tea waste for the removal of nickel(II) from aqueous solution. *J. Hazard. Mater.* **2011**, *186*, 160–168. [CrossRef] [PubMed]
112. Karvelas, E.; Liosis, C.; Benos, L.; Karakasidis, T.; Sarris, I. Micromixing Efficiency of Particles in Heavy Metal Removal Processes under Various Inlet Conditions. *Water* **2019**, *11*, 1135. [CrossRef]

113. Mourdikoudis, S.; Kostopoulou, A.; LaGrow, A.P. Magnetic Nanoparticle Composites: Synergistic Effects and Applications. *Adv. Sci.* **2021**, *8*, 2004951. [CrossRef] [PubMed]
114. Zhang, W.; Shi, X.; Zhang, Y.; Gu, W.; Li, B.; Xian, Y. Synthesis of water-soluble magnetic graphene nanocomposites for recyclable removal of heavy metal ions. *J. Mater. Chem. A* **2013**, *1*, 1745–1753. [CrossRef]
115. Elmi, F.; Hosseini, T.; Taleshi, M.S.; Taleshi, F. Kinetic and thermodynamic investigation into the lead adsorption process from wastewater through magnetic nanocomposite Fe3O4/CNT. *Nanotechnol. Environ. Eng.* **2017**, *2*, 13. [CrossRef]
116. Takafuji, M.; Ide, S.; Ihara, H.; Xu, Z. Preparation of Poly(1-vinylimidazole)-Grafted Magnetic Nanoparticles and Their Application for Removal of Metal Ions. *Chem. Mater.* **2004**, *16*, 1977–1983. [CrossRef]
117. Faghihian, H.; Moayed, M.; Firooz, A.; Iravani, M. Evaluation of a new magnetic zeolite composite for removal of Cs+ and Sr2+ from aqueous solutions: Kinetic, equilibrium and thermodynamic studies. *C. R. Chim.* **2014**, *17*, 108–117. [CrossRef]
118. Ge, L.; Wang, W.; Peng, Z.; Tan, F.; Wang, X.; Chen, J.; Qiao, X. Facile fabrication of Fe@MgO magnetic nanocomposites for efficient removal of heavy metal ion and dye from water. *Powder Technol.* **2018**, *326*, 393–401. [CrossRef]
119. Samrot, A.V.; Sahithya, C.S.; Selvarani, A.J.; Pachiyappan, S.; Kumar, S.S. Surface-Engineered Super-Paramagnetic Iron Oxide Nanoparticles For Chromium Removal. *Int. J. Nanomed.* **2019**, *14*, 8105–8119. [CrossRef]
120. Rauwel, P.; Uhl, W.; Rauwel, E. Editorial for the Special Issue on 'Application and Behavior of Nanomaterials in Water Treatment'. *Nanomaterials* **2019**, *9*, 880. [CrossRef] [PubMed]
121. Zhang, H.; Hodges, C.S.; Mishra, P.K.; Yoon, J.Y.; Hunter, T.N.; Lee, J.W.; Harbottle, D. Bio-Inspired Preparation of Clay–Hexacyanoferrate Composite Hydrogels as Super Adsorbents for Cs+. *ACS Appl. Mater. Interfaces* **2020**, *12*, 33173–33185. [CrossRef]
122. Rauwel, P.; Rauwel, E. Towards the Extraction of Radioactive Cesium-137 from Water via Graphene/CNT and Nanostructured Prussian Blue Hybrid Nanocomposites: A Review. *Nanomaterials* **2019**, *9*, 682. [CrossRef] [PubMed]
123. Perumal, S.; Atchudan, R.; Edison, T.N.J.I.; Babu, R.S.; Karpagavinayagam, P.; Vedhi, C. A Short Review on Recent Advances of Hydrogel-Based Adsorbents for Heavy Metal Ions. *Metals* **2021**, *11*, 864. [CrossRef]
124. Nicola, R.; Costişor, O.; Ciopec, M.; Negrea, A.; Lazău, R.; Ianăşi, C.; Picioruş, E.-M.; Len, A.; Almásy, L.; Szerb, E.I.; et al. Silica-Coated Magnetic Nanocomposites for Pb2+ Removal from Aqueous Solution. *Appl. Sci.* **2020**, *10*, 2726. [CrossRef]
125. Ahmad, N.; Sereshti, H.; Mousazadeh, M.; Rashidi Nodeh, H.; Kamboh, M.A.; Mohamad, S. New magnetic silica-based hybrid organic-inorganic nanocomposite for the removal of lead(II) and nickel(II) ions from aqueous solutions. *Mater. Chem. Phys.* **2019**, *226*, 73–81. [CrossRef]
126. Culita, D.C.; Simonescu, C.M.; Patescu, R.-E.; Dragne, M.; Stanica, N.; Oprea, O. o-Vanillin functionalized mesoporous silica—Coated magnetite nanoparticles for efficient removal of Pb(II) from water. *J. Solid State Chem.* **2016**, *238*, 311–320. [CrossRef]
127. Hu, H.; Wang, Z.; Pan, L. Synthesis of monodisperse Fe3O4@silica core–shell microspheres and their application for removal of heavy metal ions from water. *J. Alloy. Compd.* **2010**, *492*, 656–661. [CrossRef]
128. Morsi, R.E.; Al-Sabagh, A.M.; Moustafa, Y.M.; ElKholy, S.G.; Sayed, M.S. Polythiophene modified chitosan/magnetite nanocomposites for heavy metals and selective mercury removal. *Egypt. J. Pet.* **2018**, *27*, 1077–1085. [CrossRef]
129. Suleiman, J.S.; Hu, B.; Peng, H.; Huang, C. Separation/preconcentration of trace amounts of Cr, Cu and Pb in environmental samples by magnetic solid-phase extraction with Bismuthiol-II-immobilized magnetic nanoparticles and their determination by ICP-OES. *Talanta* **2009**, *77*, 1579–1583. [CrossRef]
130. Zhang, W.; An, Y.; Li, S.; Liu, Z.; Chen, Z.; Ren, Y.; Wang, S.; Zhang, X.; Wang, X. Enhanced heavy metal removal from an aqueous environment using an eco-friendly and sustainable adsorbent. *Sci. Rep.* **2020**, *10*, 16453. [CrossRef] [PubMed]
131. Liu, X.; Hu, Q.; Fang, Z.; Zhang, X.; Zhang, B. Magnetic Chitosan Nanocomposites: A Useful Recyclable Tool for Heavy Metal Ion Removal. *Langmuir* **2009**, *25*, 3–8. [CrossRef] [PubMed]
132. Chen, D.; Awut, T.; Liu, B.; Ma, Y.; Wang, T.; Nurulla, I. Functionalized magnetic Fe3O4 nanoparticles for removal of heavy metal ions from aqueous solutions. *e-Polymers* **2016**, *16*, 313–322. [CrossRef]
133. Huang, Z.-n.; Wang, X.-l.; Yang, D.-s. Adsorption of Cr(VI) in wastewater using magnetic multi-wall carbon nanotubes. *Water Sci. Eng.* **2015**, *8*, 226–232. [CrossRef]
134. Rivera, F.L.; Palomares, F.J.; Herrasti, P.; Mazario, E. Improvement in Heavy Metal Removal from Wastewater Using an External Magnetic Inductor. *Nanomaterials* **2019**, *9*, 1508. [CrossRef]
135. Tao, S.; Wang, C.; Ma, W.; Wu, S.; Meng, C. Designed multifunctionalized magnetic mesoporous microsphere for sequential sorption of organic and inorganic pollutants. *Microporous Mesoporous Mater.* **2012**, *147*, 295–301. [CrossRef]
136. Zheng, T.; Li, J.; Ji, Y.; Zhang, W.; Fang, Y.; Xin, F.; Dong, W.; Wei, P.; Ma, J.; Jiang, M. Progress and Prospects of Bioelectrochemical Systems: Electron Transfer and Its Applications in the Microbial Metabolism. *Front. Bioeng. Biotechnol.* **2020**, *8*, 10. [CrossRef]
137. Bajracharya, S.; Sharma, M.; Mohanakrishna, G.; Dominguez Benneton, X.; Strik, D.P.B.T.B.; Sarma, P.M.; Pant, D. An overview on emerging bioelectrochemical systems (BESs): Technology for sustainable electricity, waste remediation, resource recovery, chemical production and beyond. *Renew. Energy* **2016**, *98*, 153–170. [CrossRef]
138. Sukrampal; Kumar, R.; Patil, S.A. 3—Removal of heavy metals using bioelectrochemical systems. In *Integrated Microbial Fuel Cells for Wastewater Treatment*; Abbassi, R., Yadav, A.K., Khan, F., Garaniya, V., Eds.; Butterworth-Heinemann: Oxford, UK, 2020; pp. 49–71.
139. Rabaey, K.; Lissens, G.; Siciliano, S.D.; Verstraete, W. A microbial fuel cell capable of converting glucose to electricity at high rate and efficiency. *Biotechnol. Lett.* **2003**, *25*, 1531–1535. [CrossRef]

140. Kumar, R.; Yadav, S.; Patil, S.A. Bioanode-Assisted Removal of Hg^{2+} at the Cathode of Microbial Fuel Cells. *J. Hazard. Toxic Radioact. Waste* **2020**, *24*, 04020034. [CrossRef]
141. Zhang, Y.; Sun, J.; Hu, Y.; Li, S.; Xu, Q. Bio-cathode materials evaluation in microbial fuel cells: A comparison of graphite felt, carbon paper and stainless steel mesh materials. *Int. J. Hydrogen Energy* **2012**, *37*, 16935–16942. [CrossRef]
142. Yang, Q.; Liang, S.; Liu, J.; Lv, J.; Feng, Y. Analysis of Anodes of Microbial Fuel Cells When Carbon Brushes Are Preheated at Different Temperatures. *Catalysts* **2017**, *7*, 312. [CrossRef]
143. He, Y.-R.; Xiao, X.; Li, W.-W.; Sheng, G.-P.; Yan, F.-F.; Yu, H.-Q.; Yuan, H.; Wu, L.-J. Enhanced electricity production from microbial fuel cells with plasma-modified carbon paper anode. *Phys. Chem. Chem. Phys.* **2012**, *14*, 9966–9971. [CrossRef] [PubMed]
144. Ezziat, L.; Elabed, A.; Ibnsouda, S.; El Abed, S. Challenges of Microbial Fuel Cell Architecture on Heavy Metal Recovery and Removal From Wastewater. *Front. Energy Res.* **2019**, *7*. [CrossRef]
145. Logan, B.; Cheng, S.; Watson, V.; Estadt, G. Graphite Fiber Brush Anodes for Increased Power Production in Air-Cathode Microbial Fuel Cells. *Environ. Sci. Technol.* **2007**, *41*, 3341–3346. [CrossRef]
146. Fan, Y.; Liu, H. *Materials for Low-Temperature Fuel Cells*; Wiley: Hoboken, NJ, USA, 2014; pp. 145–166.
147. Savla, N.; Anand, R.; Pandit, S.; Prasad, R. Utilization of Nanomaterials as Anode Modifiers for Improving Microbial Fuel Cells Performance. *J. Renew. Mater.* **2020**, *8*, 1581–1605. [CrossRef]
148. Nakamura, R.; Kai, F.; Okamoto, A.; Newton, G.J.; Hashimoto, K. Self-Constructed Electrically Conductive Bacterial Networks. *Angew. Chem. Int. Ed.* **2009**, *48*, 508–511. [CrossRef]
149. Zhao, C.; Wang, Y.; Shi, F.; Zhang, J.; Zhu, J.-J. High biocurrent generation in Shewanella-inoculated microbial fuel cells using ionic liquid functionalized graphene nanosheets as an anode. *Chem. Commun.* **2013**, *49*, 6668–6670. [CrossRef] [PubMed]
150. Zhao, C.; Gai, P.; Liu, C.; Wang, X.; Xu, H.; Zhang, J.; Zhu, J.-J. Polyaniline networks grown on graphene nanoribbonscoated carbon paper with a synergistic effect for highperformance microbial fuel cells. *J. Mater. Chem. A.* **2013**, *1*, 12587. [CrossRef]
151. Hou, J.; Liu, Z.; Zhang, P. A new method for fabrication of graphene/polyaniline nanocomplex modified microbial fuel cell anodes. *J. Power Sources* **2013**, *224*, 139–144. [CrossRef]
152. Dessie, Y.; Tadesse, S.; Eswaramoorthy, R.; Adimasu, Y. Biosynthesized α-MnO2-based polyaniline binary composite as efficient bioanode catalyst for high-performance microbial fuel cell. *All Life* **2021**, *14*, 541–568. [CrossRef]
153. Mehdinia, A.; Ziaei, E.; Jabbari, A. Facile microwave-assisted synthesized reduced graphene oxide/tin oxide nanocomposite and using as anode material of microbial fuel cell to improve power generation. *Int. J. Hydrogen Energy* **2014**, *39*, 10724–10730. [CrossRef]
154. Velvizhi, G.; Venkata Mohan, S. Biocatalyst behavior under self-induced electrogenic microenvironment in comparison with anaerobic treatment: Evaluation with pharmaceutical wastewater for multi-pollutant removal. *Bioresour. Technol.* **2011**, *102*, 10784–10793. [CrossRef] [PubMed]
155. Tao, C.-H.; Liang, M.; Li, W.; Zhang, J.-L.; Ni, R.-J.; Wu, M.-W. Removal of copper from aqueous solution by electrodeposition in cathode chamber of microbial fuel cell. *J. Hazard. Mater.* **2011**, *189*, 186–192. [CrossRef]
156. Huang, L.; Chen, J.; Quan, X.; Yang, F. Enhancement of hexavalent chromium reduction and electricity production from a biocathode microbial fuel cell. *Bioprocess. Biosyst. Eng.* **2010**, *33*, 937–945. [CrossRef] [PubMed]
157. Zhang, B.; Feng, C.; Ni, J.; Zhang, J.; Huang, W. Simultaneous reduction of vanadium (V) and chromium (VI) with enhanced energy recovery based on microbial fuel cell technology. *J. Power Sources* **2012**, *204*, 34–39. [CrossRef]
158. Wang, G.; Huang, L.; Zhang, Y. Cathodic reduction of hexavalent chromium [Cr(VI)] coupled with electricity generation in microbial fuel cells. *Biotechnol. Lett.* **2008**, *30*, 1959. [CrossRef] [PubMed]
159. Wang, Z.; Lim, B.; Choi, C. Removal of Hg2+ as an electron acceptor coupled with power generation using a microbial fuel cell. *Bioresour. Technol.* **2011**, *102*, 6304–6307. [CrossRef]
160. Rodenas Motos, P.; ter Heijne, A.; van der Weijden, R.; Saakes, M.; Buisman, C.J.N.; Sleutels, T.H.J.A. High rate copper and energy recovery in microbial fuel cells. *Front. Microbiol.* **2015**, *6*, 527. [CrossRef] [PubMed]
161. Choi, C.; Cui, Y. Recovery of silver from wastewater coupled with power generation using a microbial fuel cell. *Bioresour. Technol.* **2012**, *107*, 522–525. [CrossRef] [PubMed]
162. Catal, T.; Bermek, H.; Liu, H. Removal of selenite from wastewater using microbial fuel cells. *Biotechnol. Lett.* **2009**, *31*, 1211–1216. [CrossRef] [PubMed]
163. Huang, L.; Li, T.; Liu, C.; Quan, X.; Chen, L.; Wang, A.; Chen, G. Synergetic interactions improve cobalt leaching from lithium cobalt oxide in microbial fuel cells. *Bioresour. Technol.* **2013**, *128*, 539–546. [CrossRef] [PubMed]
164. Lim, B.S.; Lu, H.; Choi, C.; Liu, Z.X. Recovery of silver metal and electric power generation using a microbial fuel cell. *Desalination Water Treat.* **2015**, *54*, 3675–3681. [CrossRef]
165. Gupta, S.; Yadav, A.; Verma, N. Simultaneous Cr(VI) reduction and bioelectricity generation using microbial fuel cell based on alumina-nickel nanoparticles-dispersed carbon nanofiber electrode. *Chem. Eng. J.* **2017**, *307*, 729–738. [CrossRef]
166. Shi, J.; Zhao, W.; Liu, C.; Jiang, T.; Ding, H. Enhanced Performance for Treatment of Cr (VI)-Containing Wastewater by Microbial Fuel Cells with Natural Pyrrhotite-Coated Cathode. *Water* **2017**, *9*, 979. [CrossRef]
167. Wu, Y.; Wang, L.; Jin, M.; Kong, F.; Qi, H.; Nan, J. Reduced graphene oxide and biofilms as cathode catalysts to enhance energy and metal recovery in microbial fuel cell. *Bioresour. Technol.* **2019**, *283*, 129–137. [CrossRef] [PubMed]
168. Li, M.; Zhou, S. α-Fe2O3/polyaniline nanocomposites as an effective catalyst for improving the electrochemical performance of microbial fuel cell. *Chem. Eng. J.* **2018**, *339*, 539–546. [CrossRef]

169. Ali, J.; Wang, L.; Waseem, H.; Djellabi, R.; Oladoja, N.A.; Pan, G. FeS@rGO nanocomposites as electrocatalysts for enhanced chromium removal and clean energy generation by microbial fuel cell. *Chem. Eng. J.* **2020**, *384*, 123335. [CrossRef]
170. Xue, A.; Shen, Z.-Z.; Zhao, B.; Zhao, H.-Z. Arsenite removal from aqueous solution by a microbial fuel cell–zerovalent iron hybrid process. *J. Hazard. Mater.* **2013**, *261*, 621–627. [CrossRef] [PubMed]
171. Qin, B.; Luo, H.; Liu, G.; Zhang, R.; Chen, S.; Hou, Y.; Luo, Y. Nickel ion removal from wastewater using the microbial electrolysis cell. *Bioresour. Technol.* **2012**, *121*, 458–461. [CrossRef] [PubMed]
172. Choi, C.; Hu, N.; Lim, B. Cadmium recovery by coupling double microbial fuel cells. *Bioresour. Technol.* **2014**, *170*, 361–369. [CrossRef] [PubMed]
173. Ge, Z.; Li, J.; Xiao, L.; Tong, Y.; He, Z. Recovery of Electrical Energy in Microbial Fuel Cells. *Environ. Sci. Technol. Lett.* **2014**, *1*, 137–141. [CrossRef]
174. Fan, Y.; Hu, H.; Liu, H. Enhanced Coulombic efficiency and power density of air-cathode microbial fuel cells with an improved cell configuration. *J. Power Sources* **2007**, *171*, 348–354. [CrossRef]
175. Liu, H.; Cheng, S.; Logan, B.E. Production of Electricity from Acetate or Butyrate Using a Single-Chamber Microbial Fuel Cell. *Environ. Sci. Technol.* **2005**, *39*, 658–662. [CrossRef] [PubMed]
176. Tang, X.; Guo, K.; Li, H.; Du, Z.; Tian, J. Microfiltration membrane performance in two-chamber microbial fuel cells. *Biochem. Eng. J.* **2010**, *52*, 194–198. [CrossRef]

processes

MDPI

Article

Screening of Ion Exchange Resins for Hazardous Ni(II) Removal from Aqueous Solutions: Kinetic and Equilibrium Batch Adsorption Method

Anna Wołowicz * and Monika Wawrzkiewicz

Department of Inorganic Chemistry, Institute of Chemical Sciences, Faculty of Chemistry, Maria Curie-Sklodowska University in Lublin, M. Curie-Sklodowska Sq. 2, 20-031 Lublin, Poland; m.wawrzkiewicz@poczta.umcs.lublin.pl
* Correspondence: anna.wolowicz@poczta.umcs.lublin.pl; Tel.: +48-81-537-57-38

Abstract: The development of new, cheaper, and more effective technologies to decrease the amount of wastewater containing heavy metals and to improve the quality is indispensable. Adsorption has become one of the alternative treatment methods. A small number of studies focusing on the batch technique for nickel ion removal by the new generation ion exchangers are described in the literature. In this paper, the Ni(II) removal from aqueous solutions using the ion exchange resins of different types was investigated. The experiments were conducted at different HCl and HCl/HNO_3 concentrations, and the initial concentration was 100 mg Ni(II)/L. The investigation of the Ni(II) desorption from the chosen resins were carried out. The Ni(II) removal efficiency and the rate of removal are shown on the kinetic curves and the rate constants as well as kinetic parameters were collected and compared. The isotherm parameters were calculated and Fourier-transform infrared spectroscopy with the attenuated total reflection spectra was performed to determine the nature of adsorption. The experimental results showed that the Ni(II) percentage removal is high and Lewatit MonoPlus TP220 could be an alternative for the treatment of nickel(II) containing wastewaters.

Keywords: nickel removal; adsorption; ion exchangers; water pollution; Lewatit MonoPlus TP220

Citation: Wołowicz, A.; Wawrzkiewicz, M. Screening of Ion Exchange Resins for Hazardous Ni(II) Removal from Aqueous Solutions: Kinetic and Equilibrium Batch Adsorption Method. *Processes* **2021**, *9*, 285. https://doi.org/10.3390/pr9020285

Academic Editor: Andrea Petrella
Received: 11 January 2021
Accepted: 29 January 2021
Published: 2 February 2021

1. Introduction

Nickel is widely distributed in the environment, and can be found in water, air, soil, or food because of its natural occurrence in nature and anthropogenic origin (Figure 1). It is the 24th most abundant metal in the earth's crust and accounts for about 3% of the earth's composition [1,2]. The natural sources of atmospheric nickel include dusts from volcanic emissions, meteoric dust, weathering of soils and rocks, forest fires, and sea salt spray [3,4]. From 30% to 50% of natural Ni sources are generated by soil particles moved by wind, blown from eroded areas [5], whereas almost 90% of the global anthropogenic Ni emissions are generated by oil combustion [6]. About 6–20 ng/m^3 of nickel is present in ambient air, whereas air contaminated by anthropogenic sources could include 150 ng/m^3 of nickel [4]. Solubilization of nickel compounds from soils as well as biological cycles are the main sources of nickel in water [3]. The content and mobility of nickel in soils depends on its compound solubility, pH, and soil types [3]. At pH < 6.5, nickel compounds are soluble in water, whereas at above pH > 6.7, insoluble hydroxides are usually present. Nickel salts such as chloride, nitrate, and sulfate (salts of strong acids, organic acids) are soluble in water, whereas metallic nickel, nickel sulfides, and nickel oxides, as well as nickel salts of weak inorganic acids, are poorly water-soluble [4]. Moreover, the use of pesticides and fertilizer increases the nickel content in soils and could be a source of heavy metals in food.

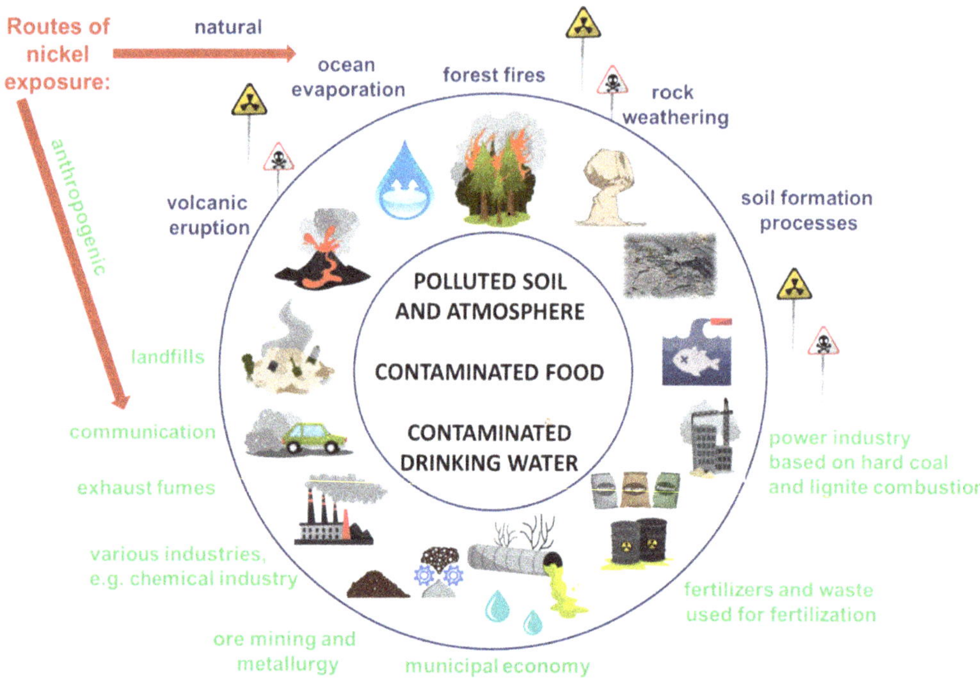

Figure 1. Nickel sources in the environment with division onto natural and anthropogenic origin.

The nickel content in farm soils is in the range from 3 to 1000 mg Ni/kg, soil but in the soils collected near to metal refineries it could be much higher; in the range from 24,000 to 53,000 mg/kg [3]. The nickel contents in different areas of the environment are presented in Table 1.

Table 1. Nickel contents in the environment as a result of its natural and anthropogenic emission.

Ni Occurrence		Concentration of Ni	References
water	Baltic water	0.09–1.08 μg/L	[2]
	river water	0.7 μg/L	
	bottled mineral waters	0.71–3.20 μg/L	
	drinking water from Stalowa Wola (an area affected by industrial emissions)	17 μg/L	
	uncontaminated water	300 ng/L	[3]
air	ambient air	6–20 ng/L	
	air (anthropogenic sources)	150 ng/L	
soil	farm soils	3–1000 mg/L	[2]
	soil	0.2–450 mg/kg	
	soil near metal refineries dried sludge	24,000–53,000 mg/kg	
	average content of nickel in Poland	6.5 mg/kg	
	average content of nickel in the world	13–37 mg/kg	
	soil affected by industrial emissions from Stalowa Wola	17.20 mg/kg	

Table 1. *Cont.*

	Ni Occurrence	Concentration of Ni	References
	soil affected by the Bolesław Mining and Metallurgical Plant	19.62 mg/kg	[6]
fertilizer	fertilizer based on dolomite	7.6–396.0 mg/kg	[6]
fertilizer	11 types of fertilizer [1]	0.4–295.1 mg/kg	[7,8]

Where: [1] Chloride potassium salt (0.4 mg/kg), Salmag (0.6 mg/kg), Calcium sulfate tetraurea (1.5 mg/kg), Phosphogypsum (4.3 mg/kg), Triple superphosphate (6.5 mg/kg), Granulate (26 mg/kg), Polifoska B (33 mg/kg), Polifoska 8 (35.1 mg/kg), Polifoska 6 (38 mg/kg), Polimag 405 (295.1 mg/kg).

Due to its properties, Nickel (Ni) can be utilized in various branches of industry and applied in many processes, such as electroplating, mineral processing, production of stainless steel, batteries, metallic alloys, coins, ceramic coloring, and paint [2,9]. The details of nickel properties, global uses, as well as nickel consumer markets are presented in Figure 2. As a consequence, huge amounts of nickel-containing wastes, e.g., spent batteries, catalysts, waste electrical and electronic equipment as well as wastewaters and electrolytes are generated and leach into the environment [9,10].

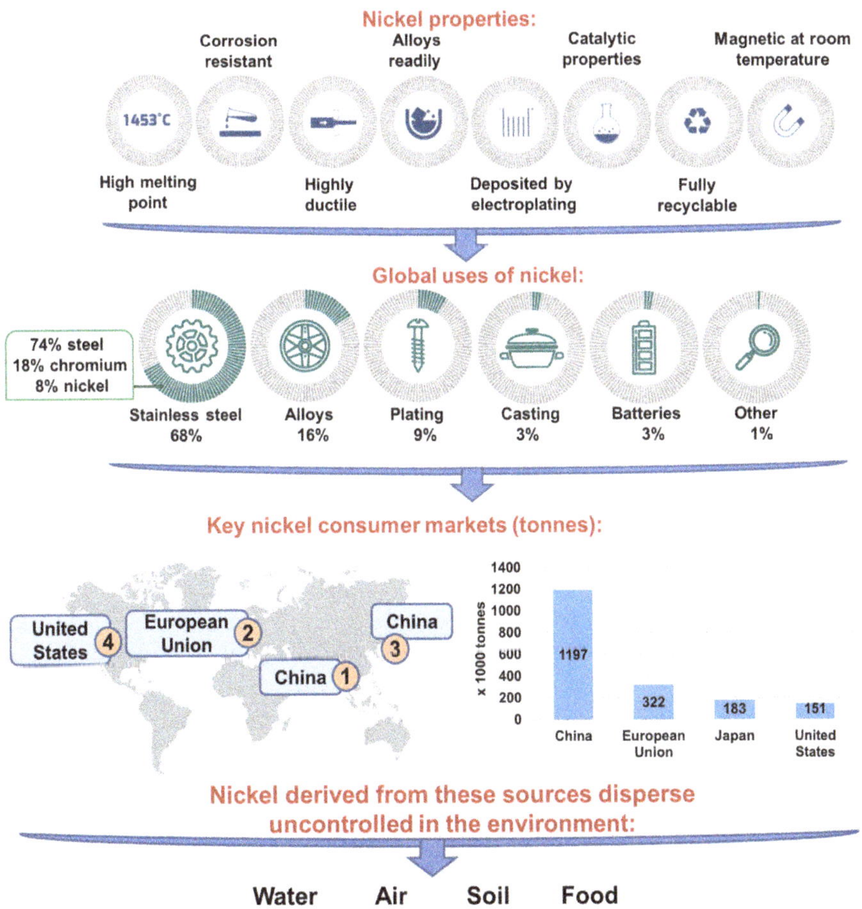

Figure 2. Nickel properties, global uses, and consumer markets.

Drinking of contaminated water, inhalation of particulates from the atmosphere, eating of contaminated food, due to the toxic and carcinogenic properties of nickel results in harmful effects to humans as well as to other living organisms [2]. This metal is capable of bioaccumulation in the aquatic environment and biomagnifications along the food chain [9,11,12]. Nickel causes kidney and lung diseases, chronic asthma, cough, pulmonary fibrosis, gastrointestinal distress (nausea, vomiting, diarrhea), skin dermatitis, nickel-induced apoptosis, allergy, headaches, cardiovascular diseases, lung and nasal cancer, as well as epigenetic effects [10–14] (Figure 3). A tolerable daily intake of nickel is equal to 2.8 μg/kg body weight (b.w.) [15]. As is indicate in Figure 3, the maximum contaminant level (MCL) of heavy metals such as lead (Pb), cadmium (Cd), mercury (Hg), arsenic (As), chromium (Cr), zinc (Zn) and nickel (Ni) that is allowed in drinking water is in the range from 0.00003 to 0.8 mg/L (establish by the United States Environmental Protection Agency, USEPA) [16]. As was suggested by the World Health Organization, the permissible nickel concentration in drinking water and wastewaters should not exceed 0.02 mg/L and 900 mg/L, respectively [2].

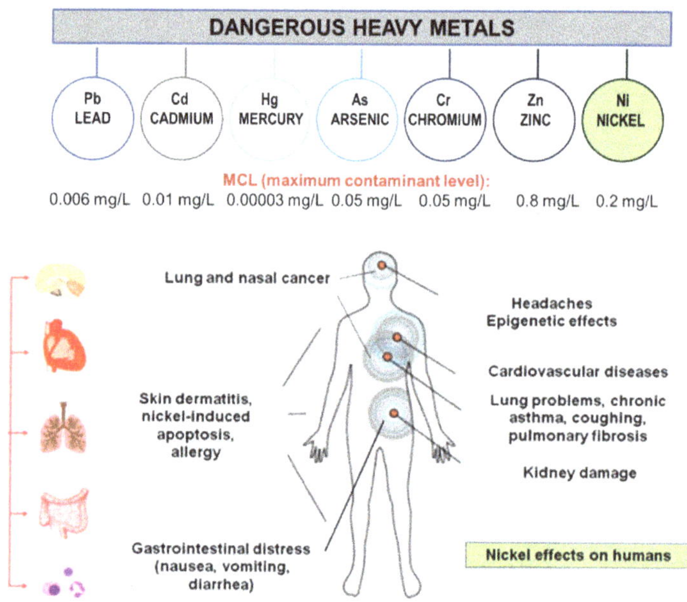

Figure 3. The maximum contaminant level (MCL) of heavy metals that is allowed in drinking water established by the United States Environmental Protection Agency (USEPA) and the effect of nickel on the human body.

Taking into account nickel deficit due to its large demand by different industrial sectors, its widespread dispersion and persistence in the environment, and toxic effects on living organisms, its recycling and reuse is essential from the economic (new sources of valuable metal) as well as ecological (environment protection, wastes reduction) points of view. Despite of some technologies proposed for reclamation of solid matrices containing nickel, e.g., phytoremediation, solvent extraction, immobilization [17,18], etc., the most effective way to avoid nickel diffusion in the environment is its removal from industrial effluents using efficient treatment methods [2,19,20]. Nickel is removed by using various physicochemical methods such as coagulation, flocculation [21], electrocoagulation [22], co-precipitation [23], reverse osmosis, electrodialysis [24], ultrafiltration, complexation [25], membrane separation [26], adsorption [27], and ion exchange [28]. These techniques have notable limitations, such as incomplete removal, low efficiency, high operating and capital costs, sensitivity to operating conditions, generation of by-products, and excess sludge, which require further treatment; therefore, only few of them are suitable for application on a large-scale [2,19,20,29–34]. A comparison of treatment methods applied for nickel removal both with their advantages and disadvantages are presented in Figure 4. Adsorption, due to its advantageous such as cost-effectiveness, efficiency, easiness of application, effectiveness, high adsorption capacities of polymeric adsorbents, applicability for low pollutant concentrations, and non-toxic by products, is already applied in nickel removal from wastewaters [2,29,30,33–35]. A large amount of adsorbents, both inorganic and organic, were studied for Ni(II) ion removal from aqueous solutions, such as carbonaceous materials (coal, lignite, commercial activated carbon, activated carbon from waste materials), industrial and agricultural wastes (fly ash, sugarcane bagasse, red mud, sludge, peels, brans, barks, coir pith, tea), polymeric adsorbents (biopolymers, synthetic polymers), mineral adsorbents (clay, zeolite, siliceous materials) and bio-adsorbents (algae, fungi, bacteria) [2,33,35–41].

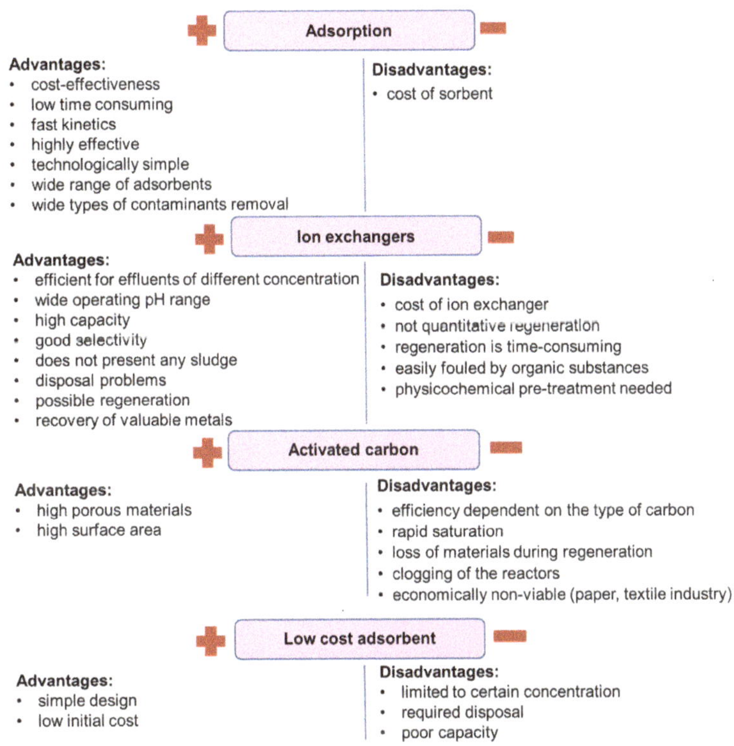

Figure 4. Comparison of nickel containing wastewaters treatment methods.

The aim of this study was the applicability of various adsorbents, with particular consideration of new generation ion exchangers: Purolite S984 (S984), Lewatit MonoPlus TP220 (TP220), Purolite A830 (A830), Lewatit MonoPlus SR7 (SR7), Purolite A400TL (A400TL), Dowex PSR2 (PSR2), Dowex PSR3 (PSR3) and Lewatit AF5 (AF5) for nickel removal from acidic solutions (0.1 M–6.0 M HCl and 0.9–0.1 M HCl/0.1–0.9 M HNO_3 systems).

The adsorption efficiency of Ni(II) onto selected adsorbents was studied in this paper, as well as previously published ones [42–45], to choose the most efficient adsorbent for Ni(II) removal. The equilibrium studies, calculations with kinetic and isotherm models, desorption studies, FTIR-ATR spectra analysis of TP220 after Ni(II) adsorption, and proposed mechanism of Ni(II) adsorption are presented in this paper.

Nickel is frequently recovered from nickel-based solid waste (spent batteries, used catalysts, alloy scraps) as well as from high-Ni content solutions generated during technological processes. Nickel could be effectively removed from aqueous solutions using physicochemical techniques, but in the case of solids wastes, such materials must be hydrometallurgy treated (leached into a solution and subsequently recovered from it) or pyrometallurgy treated (thermal treatment). During the leaching procedure, HCl, aqua regia, HNO_3, H_2SO_4, H_2SO_4—H_2O_2 of different concentrations [11] were used; therefore, the compositions of the solutions under discussion were selected in such a way to reflect the real wastewaters.

2. Materials and Methods

2.1. Materials

The stock (10,000 mg Ni(II)/L) and working solutions used for kinetic studies (100 mg Ni(II)/L), isotherm studies (100–10,000 mg Ni(II)/L) and desorption (1 and 2 M HNO_3, HCl, H_2SO_4, NH_4OH, NaOH and NaCl) were prepared using the chemical compounds of analytical grade purchased from the POCh S.A. company (Gliwice, Poland). The working solutions used for sorption were prepared by dilution of the stock solutions obtained by weighing a proper amount of solid $NiCl_2 \times 6H_2O$ and dissolving it in 0.1 M HCl and distilled water. The required volumes of 36–38% HCl as well as 36–38% HCl and 65% HNO_3 were also added to obtain the desired acids concentrations (the chloride solutions: 0.1; 1.0; 3.0 and 6.0 M HCl and the chloride–nitrate(V) solutions: 0.1–0.9 M HCl/0.9–0.1 M HNO_3). The presence of nickel complexes in the hydrochloric acid solutions, i.e., the chemical speciation of nickel (HCl–Ni), was modelled using the HYDRA–MEDUSA chemical equilibrium software for Windows (version: August 2019). In HYDRA (Hydrochemical Equilibrium Constant Database), the components for presented systems such as nickel species and their formation constants were obtained, then with MEDUSA (Make Equilibrium Diagrams Using Sophisticated Algorithms), the diagram for nickel in HCl solutions was obtained. During preparation of the solutions for isotherm studies similar procedure as in the kinetic once was applied whereas the desorption solutions were prepared by adding the proper volume of mineral acid, ammonium hydroxide, sodium hydroxide or sodium chloride into the volumetric flask and diluted or dissolved by distilled water. The concentration of eluting agent solutions was 1.0 or 2.0 M. The adsorbents for nickel removal from the acidic solutions, i.e., S984, TP220, A830, SR7, A400TL, PSR2, PSR3 and AF5, were cleaned by decantation before use to prevent from mechanical contamination, and then rinsed several times with distilled water or treated with 1 M hydrochloric acid to convert from the free base form to the chloride form, and then washed with distilled water. The commercial adsorbents characteristics are given in Table 2.

Table 2. Characteristics of the adsorbents applied for screening test for Ni(II) removal [46,47].

Name	Type	Matrix	Structure	Functional Groups	Mean Bead Size (mm)	Total Capacity (val/L)	Water Retention (%)
S984	Chelating ion exchanger	Cross-linked polyacrylic	Macroporous	Polyamine	–	2.7	44–55
TP220	Chelating/Weakly basic anion exchanger	Cross-linked polystyrene	Macroporous	Bis-picolylamine, bis(2-pyridyl-methyl)amine	0.62 (±0.05)	2.2	48–60

Table 2. *Cont.*

Name	Type	Matrix	Structure	Functional Groups	Mean Bead Size (mm)	Total Capacity (val/L)	Water Retention (%)
A830	Weakly basic anion exchanger	Cross-linked polyacrylic	Macroporous	Complex amine	0.3–1.2	2.75	47–53
SR7	Strongly basic anion exchanger	Cross-linked polystyrene	Macroporous	Quaternary ammonium, type 3	0.57–0.67	0.6	59–64
A400 TL	Strongly basic anion exchanger	Cross-linked polyacrylic	Microporous	Quaternary ammonium, type 1	0.425–0.85	1.3	48–54
PSR2	Strongly basic anion exchanger	Cross-linked polystyrene	Microporous	Quaternary ammonium, type, tri-n-butyl amine	0.3–1.2	0.65	40–48
PSR3	Strongly basic anion exchanger	Cross-linked polystyrene	Macroporous	Quaternary ammonium, type, tri-n-butyl amine	0.3–1.2	0.6	50–65
AF5	Adsorbent without functional group	Carbonaceous	Microporous	–	0.4–0.8	–	48–60

Prices of ion exchangers are affected by their types, quality, and spherical bead size. Usually, the price of cation exchange resins (strong and weak acid) ranges from USD 1.4 to USD 7.1 per 1 L, whereas anion exchange resins (strong and weak base) range from USD 4.6 to USD 7.1 per 1 L. Type 2 resins are generally more expensive than type 1 resins. The chelating ion exchange resins price range from USD 17.7 to USD 70.7 and above per 1L [48,49].

2.2. *Batch Adsorption Studies*

In the batch adsorption experiments, the effects of the phase contact times, acid (HCl, HNO_3), and Ni(II) concentrations were investigated as factors determining Ni(II) adsorption. The volume of the liquid phase was 50 mL, and the mass of adsorbent was equal to 0.5 ± 0.0005 g. All adsorption experiments were performed at 25 °C using the laboratory shaker Elpin 358+ (Poland) at 180 rpm (rotations per minute), amplitude 8. After the separation of Ni(II) solutions from the adsorbent phase by filtration (qualitative medium filter paper) using Atomic Absorption Spectroscopy (ASA) absorption, a Varian AA240FS spectrometer with SIPS autosampler (Varian, Australia) was used for determination of the Ni(II) concentration after sorption (measurement parameters: wavelength 232.0 nm; lamp current 4 mA; slit width 0.2 nm; and the air/acetylene flow 13.5/2 L/min).

2.2.1. Kinetic Studies

Using the chosen adsorbents, the Ni(II) uptake was examined as a function of time (1–240 min.) and acid concentrations (0.1 M, 1.0 M, 3.0 M and 6.0 M HCl, as well as 0.1 M HCl/0.9 M HNO_3, 0.2 M HCl/0.8 M HNO_3, 0.5 M HCl/0.5 M HNO_3, 0.8 M HCl/0.2 M HNO_3, 0.9 M HCl/0.1 M HNO_3) in order to explore the rate-controlling mechanism. The amount of Ni(II) sorbed by the adsorbents at time t was calculated from the equation:

$$q_t = \frac{(C_0 - C_t)}{m} \cdot V \tag{1}$$

where C_0 and C_t (mg/L)—the Ni(II) concentration in the solution before and after the sorption time t, respectively; V (L)—the volume of the Ni(II) solution; and m (g)—the mass of the adsorbent.

The pseudo-first-order (PFO), pseudo-second-order (PSO), and intraparticle diffusion (IPD) kinetic models [50,51] were applied for description of the experimental data as follows:

$$\frac{dq_t}{dt} = k_1(q_e - q_t) \tag{2}$$

$$\frac{dq_t}{dt} = k_2(q_e - q_t)^2 \tag{3}$$

$$q_t = k_i t^{1/2} \tag{4}$$

wher q_e and q_t (mg/g)—the Ni(II) amounts sorbed at the equilibrium and at any time t; k_1 (1/min) and k_2 (g/mg min)—the rate constants of sorption determined from PFO and PSO equations, respectively; k_i (mg/g min$^{0.5}$)—the intraparticle diffusion rate constant.

2.2.2. Equilibrium Studies

Series of Ni(II) solutions of the increasing metal ions concentrations from 100 to 10,000 mg Ni(II)/L in 0.1 M HCl were prepared. The equilibration time was established as 24 h. The amounts of Ni(II) sorbed at equilibrium, denoted as the sorption capacities (q_e) of selected adsorbents, were calculated from the equation:

$$q_e = \frac{(C_0 - C_e)}{m} \cdot V \tag{5}$$

where C_0 and C_e (mg/L)—the Ni(II) concentrations in the solution before and after the sorption at equilibrium, respectively; V (L)—the volume of Ni(II) solution; and m (g)—the mass of the adsorbent.

In order to explain the relationship between the Ni(II) concentration in the solution and sorbent phase at equilibrium, four isotherm models were chosen to describe the experimental data, i.e., the Langmuir, Freundlich, Temkin and Dubinin–Raduskievich models [52]. The non-linear forms of the above-mentioned models are as follows:

$$q_e = k_F C_e^{1/n} \tag{6}$$

$$q_e = \frac{k_L Q_0 C_e}{1 + C_e k_L} \tag{7}$$

$$q_e = \frac{RT}{b_T} lnAC_e \tag{8}$$

$$q_e = q_m e^{k_{DR}\varepsilon^2} \tag{9}$$

where k_F (mg$^{1-1/n}$ L$^{1/n}$/g) and n—the Freundlich constants related to the adsorption capability and adsorption intensity, respectively; k_L (L/mg)—the constant parameters of adsorption equilibrium; Q_0 (mg/g)—the monolayer adsorption capacity; b_T (J g/mol mg) —Temkin constant related to the heat of adsorption; A (L/mg)—the Temkin isotherm equilibrium binding constant; q_m (mg/g)—the maximum adsorption capacity; k_{DR} (mol^2 J^2) —constant related to the adsorption energy; ε (J/mol)—the adsorption potential calculated as $\varepsilon = RTln\left[1 + \frac{1}{C_e}\right]$; R—the gas constant (8.314 J/mol K); and T (K)—the temperature.

2.2.3. Error Analysis

All adsorption experiments were performed in triplicates. The mean values of the results were used for data evaluation. The standard deviation did not exceed 3–5% in all cases. Non-linear methods for the calculation of kinetic and equilibrium adsorption parameters were applied using the software Microsoft Excel 2013 with Solver add-in. Based on the values of Marquardt's percent standard deviation (*MPSD*), the determination coefficient (R^2) and the adjusted R-squared (R^2_{adj}) of the best fitted model was proposed. The above-mentioned parameter can be determined using the following equations [53,54]:

$$MPSD = \sum_{i=1}^{n}\left(\frac{q_{e\,exp} - q_{e\,cal}}{q_{e\,exp}}\right)_i^2 \tag{10}$$

$$R^2 = 1 - \frac{\sum\left(q_{e\,exp} - q_{e\,cal}\right)^2}{\sum\left(q_{e\,exp} - q_{e\,mean}\right)^2} \tag{11}$$

$$R^2_{adj} = 1 - \left[\frac{(1-R^2)(n-1)}{n-k-1}\right] \tag{12}$$

where $q_{e\,exp}$ (mg/g)—the amount of Ni(II) sorbed at equilibrium; $q_{e\,cal}$ (mg/g)—the amount of Ni(II) sorbed calculated from the non-linear models; $q_{e\,mean}$ (mg/g)—the measured by the means of $q_{e\,exp}$ values; n—the points number in data sample; and k—the number of independent regressors.

2.2.4. FTIR-ATR Analysis

The Fourier-transform infrared spectroscopy with attenuated total reflection (FTIR-ATR) technique was used for the recorded spectra of adsorbents under discussion using the Agilent Cary 630 FTIR spectrometer. The above-mentioned spectra were recorded in the frequency range from 400 to 4000 cm^{-1} for the adsorbents before and after the Ni(II) adsorption. The FTIR-ATR technique was used to confirm and identify the presence of functional groups in the adsorbents used, as well as to provide information about possible Ni(II) interactions with the functional groups.

2.2.5. Batch Desorption Experiments

Regeneration of the adsorbent with the highest sorption capacity for Ni(II) ions was performed in three sorption–desorption cycles. The sorption–desorption cycles were performed using the batch technique (laboratory shaker Elpin 358+, Poland) and applying the following parameters: m = 0.5 ± 0.0005 g, V = 50 mL, T = 25 °C, amplitude and rotary of shaking: 8 and 180 rpm, respectively. The solutions of the compositions 100 mg Ni(II)/L in 0.1–6.0 M HCl and 0.1–0.9 M HCl/0.9–0.1 M HNO_3 were used in each sorption step, and then Ni(II) concentration in the sorbent phase was calculated. Regeneration was carried out using the samples of adsorbent uploaded with Ni(II) ions, which were shaken for 2 h with 50 mL of the eluting solutions such as HNO_3, HCl, H_2SO_4, NH_4OH, NaOH, NaCl of the 1.0 M and 2.0 M concentrations. The amounts of Ni(II) desorbed form the adsorbent phase were determined in the liquid phase using the AAS measurements. Desorption of Ni(II) ions as a percentage (D) was calculated using the equation:

$$D = \frac{m_{des}}{m_{ads}} \times 100\% \tag{13}$$

where m_{des} (mg)—the mass of Ni(II) desorbed; m_{ads} (mg)—the mass of Ni(II) adsorbed.

3. Results and Discussion

3.1. Effect of Phase Contact Time and Acids Concentration on Ni(II) Adsorption

A series of shaking time studies for Ni(II) ions was carried out with 100 mg/L initial metal concentration at 25 °C and different HCl and HNO_3 acids concentrations (HCl (M): 0.1; 1.0; 3.0; 6.0; HCl (M)/HNO_3 (M): 0.1/0.9; 0.2/0.8; 0.5/0.5; 0.8/0.2; 0.9/0.1). The effect of phase contact times ranging from 1 min to 240 min on the Ni(II) adsorption was also examined. The adsorption of 100 mg/L concentration of Ni(II) onto selected adsorbents for 4 h was studied in this paper and compared to the results obtained previously [42–45] to choose the most efficient adsorbent for Ni(II) removal. The results are compared in Table 3, and presented in Figure 5 (chosen examples).

Table 3. Adsorption capacities obtained during the Ni(II) adsorption, q_e (mg/g) for 240 min of phase contact time from the chloride solutions (where: * references this paper).

Adsorbent	HCl (M)				Ref.	HCl (M)/HNO_3 (M)					Ref.
	0.1	1.0	3.0	6.0		0.1/0.9	0.2/0.8	0.5/0.5	0.8/0.2	0.9/0.1	
S984	4.95	4.83	5.52	4.74	[42]	5.43	4.93	4.75	5.27	6.41	[42]
TP220	6.24	4.93	4.68	4.89	[43]	4.88	4.93	4.91	5.00	4.95	[43]
A830	4.60	4.85	4.53	5.02	*	4.59	4.49	4.36	4.40	4.53	*
SR7	4.56	4.74	4.87	4.82	*	4.55	4.44	4.25	4.34	4.32	*
A400TL	4.72	4.79	4.91	4.76	*	4.17	4.40	4.30	3.92	4.02	*
PSR2	3.70	3.29	4.36	4.14	[44]	4.78	4.75	4.76	4.69	4.78	*
PSR3	4.73	4.82	4.53	3.92	[44]	4.69	4.78	4.57	4.60	4.64	*
AF5	4.89	4.76	4.73	4.72	[45]	4.83	4.90	4.84	5.00	4.88	[45]

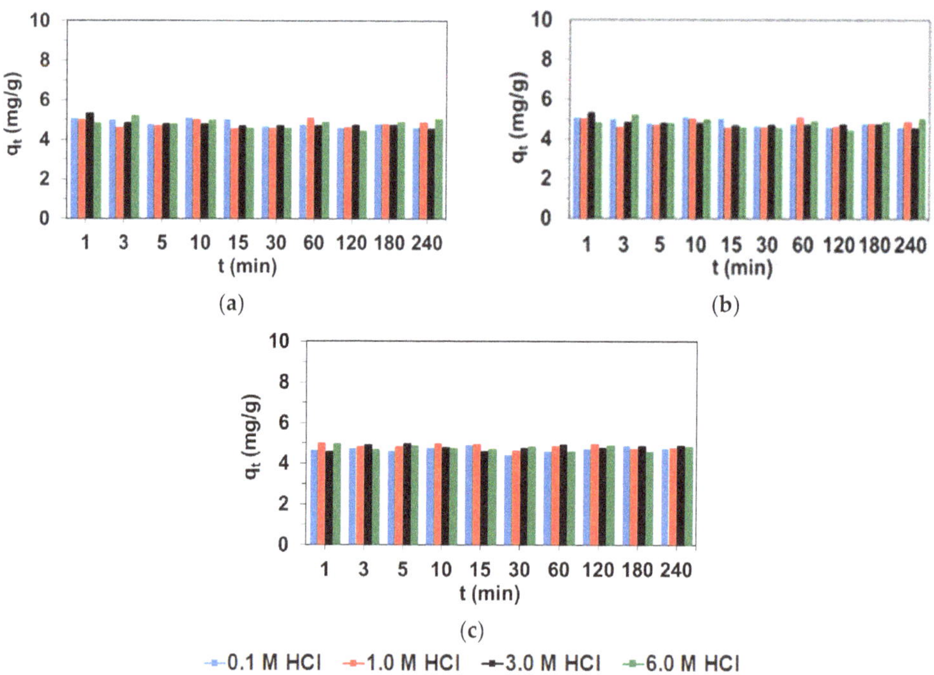

Figure 5. *Cont.*

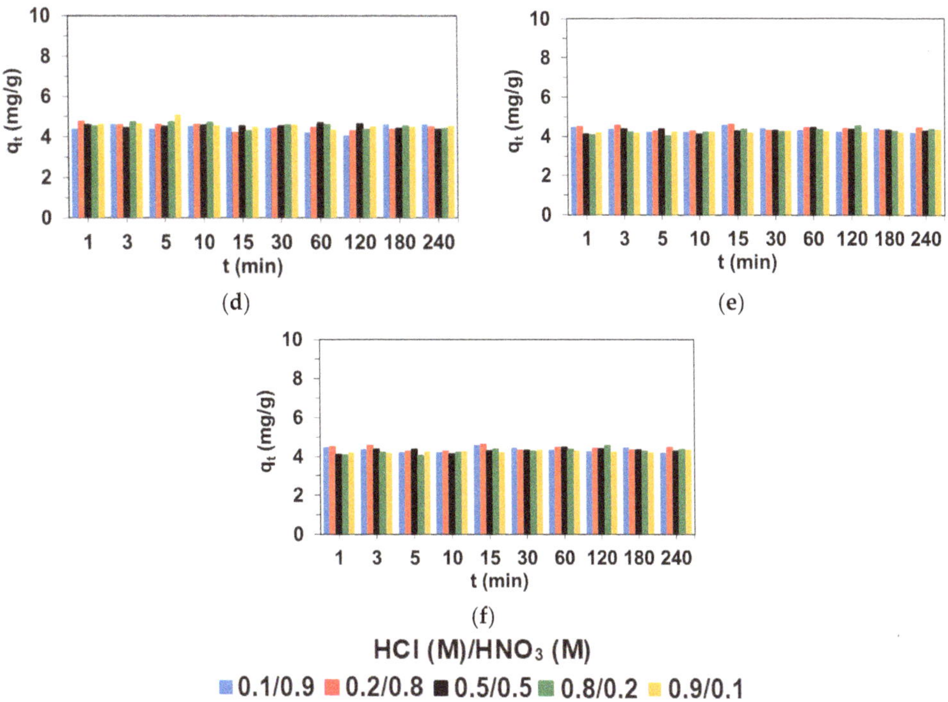

Figure 5. Effect of phase contact time on Ni(II) adsorption on (**a**) A830; (**b**) SR7; (**c**) A400TL from the chloride 0.1–6.0 M HCl–100 mg Ni(II)/L and (**d–f**) 0.1–0.9 M HCl/0.9–0.1 M HNO_3–100 mg Ni(II)/L.

It was observed that, for seven adsorbents (S984, A830, SR7, A400TL, PSR2, PSR3 and AF5), Ni(II) ion adsorption did not increase gradually at the beginning of the adsorption, but the amount of Ni(II) adsorbed at time *t*, as well as the adsorption capacities, reached similar values in the phase contact time ranging from 1 min to 240 min. On the other hand, in the case of TP220 ion exchanger (adsorption from the diluted HCl solutions), the amount of Ni(II) adsorbed at time *t* increased with the increasing phase contact time, and at 60 min of phase contact time, a sharp increase in q_t values was observed. The adsorption capacity towards Ni(II) was the highest for TP 220 and 0.1 M HCl. Moreover, comparing the adsorption capacities collected in Table 2 can conclude that TP220 showed the highest Ni(II) adsorption ability in most cases. At 240 min of phase contact time, the Ni(II) percentage removal (%*R*):

$$\%R = \frac{(C_0 - C_t)}{C_t} 100\% \quad (14)$$

where C_0 (mg/L) and C_t (mg/L) are the Ni(II) concentrations in the solution before and after the sorption at time *t*, respectively,. This value was the highest for TP220 and equal to %*R* = 62.4%, whereas for the other adsorbents these values were in the range from 32.9 to 55.2% for the HCl system. Additionally, the highest adsorption ability towards Ni(II), similar to TP220, was exhibited by the S984 adsorbent from the HCl/HNO_3 systems (the percentage removal was approximately 50–65%).

The time required to reach system equilibrium was above 240 min for TP220 [43], whereas in the case of other adsorbents the q_t values remained at a similar level at 1 min as well as at 240 min of phase contact time. As pointed out by Uzun and Guzel [55], the time required to reach equilibrium (t_{eq}) during the Ni(II), Mn(II), Fe(II), Cu(II) adsorption on Merck 2514 activated carbon was equal to 134 h for Ni(II) (m = 1 g, V = 50 mL, C_0 = 200 mg/L, 140 rpm) whereas during the Ni(II) adsorption on raw and modified Filtrasorb-400 by potassium bromate t_{eq} = 120 min (V = 400 mL, 1000 rpm) [56]. Adsorption of Ni(II) or separation was also carried out on ion exchange, resins e.g., Amberlite IRA-900 and a novel silica-based anion exchanger, AR-01 (Co(II), Ni(II) and Cu(II) from the nitrite solution) [57], anion exchange Amberjet™ 4200 Cl and cation exchange Amberjet™ 1200 H [58], anion exchanger D301R [59], and chelating Purolite S950 resin [60], but the time required to reach the system equilibrium was not determined. Adsorption of Ni(II) in the presence of complexing agents on the polyacrylate anion exchangers Amberlite IRA 458, Amberlite IRA 958 and Amberlite IRA 67 showed that the curve kinetics reached the plateau at the time greater than 30 min [61]. Effect of acid concentration were not observed in the HCl or HCl/HNO_3 systems. The q_t values were similar to those of the adsorption capacities in all examined aqueous solutions. Adsorption of Co(II), Ni(II), and Cu(II) from the nitrite solutions by anion exchangers was insignificant at the nitrite concentrations below 0.1 M, and then very slight increase in Ni(II) adsorption with the increasing concentration of nitrate ions was observed [57]. The experimental results obtained during the Ni(II) adsorption on S984, A830, SR7, A400TL, PSR2, PSR3 and AF5 adsorbents were modelled using the pseudo-first-order (PFO), pseudo-second-order (PSO), and intraparticle diffusion (IPD) kinetic models (Equations (2)–(4)). Taking into account the shape of kinetic curves and the arrangement of the experimental points, it was possible to calculate kinetic parameters at 0.1 M HCl using TP220 which exhibited the highest $\%R$ and adsorption capacity. The kinetic parameters were calculated and based on error analysis such as Marquardt's percent standard deviation, determination coefficient, and adjusted R-squared values; then, the best fitted model was proposed. The obtained kinetic parameters as well as the errors values are collected in Table 4, whereas the fitting plots are presented in Figure 6. The mechanism of Ni(II) adsorption depends on the structure and functional groups of the adsorbent as well as on the physicochemical characteristics of the solute and adsorbent. The pseudo-first-order and pseudo-second-order kinetic models were based on the different assumption: PFO on the physisorption process, and PSO on the chemisorption process [50,51].

Table 4. Kinetic parameters for the Ni(II) adsorption from the 0.1 M HCl chloride solutions by TP220.

Parameters	Pseudo-First Order	Pseudo-Second Order	Intraparticle Diffusion Model
		$q_{e\,exp}$ = 6.24 mg/g	
$q_{e\,cal}$ (mg/g)	5.20	5.31	5.52
k_1 (1/min)	2.75	-	-
k_2 (g/mg min)	-	1.22	
k_i (mg/g $min^{0.5}$)	-	-	0.05
MPSD	0.0748	0.0631	-
R^2	0.9131	0.9280	0.6869
R^2_{adj}	0.8883	0.9074	0.5974

The PFO kinetic model did not adjust during the whole range of phase contact times, whereas the PSO could be applied in the whole range of adsorption. From a physicochemical point of view, the PFO model was expected when adsorption was limited by the molecules transport from the solution to the adsorbent (film diffusion) or when this transport was not the rate-limiting step [62]. The PSO kinetic equation can be used as an estimation of the adsorption capacity based on the extrapolation of the kinetic data q_t (q_t > 80% q_e) [63]. Taking into account the kinetic parameters obtained for PFO and PSO, it was found that the determination coefficient as well as the adjusted R-squared values were slightly higher for the PSO model (R^2 = 0.9280, R^2_{adj} = 0.9074) than for PFO (R^2 = 0.9131, R^2_{adj} = 0.8883), whereas the MPSD was lower for the PSO model. Comparing the q_{exp} values (q_{exp} = 6.24 mg/g) with the calculated q_e results (PFO q_{exp} = 5.20 mg/g, PSO q_{exp} = 5.31 mg/g), a slightly higher compatibility between q_{exp} and q_e was observed in the case of the PSO model. Taking into account the above-mentioned parameters, the error analysis, as well as the fitting plot (Figure 5), it can be concluded that neither PFO nor

PSO kinetic models can be applied for describing Ni(II) adsorption on TP220; however, the better-fitting was found in the case of the PSO model. Taking into account the IPD model, which is frequently used to predict the rate controlling step, it was found that the intraparticle diffusion rate constant was equal to 0.05 mg/g min$^{0.5}$. Comparing the R^2 value obtained for IPD, which was equal to 0.6869, to the other values obtained for PFO and PSO, it was much smaller; therefore, the intraparticle diffusion may not be the only rate-limiting step—the film diffusion may also be limiting as well. The kinetics of Ni(II) adsorption on Amberlite IRA 458, Amberlite IRA 958, and Amberlite IRA 67 with the non-biodegradable complexing agents followed the pseudo second-order rate expression (R^2 = 0.999 for PSO, R^2 = 0.663 (IRA 458), R^2 = 0.952 (IRA 958), R^2 = 0.742 (IRA 67) for PFO, R^2 = 0.526 (IRA 458), R^2 = 0.490 (IRA 958), R^2 = 0.793 (IRA 67) for IPD) [61], similar to the Ni(II) adsorption on the above-mentioned ion exchangers with a biodegradable complexing agent, e.g., IDS (sodium salt of *N*-(1,2-dicarboxyethyl)-D,L-aspartic acid) (R^2 in the range from 0.999 to 1.000 for PSO, from 0.592 to 0.822 for PFO as well as from 0.637 to 0.682 for IPD) [40].

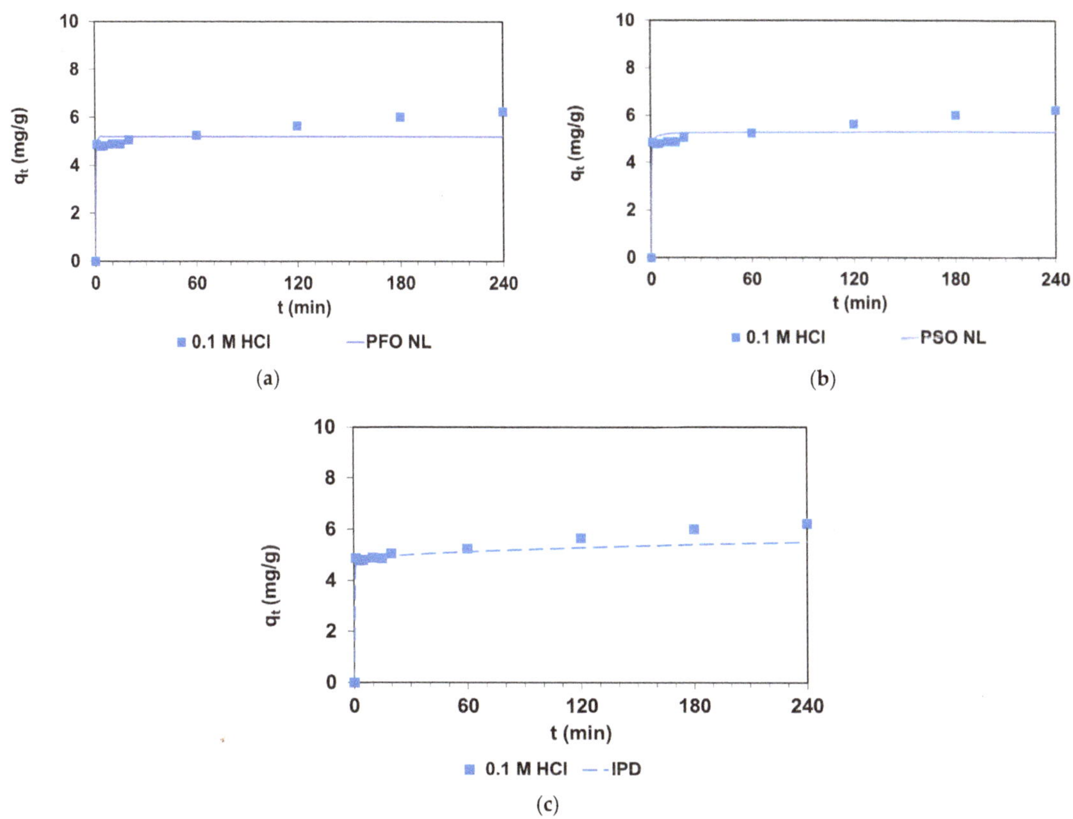

Figure 6. Influence of phase contact time on the Ni(II) uptake (C_0 = 100 mg/L) by TP220 with non-linear (NL) fitting of (**a**) PFO, (**b**) PSO and linear fitting of (**c**) IPD kinetic models to the experimental data.

3.2. Equilibrium Studies

Determination of adsorption capacities (q_e) towards a given pollutant such as Ni(II) ions, is a key step of the adsorption process. Determination of these values makes it possible to assess the suitability of a given material as a potential adsorbent, which may imply its use in industrial installations. The magnitude of these values is influenced not only by the properties of the adsorbent, such as chemical composition, specific surface area, type of functional groups, and porosity, but also the properties of the adsorbate and the nature of the interactions between the adsorbent and the adsorbate at equilibrium. The dependence of q_e vs. C_e in the selected investigated systems is presented in Figure 7. The shape of the presented isotherms corresponds to type II of the IUPAC adsorption isotherms classification that reflects adsorption on the macroporous materials, and weak and strong interactions in the adsorbate–adsorbent are considered.

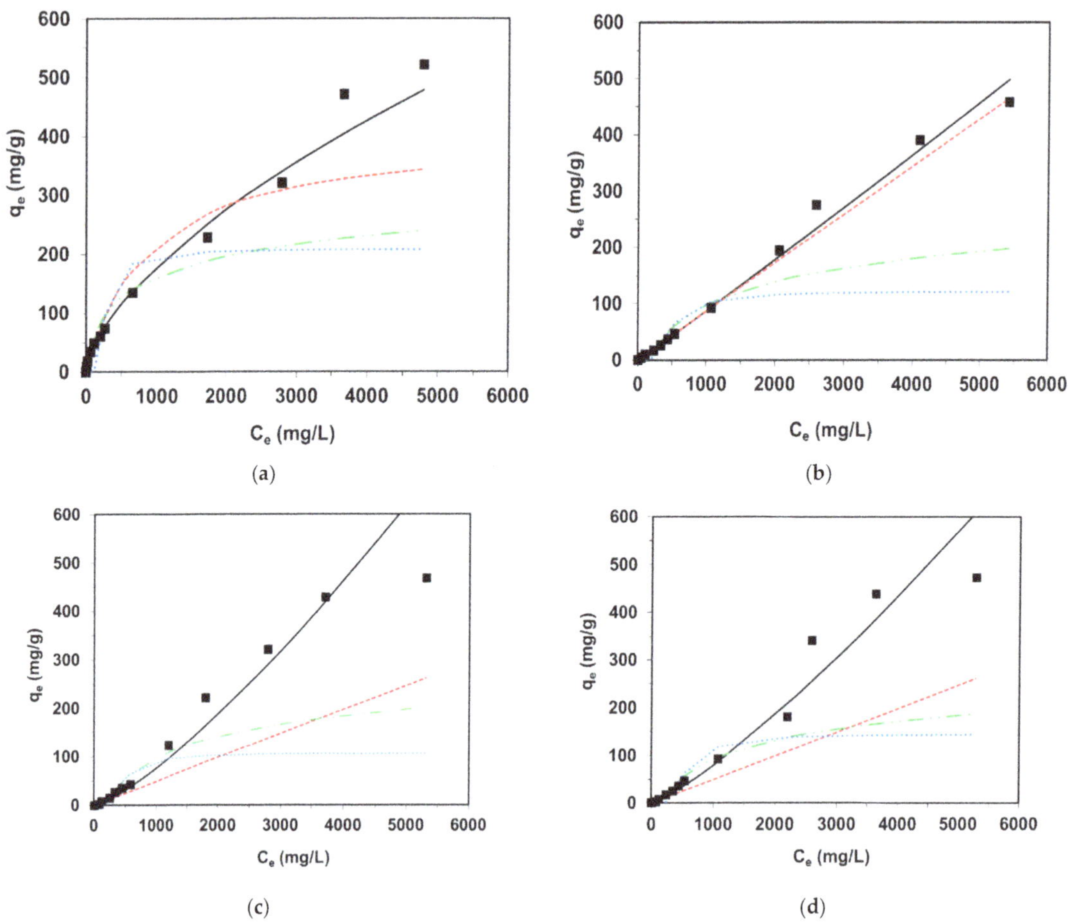

Figure 7. Equilibrium sorption isotherms of Ni(II) on (**a**) TP220, (**b**) A830, (**c**) SR7, and (**d**) AF5 adsorbents, and the fitting of experimental points to the Freundlich, Langmuir, Temkin and Dubinin–Radushkevich isotherm models using the non-linear regression.

The four most frequently used isotherm models (Freundlich, Langmuir, Temkin, and Dubinin–Radushkevich) were chosen for description of the Ni(II) sorption on the chelating (TP220, S984), weakly (A830), and strongly (SR7, PSR2, PSR3, A400TL) basic resins, as well as activated carbon (AF5). Characteristic parameters of the Freundlich, Langmuir, Temkin, and Dubinin–Radushkevich isotherm models were calculated using the non-linear regression and are listed in Table 5.

Table 5. Parameters of Frendlich, Langmuir, Temkin, and Dubinin–Radushkevich isotherm models calculated for the adsorption systems.

Model	Parameters	Adsorbents							
		S984	TP220	A830	SR7	A400TL	PSR2	PSR3	AF5
Freundlich	k_F	0.0479	2.317	0.070	0.0096	0.0260	0.0363	0.0376	0.0169
	$1/n$	1.081	0.629	1.031	1.299	1.165	1.118	1.108	1.224
	$MPSD$	0.272	0.079	0.097	0.905	0.248	0.368	0.212	0.582
	R^2	0.982	0.986	0.987	0.927	0.969	0.881	0.962	0.929
	R^2_{adj}	0.977	0.983	0.984	0.909	0.961	0.852	0.952	0.911
Langmuir	k_L	2.07×10^{-6}	0.0011	2.27×10^{-6}	1.30×10^{-6}	1.72×10^{-6}	1.30×10^{-6}	1.91×10^{-6}	1.30×10^{-6}
	Q_0	38,371.1	408.98	38,322.8	38,322.8	40,472.9	38,682.4	38,604.4	38,322.7
	$MPSD$	0.474	1.002	0.123	3.119	0.944	0.737	0.538	2.153
	R^2	0.986	0.901	0.989	0.965	0.978	0.892	0.970	0.947
	R^2_{adj}	0.983	0.877	0.986	0.956	0.972	0.865	0.962	0.934
Temkin	b_T	48.05	49.74	41.78	39.72	48.29	46.19	45.56	43.75
	A	0.0061	0.0256	0.0052	0.0048	0.0055	0.0054	0.0053	0.0052
	$MPSD$	1.921	2.089	1.797	1.672	1.942	1.894	1.837	1.756
	R^2	0.900	0.858	0.906	0.926	0.882	0.850	0.908	0.877
	R^2_{adj}	0.875	0.823	0.883	0.907	0.852	0.813	0.884	0.846
Dubinin–Radushkevich	q_m	183.86	208.12	121.59	107.8	90.41	141.33	165.63	144.56
	k_{DR}	0.0048	0.0086	0.0282	0.0263	0.0216	0.0374	0.0458	0.038
	E	3.237	7.605	4.178	4.359	4.811	3.655	3.304	3.624
	$MPSD$	4.391	4.889	4.266	4.602	4.708	4.533	4.297	4.389
	R^2	0.773	0.706	0.680	0.689	0.606	0.698	0.767	0.711
	R^2_{adj}	0.717	0.632	0.600	0.612	0.508	0.623	0.709	0.639

Where: k_F ($mg^{1-1/n}$ $L^{1/n}$/g), k_L (L/mg), Q_0 (mg/g), b_T (J g/mol mg), A (L/mg), q_m (mg/g), k_{DR} (mol^2 J^2), and E (J/mol).

Analyzing the data presented in Table 5 and Figure 7, the smallest *MPSD* values and the highest R^2 and R^2_{adj} values—among the applied isotherm models—were calculated using the Freundlich equation. The model takes into consideration the multilayer adsorption in the system and proceeds on a heterogeneous surface of unequal energetic active sites with a different binding energy. The Freundlich model is based on physisorption. The highest k_F value was calculated for the TP220 chelating resin, and was found to be 2.317 $mg^{1-1/n}$ $L^{1/n}$/g. The parameter *1/n* describing the intensity of adsorption was 0.629 for the Ni(II)—TP220 system and ranged from 0 to 1, which points to favorable adsorption, whereas in the case of other systems, this value was greater than 1 (unfavorable adsorption). The %*R* values were calculated to be 62.4% in 100 mg Ni(II)/L—0.1 M HCl—TP220 and confirmed the favorable uptake of Ni(II) ions by TP220.

For the Langmuir adsorption model assuming the monolayer adsorption, the MPSD values were greater than for the Freundlich model.

Analyzing of the distribution of experimental points in Figure 6, as well the results of error analysis, it can be stated that the Langmuir model describing the monolayer adsorption cannot be applied for description of the equilibrium sorption data of Ni(II) on the different types of resins.

The Temkin isotherm model was firstly applied for hydrogen adsorption onto platinum electrodes in an acidic medium and assumes that the adsorption heat of molecules in the surface layer declines linearly rather than logarithmically. The b_T and A values were in the range of 39.72–49.74 J g/mol mg and 0.0048–0.0256 L/mg, respectively. However, the R^2 (from 0.850 to 0.926) and R^2_{adj} (from 0.813 to 0.907) values obtained for the Temkin isotherm model were smaller than for the Freundlich model.

The Dubinin–Radushkevich isotherm model was proposed for the adsorption process related to micropore volume filling as opposed to the layer-by-layer adsorption on the pore [64]. Moreover, the k_{DR} parameter enables estimation of the mean free energy E of adsorption. This value, on the other hand, allows the assessing of the type of the adsorption. When the magnitude of E is smaller than 8 kJ/mol, the adsorption process has physical character, and when E is between 8 kJ/mol and 16 kJ/mol, the process is a chemical reaction.

In the investigated systems, the q_m parameters indicating the maximum sorption capacities for the resins and activated carbon were equal to 90.41–208.12 mg/g and 144.56 mg/g, respectively, and did not match well the experimental data. The mean free energies were in the range of 3.237–7.605 J/mol, which revealed the physical character of Ni(II) ion binding by the adsorbents. However, these data cannot properly reflect the Ni(II) ion retention by the resins, and AF5 as the highest values of *MPSD* and the lowest values of R^2 and R^2_{adj} were obtained for the Dubinin–Radushkevich model compared with the others.

Based on the obtained equilibrium data, Figure 8 presents a possible mechanism of Ni(II) retention by the resins of various basicity values of the functional groups in the acidic medium.

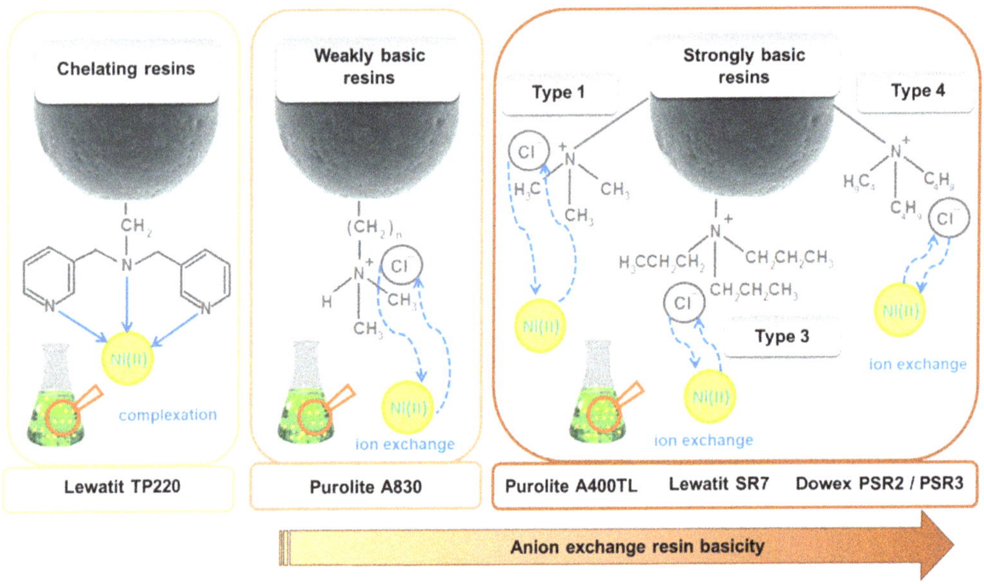

Figure 8. Mechanism of Ni(II) adsorption on the applied sorbents.

In order to assess and explain the adsorption capacity of the adsorbents, it is necessary to consider the complexes that can be formed in the hydrochloric acid solutions, as well as the nature and physical properties of the adsorbents under discussion. Based on the literature, spectrophotometric and anion-exchange measurements [43,61,65,66] in the hydrochloric acid system nickel ions form complexes with the chloride and hydroxide ions. Most of nickel exist as Ni^{2+} up to 5 M HCl [66,67], but in aqueous solutions at chloride concentrations up to 0.66 M, the formation of nickel complexes such as $NiCl^+$ and $NiCl_2$ starts ($Ni^{2+} + Cl^- = NiCl^+$, equilibrium constant reported by the Russian Academy of Science, K = 2.1265; $Ni^{2+} + 2Cl^- = NiCl_{2,aq}$ K = 5.9237) [55]. The $NiCl^+$ concentration gradually increased with the increasing concentration of HCl, whereas the concentration of Ni^{2+} decreased (Figure 9) [66]. With the hydrochloric acid concentration increase, the absorption spectrum general shift to longer wavelengths and octahedral complexes such as $[NiCl \times 5H_2O]^+$ and $[NiCl_2 \times 4H_2O]$ can be formed (substitution of water molecules by the chloride ions in the first coordination sphere of the $[Ni \times 6H_2O]^{2+}$ ion). The formation of anionic nickel(II) chloro-complexes $[NiCl_4]^{2-}$ in the concentrated HCl solutions is not clearly stated [55]. The nickel ion reacts also with the hydroxide ion to form the complexes of the following composition: $NiOH^+$, $Ni(OH)_{2aq}$, $Ni(OH)_3^-$, Ni_2OH^{3+}, and $Ni_4(OH)_4^{4+}$, but at the high concentration of HCl the hydroxide ion concentration is so low that the formation of nickel hydroxide complexes can be neglected. The mass and charge balance equations obtained for the strong HCl solution by Lee and Nam [67] were the following: $[Cl]_{total} = 2[NiCl_2]_{total} + [HCl]_{total} = [Cl^-] + [NiCl^+] + 2[NiCl_{2aq}]$; $[Ni]_{total} = [NiCl_2]_{total} = [Ni^{2+}] + [NiCl^+] + [NiCl_{2aq}]$ and $[H^+] = [HCl]_{total}$.

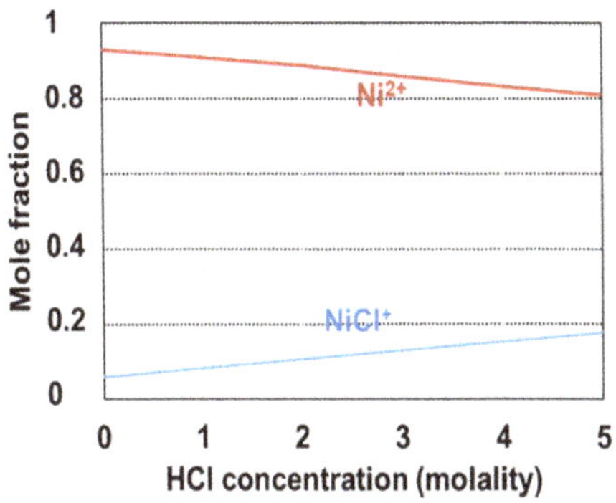

(a)

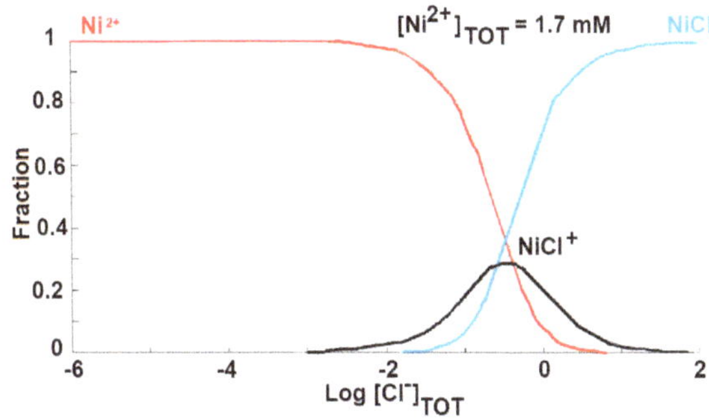

(b)

Figure 9. Distribution of nickel species in the HCl solution (**a**) with $NiCl_2$ of 1.0 molality, (**b**) with Ni^{2+} of 1.7 mM concentration depending on the total concentration of Cl^- ions.

Moreover, Table 6 makes a comparison of the sorptive properties of investigated materials with the data published in the literature possible.

Table 6. Equilibrium parameters of Ni(II) sorption on various adsorbents based on the literature review.

Sorbent	Isotherm Model	Equilibrium Parameters	Ref.
Activated carbons from the doum seed	Freundlich	k_F = 0.36–0.98 L/g, T = room temperature, pH = 7, a.d. = 5 g/L	[36]
Expanded graphite decorated with manganese oxide nanoparticles	Langmuir	q_e = 0.0065 mg/g, T = 25 °C, a.d. = 0.5 g/L,	[37]
Lewatit TP207 (chelating iminodiacetic acid groups in PS-DVB matrix)	-	q_e = 1.23 mg/g, T = 25 °C, a.d. = 0.27 g/L, pH = 6	[38]
Biochars produced from the wheat straw pellets (WSP550, WSP700) and rice husk (RH550, RH700) at 550 and 700 °C	Freundlich	WSP700: q_e = 25.1 mg/g WSP550: q_e = 12.6 mg/g RH700: q_e = 10.15 mg/g RH550: q_e = 6.87 mg/g T = 20 °C, a.d. = 0.1 g/20 mL, pH = 5	[39]
Amberlite IRA 458 (quaternary ammonium groups in the PA-DVB matrix) Amberlite IRA 958 (quaternary ammonium groups in the PA-DVB matrix) Amberlite IRA 67 (tertiary amine groups in the PA-DVB matrix)	Langmuir	q_e = 16.72 mg/g q_e = 13.22 mg/g q_e = 10.03 mg/g T = room temperature, a.d. = 0.5 g/50 mL, pH = 4–8	[40]
Modified carboxymethyl cellulose hydrogel	Freundlich	k_F = 4.614 L/g, T = 30 °C, a.d. = 100 mg/L, pH = 5	[41]

Where: T—temperature; a.d.—the adsorbent dose; PS-DVD—polystyrene-divinylbenzene; PA-DVB—poly-acrylate-divinylbenzene.

3.3. FTIR-ATR Analysis of Pure and Loaded TP220 by Ni(II)

The Fourier-transform infrared spectra with the ATR mode (FTIR-ATR) for TP220 before and after the Ni(II) adsorption are shown in Figure 10. The obtained FTIR-ATR spectra of TP220 before the adsorption (pure, without metal ions) show the characteristic bands for the bis(picolylamine) functional groups as well as for the polystyrene matrix which was previously described in the paper by Wołowicz and Hubicki [43], Kołodyńska et al. [68], Zagorodni et al. [69], Lazar et al. [70], Ghosh et al. [71], and Traboulsi et al. [72]. The spectra confirm the ion exchange resin structure as well as the composition. In the higher frequency region, big, broad bands between 3700 cm^{-1} and 3100 cm^{-1} with the maximum located at about 3421 cm^{-1} were observed, reflecting the stretching vibrations of the –OH group (residual hydration water), but this peak is not the structural part of the resin. The peak assigned to the asymmetric stretching vibrations of –C–H of the benzene ring at 3048 cm^{-1} and the bands assigned to the asymmetric and symmetric stretching vibrations of the $-CH_2$ group at 2800–3000 cm^{-1} confirmed the polystyrene structure of TP220. The peaks characteristic of the styrene–divinylbenzene backbone were described in detail in ref 72. There were found to be the stretching vibrations of –C–H of benzene rings (3083, 3059, 3025, 3001 cm^{-1}), asymmetric stretching vibrations of $-CH_2$ and $-CH_3$ in $-CH_3$–N (2915 cm^{-1}), symmetric stretching vibrations of $-CH_2$ and $-CH_3$ in $-CH_3$–N (2849 cm^{-1}), stretching vibrations of –C=C– of benzene rings, deformation vibrations of $-CH_2-N^+(CH_3)_3 \ldots OH^-$ (1602, 1582 cm^{-1}), stretching vibrations of –C=C– of benzene rings, –C–H asymmetric deformation vibrations of CH_3 (1493 cm^{-1}), C–H deformation vibrations of $-CH_2$ and $-CH_3$, –C=C– stretching vibrations, scissoring vibrations of $-CH_2$ (1452 cm^{-1}), C–H deformation vibrations of the aliphatic group >CH– (1372 cm^{-1}), –C–H plane deformation vibrations of benzene rings (1151, 1155, 1068, 1028 cm^{-1}), as well as C–H out-of-plane deformation vibrations of monosubstituted benzene rings (905, 840, 755 cm^{-1}) [72].

Peaks characteristic of bis-picolylamine (bis(2-pyridyl-methyl)amine) functional groups are also present, such as skeletal vibrations of pyridine, in-plane and out-of-plane C–C–H deformations in pyridine rings, stretching vibrations of C=N and C–N bonds, as well as stretching vibrations of aliphatic amino groups, (1590, 1471, 1437, 995–699 cm^{-1}) [43,49,69]. After the adsorption of Ni(II) on TP220, it was found that some of the bands reduced its intensity and the maxima of some peaks were moved to the major wave length. Such changes are observed in the region in which the functional groups were confirmed (C=N and C–N bond, pyridine, and aliphatic amine) indicating that the functional groups play a part in the mechanism of sorption and a coordinate bond with metals is formed [49].

3.4. Desorption of Ni(II)

From the economic point of view, the high adsorption in a short time towards pollutions as well as the sorbent desorption possibility and its reuse are important factors influencing the total cost of adsorption removal. As was pointed out by Zong et al. [49], the price of Dowex M4195 as well as TP220 is 134.6 USD/kg, therefore their regeneration and reuse could reduce the cost. Therefore, the desorption possibility and TP220 reuse were analyzed. As described in the experimental section, 50 mL of the eluting agents (HNO_3, HCl, H_2SO_4, NH_4OH, NaOH, NaCl) of the 1.0 M and 2.0 M concentrations were applied in the desorption test. The loaded anion exchange resin (sorption from 0.1 M HCl—100 mg Ni(II)/L, t = 4 h) was contacted with the above-mentioned solutions for 2 h and then the percentage desorption of Ni(II) was calculated based on Equation (13). Three cycles of sorption–desorption were made, and the percentage of sorption after the first (%*S1*), second (%*S2*), and third (%*S3*), as well as the percentage of desorption after the first (%*D1*), second (%*D2*), and third (%*D3*) cycles were calculated. In %*D2* and %*D3*, the amounts of not-desorbed Ni(II) were taken into account during calculations. The results of Ni(II) sorption and desorption after the first cycles are depicted in Figure 11a. It was found that the eluting agents are characterized by diverse behavior towards loaded Ni(II). The %*D1* was the smallest in the case of the ammonium hydroxide, sodium hydroxide, and sodium chloride solutions. In this case, the %*D1* was close to 2–4%. With the NH_4OH and NaOH concentration increase, the %*D1* was the highest but still very low, i.e., %*D1* = 2.06% for 1 M NH_4OH and %*D1* = 4.23% for 2 M NH_4OH.

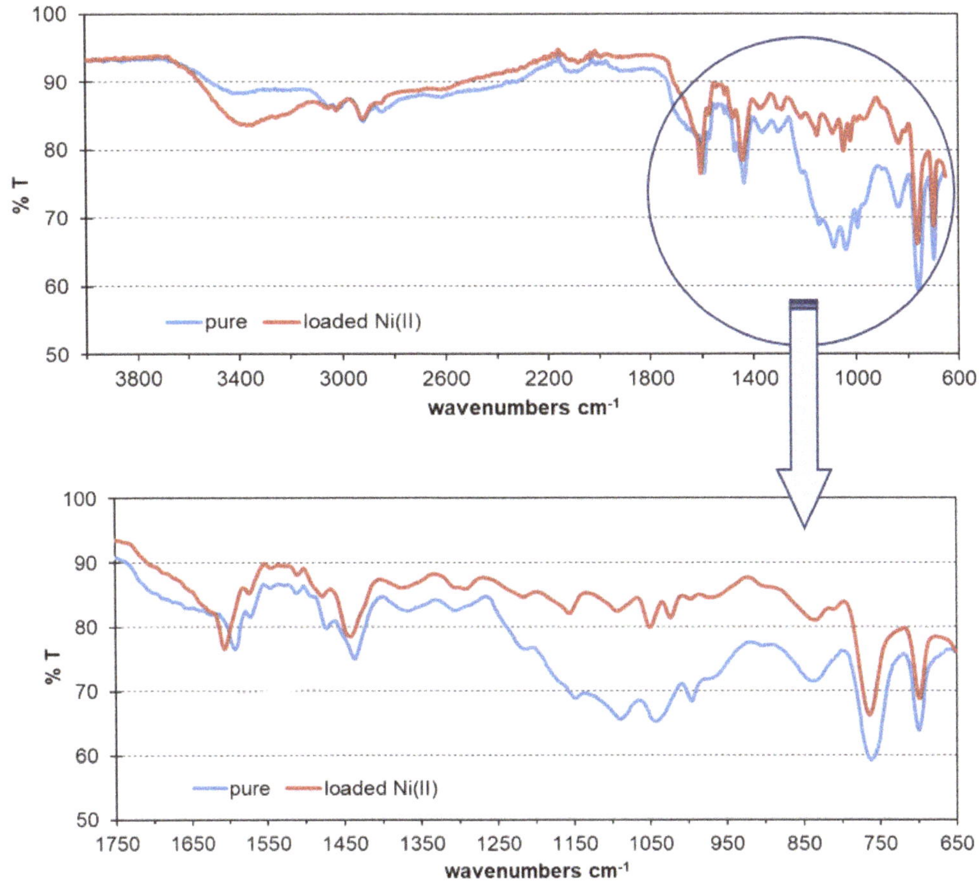

Figure 10. FTIR spectra for TP220 before and after the adsorption of Ni(II).

Much higher desorption yield was obtained using the acids such hydrochloric, nitric(V), and sulfuric(VI) onces acids. In all cases, the desorption yield after the first cycle was close to 22–23%; therefore, acids were selected for desorption in the second and third cycles (Figure 11b). It was found that the ability of TP220 in the second and third cycles was slightly reduced, which resulted from the incomplete regeneration of anion exchange resins, but the adsorption was still high. Moreover, in the second and third desorption cycles, the desorption yield decreased, i.e., %*D1* = 23.8%, %*D2* = 6.7%, %*D3* = 4.5% for 2 M HNO_3. The small desorption yield and its decrease with the increasing number of cycles indicated strong interactions between the Ni(II) and functional groups of TP220, and proved the coordinate mechanism of adsorption.

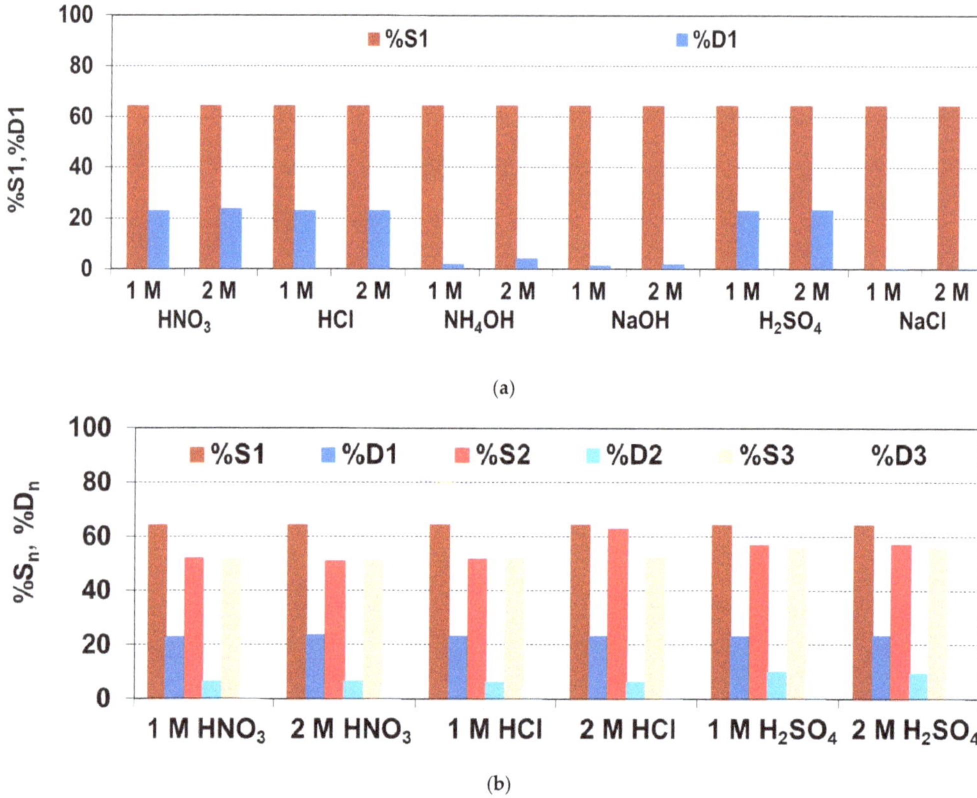

Figure 11. Comparison of %sorption (%*S*) and %desorption (%*D*) for Ni(II) on/from TP220 (**a**) in the first cycle and (**b**) in three cycles of sorption–desorption for the best eluting agents.

4. Conclusions

The adsorbents of different matrices, structures and functional groups, such as S984, TP220, A830, SR7, A400TL, PSR2, PSR3 and AF5, were selected for Ni(II) removal from the acidic solutions of different composition such as HCl, HCl/HNO_3 systems. It was found that Lewatit MonoPlus TP220 shows the highest adsorption ability towards Ni(II) (%*R* = 62.4%) compared to the other adsorbent; %*R* was in the range from 32.9% to 55.2% for the 0.1 M HCl system. In the HCl/HNO_3 systems, TP220 as well as S984 exhibited the highest percentage removal (48.8% for TP220, 54.3% for S984, 0.1 M HCl/0.9 M HNO_3). The adsorption mechanism of Ni(II) on the adsorbents was examined, studying the kinetic and isotherm models. The pseudo-first-order, pseudo-second-order, and intraparticle diffusion kinetic models as well as the Langmuir, Freundlich, Temkin, Dubinin–Radushkevich isotherm models were applied to provide the best fit to the data. Taking into account the correlation coefficient and error analysis, the PSO model showed the best parameters but the fitting was not satisfactory. Intraparticle diffusion, as well as film diffusion, could be the rate limiting steps. Moreover, the Freundlich model confirmed favorable uptake of Ni(II) ions by TP220. The desorption studies (%*D1* = 22–23%) similar to the FTIR-ATR analysis of TP220 before and after the Ni(II) adsorption demonstrated that TP220 interacts with Ni(II), the interactions are strong, the coordinate mechanism takes place, and the functional groups are involved in the Ni(II) removal. TP220 as a new generation anion exchanger is characterized by high adsorption ability of Ni(II) from acidic solutions and may be considered as a good candidate for wastewater treatment from nickel-containing wastes. The preliminary research presented in this paper may become the basis for the development of an effective method of nickel ion removal on an industrial scale. However, additional tests in a column system regarding the kinetics of the process, desorption (100 cycles of sorption–desorption), and the mechanical strength of the ion exchanger, as well as a thorough economic analysis of the process, are required. It is also necessary to investigate the selectivity of the TP220 exchanger towards heavy metals contained in real wastewaters.

Author Contributions: Conceptualization, A.W.; methodology, A.W.; software, A.W. and M.W.; validation, A.W. and M.W.; formal analysis, A.W.; investigation, A.W.; resources, A.W. and M.W.; data curation, A.W.; writing—original draft preparation, A.W. and M.W.; writing—review and editing, A.W. and M.W.; visualization, A.W. and M.W.; supervision, A.W. All authors have read and agreed to the published version of the manuscript.

Funding: This research received no external funding.

Institutional Review Board Statement: Not applicable.

Informed Consent Statement: Not applicable.

Data Availability Statement: Not applicable.

Conflicts of Interest: The authors declare no conflict of interest.

Abbreviations

MCL (mg/L)	maximum contaminant level
HYDRA	Hydrochemical Equilibrium Constant Database
MEDUSA	Make Equilibrium Diagrams Using Sophisticated Algorithms
q_t (mg/g)	the amount of Ni(II) sorbed by the adsorbents at time t
q_e (mg/g)	the amount of Ni(II) sorbed at the equilibrium
C_0 (mg/L)	the Ni(II) concentration in the solution before sorption
C_t (mg/L)	the Ni(II) concentration in the solution after sorption
C_e (mg/L)	the Ni(II) concentrations in the solution after sorption at equilibrium
t (min)	the phase contact time
V (L)	the volume of the Ni(II) solution
m (g)	the mass of the adsorbent
PFO	the pseudo-first-order kinetic model
k_1 (1/min)	the rate constant of sorption determined from PFO equation
PSO	the pseudo-second-order kinetic model
k_2 (g/mg min)	the rate constant of sorption determined from PSO equation
IPD	the intraparticle diffusion kinetic model
k_i (mg/g min$^{0.5}$)	the intraparticle diffusion rate constant
k_F (mg$^{1-1/n}$ L$^{1/n}$/g)	the Freundlich constant related to the adsorption capability
n	the Freundlich constant related to adsorption intensity
k_L (L/mg)	the constant parameter of adsorption equilibrium
Q_0 (mg/g)	the monolayer adsorption capacity
b_T (J g/mol mg)	Temkin constant related to the heat of adsorption
A (L/mg)	the Temkin isotherm equilibrium binding constant
q_m (mg/g)	the maximum adsorption capacity
k_{DR} (mol^2 J^2)	the constant related to the adsorption energy
ε (J/mol)	the adsorption potential
R (J/mol K)	the gas constant
T (K)	the temperature
MPSD	Marquardt's percent standard deviation
R^2	the determination coefficient
R^2_{adj}	the adjusted R-squared

$q_{e\,exp}$ (mg/g)	the experimental amount of Ni(II) sorbed at equilibrium
$q_{e\,cal}$ (mg/g)	the amount of Ni(II) sorbed calculated from the non-linear models
$q_{e\,mean}$ (mg/g)	the measured by the means of $q_{e\,exp}$ values
n	the number of points in the data sample
k	the number of independent regressors
FTIR-ATR	the Fourier-transform infrared spectroscopy with the attenuated total reflection
D (%)	the percentage values of Ni(II) desorbed from adsorbent
m_{des} (mg)	the mass of Ni(II) desorbed
m_{ads} (mg)	the mass of Ni(II) adsorbed.

References

1. Iyaka, Y.A. Nickel in soils: A review of its distribution and impacts. *Sci. Res. Essays* **2011**, *6*, 6774–6777. [CrossRef]
2. Vakili, M.; Rafatullah, M.; Yuan, J.; Zwain, H.M.; Mojiri, A.; Gholami, Z.; Gholami, F.; Wang, W.; Giwa, A.S.; Yu, Y.; et al. Nickel ion removal from aqueous solutions through the adsorption process: A review. *Rev. Chem. Eng.* **2020**. [CrossRef]
3. Harasim, P.; Filipek, T. Nickel in the environment. *J. Elem.* **2015**, *20*, 525–534. [CrossRef]
4. Duda-Chodak, A.; Błaszczyk, U. The impact of nickel on human health. *J. Elementol.* **2008**, *13*, 685–696.
5. Nieminen, T.M.; Ukonmaanaho, L.; Rausch, N.; Shotyk, W. Biogeochemistry of Nickel and Its Release into the Environment. *Nickel Surpris. Impact Nat.* **2007**, *2*, 1–29. [CrossRef]
6. Pacyna, J.M.; Pacyna, E.G. An assessment of global and regional emissions of trace metals to the atmosphere from anthropogenic sources worldwide. *Environ. Rev.* **2001**, *9*, 269–298. [CrossRef]
7. Gambuś, F.; Wieczorek, J. Pollution of fertilizers with heavy metals. *Ecol. Chem. Eng. A* **2012**, *19*, 353–360. [CrossRef]
8. Łukowski, A.; Wiater, J. The influence of mineral fertilization on heavy metal fraction contents in soil. Part II. Copper and nickel. *Pol. J. Environ. Stud.* **2009**, *18*, 645–650.
9. Coman, V.; Robotin, B.; Ilea, P. Nickel recovery/removal from industrial wastes: A review. *Resour. Conserv. Recycl.* **2013**, *73*, 229–238. [CrossRef]
10. Barakat, M. New trends in removing heavy metals from industrial wastewater. *Arab. J. Chem.* **2011**, *4*, 361–377. [CrossRef]
11. Renu; Agrawal, M.; Singh, K. Heavy metal removal from wastewater using various adsorbents: A review. *J. Water Reuse Desalin.* **2016**, *7*, 387–419. [CrossRef]
12. Tripathi, A.; Ranjan, M.R. Heavy Metal Removal from Wastewater Using Low Cost Adsorbents. *J. Bioremediat. Biodegrad.* **2015**, *6*, 315–320. [CrossRef]
13. Borba, C.; Guirardello, R.; Silva, E.; Veit, M.; Tavares, C. Removal of nickel(II) ions from aqueous solution by biosorption in a fixed bed column: Experimental and theoretical breakthrough curves. *Biochem. Eng. J.* **2006**, *30*, 184–191. [CrossRef]
14. Genchi, G.; Carocci, A.; Lauria, G.; Sinicropi, M.S.; Catalano, A. Nickel: Human Health and Environmental Toxicology. *Int. J. Environ. Res. Public Health* **2020**, *17*, 679. [CrossRef]
15. EFSA Panel on Contaminants in the Food Chain (CONTAM). Scientific Opinion on the risks to public health related to the presence of nickel in food and drinking water. *EFSA J.* **2015**, *13*, 4002. [CrossRef]
16. Khulbe, K.C.; Matsuura, T. Removal of heavy metals and pollutants by membrane adsorption techniques. *Appl. Water Sci.* **2018**, *8*, 19. [CrossRef]
17. Liu, L.; Li, W.; Song, W.; Guo, M. Remediation techniques for heavy metal-contaminated soils: Principles and applicability. *Sci. Total Environ.* **2018**, *633*, 206–219. [CrossRef]
18. Kumar, D.; Khan, E.A. 12—Remediation and detection techniques for heavy metals in the environment. In *Heavy Metals in the Environment: Impact, Assessment, and Remediation*; Candice Janco: Cambridge, MA, USA, 2020; pp. 205–222.
19. Geng, H.; Xu, Y.; Zheng, L.; Gong, H.; Dai, L.; Dai, X. An overview of removing heavy metals from sewage sludge: Achievements and perspectives. *Environ. Pollut.* **2020**, *266*, 115375. [CrossRef]
20. Azimi, A.; Azari, A.; Rezakazemi, M.; Ansarpour, M. Removal of Heavy Metals from Industrial Wastewaters: A Review. *ChemBioEng Rev.* **2017**, *4*, 37–59. [CrossRef]
21. Beltrán-Heredia, J.; Martín, J.S. Removing heavy metals from polluted surface water with a tannin-based flocculant agent. *J. Hazard. Mater.* **2009**, *165*, 1215–1218. [CrossRef]
22. Al-Shannag, M.; Al-Qodah, Z.; Bani-Melhem, K.; Qtaishat, M.R.; Alkasrawi, M. Heavy metal ions removal from metal plating wastewater using electrocoagulation: Kinetic study and process performance. *Chem. Eng. J.* **2015**, *260*, 749–756. [CrossRef]
23. Lakhdhar, I.; Belosinschi, D.; Mangin, P.; Chabot, B. Development of a bio-based sorbent media for the removal of nickel ions from aqueous solutions. *J. Environ. Chem. Eng.* **2016**, *4*, 3159–3169. [CrossRef]
24. Mohsen-Nia, M.; Montazeri, P.; Modarress, H. Removal of Cu2+ and Ni2+ from wastewater with a chelating agent and reverse osmosis processes. *Desalination* **2007**, *217*, 276–281. [CrossRef]
25. Samper, E.; Rodríguez, M.; De La Rubia, M.; Rico, D.P. Removal of metal ions at low concentration by micellar-enhanced ultrafiltration (MEUF) using sodium dodecyl sulfate (SDS) and linear alkylbenzene sulfonate (LAS). *Sep. Purif. Technol.* **2009**, *65*, 337–342. [CrossRef]

26. Moghbeli, M.; Khajeh, A.; Alikhani, M. Nanosilica reinforced ion-exchange polyHIPE type membrane for removal of nickel ions: Preparation, characterization and adsorption studies. *Chem. Eng. J.* **2017**, *309*, 552–562. [CrossRef]
27. Anirudhan, T.S.; Radhakrishnan, P.G. Uptake and desorption of nickel(II) using polymerised tamarind fruit shell with acidic functional groups in aqueous environments. *Chem. Ecol.* **2010**, *26*, 93–109. [CrossRef]
28. Fil, B.A.; Boncukcuoğlu, R.; Yilmaz, A.E.; Bayar, S. Adsorption of Ni(II) on ion exchange resin: Kinetics, equilibrium and thermodynamic studies. *Korean J. Chem. Eng* **2020**, *29*, 1232–1238. [CrossRef]
29. Crini, G.; Lichtfouse, E. Advantages and disadvantages of techniques used for wastewater treatment. *Environ. Chem. Lett.* **2019**, *17*, 145–155. [CrossRef]
30. Sharma, S.; Bhattacharya, A. Drinking water contamination and treatment techniques. *Appl. Water Sci.* **2017**, *7*, 1043–1067. [CrossRef]
31. Kanamarlapudi, S.L.R.K.; Chintalpudi, V.K.; Muddada, S. Application of biosorption for removal of heavy metals from wastewater. *InTech* **2018**, *4*, 69–116.
32. Couto, C.F.; Lange, L.C.; Amaral, M.C.S. A critical review on membrane separation processes applied to remove pharmaceutically active compounds from water and wastewater. *J. Water Process. Eng.* **2018**, *26*, 156–175. [CrossRef]
33. Fu, F.; Wang, Q. Removal of heavy metal ions from wastewaters: A review. *J. Environ. Manag.* **2011**, *92*, 407–418. [CrossRef]
34. Rubio, J.; Souza, M.L.; Smith, R.W. Overview of flotation as a wastewater treatment technique. *Miner. Eng.* **2002**, *15*, 139–155. [CrossRef]
35. Ahmad Panahi, H.; Samadi Zadeh, M.; Tavangari, S.; Moniri, E.; Ghassemi, J. Nickel adsorption from environmen-tal samples by ion imprinted aniline-formaldehyde polymer. *Iranian J. Chem. Eng.* **2012**, *31*, 35–44.
36. El-Sadaawy, M.; Abdelwahab, O. Adsorptive removal of nickel from aqueous solutions by activated carbons from doum seed (Hyphaenethebaica) coat. *Alex. Eng. J.* **2014**, *53*, 399–408. [CrossRef]
37. Do, Q.C.; Choi, S.; Kim, H.; Kang, S. Adsorption of Lead and Nickel on to Expanded Graphite Decorated with Manganese Oxide Nanoparticles. *Appl. Sci.* **2019**, *9*, 5375. [CrossRef]
38. Silva, R.M.P.; Manso, J.P.H.; Rodrigues, J.R.; Lagoa, R. A comparative study of alginate beads and an ion-exchange resin for the removal of heavy metals from a metal plating effluent. *J. Environ. Sci. Health Part A* **2008**, *43*, 1311–1317. [CrossRef]
39. Shen, Z.; Zhang, Y.; McMillan, O.; Jin, F.; Al-Tabbaa, A. Characteristics and mechanisms of nickel adsorption on biochars produced from wheat straw pellets and rice husk. *Environ. Sci. Pollut. Res.* **2017**, *24*, 12809–12819. [CrossRef]
40. Kołodyńska, D. Polyacrylate anion exchangers in sorption of heavy metal ions with the biodegradable complexing agent. *Chem. Eng. J.* **2009**, *150*, 280–288. [CrossRef]
41. Anah, L.; Astrini, N. Isotherm adsorption studies of Ni(II) ion removal from aqueous solutions by modified carboxymethyl cellulose hydrogel. *IOP Conf. Ser. Earth Environ. Sci.* **2018**, *160*, 012017. [CrossRef]
42. Wołowicz, A.; Hubicki, Z. Polyactrylate ion exchangers in sorption of noble and base metal ions from single and tertiary component solutions. *Solvent Extr. Ion Exch.* **2014**, *32*, 189–205. [CrossRef]
43. Wołowicz, A.; Hubicki, Z. The use of the chelating resin of a new generation Lewatit MonoPlus TP-220 with the bis-picolylamine functional groups in the removal of selected metal ions from acidic solutions. *Chem. Eng. J.* **2012**, *197*, 493–508. [CrossRef]
44. Wołowicz, A.; Hubicki, Z. Sorption Behavior of Dowex PSR-2 and Dowex PSR-3 Resins of Different Structures for Metal(II) Removal. *Solvent Extr. Ion Exch.* **2016**, *34*, 375–397. [CrossRef]
45. Wołowicz, A.; Hubicki, Z. Carbon-based adsorber resin Lewatit AF 5 applicability in metal ion recovery. *Microporous Mesoporous Mater.* **2016**, *224*, 400–414. [CrossRef]
46. Wołowicz, A. Zinc(II) removal from model chloride and chloride–nitrate(V) solutions using various sorbents. *Physicochem. Probl. Miner. Process.* **2019**, *55*, 1517–1534.
47. Wołowicz, A.; Hubicki, Z. Enhanced removal of copper(II) from acidic streams using functional resins: Batch and column studies. *J. Mater. Sci.* **2020**, *55*, 13687–13715. [CrossRef]
48. How Much Does It Cost to Buy, Maintain, and Dispose of Ion Exchange Resins? Available online: Samcotech.com/how-much-does-it-cost-to-buy-maintain-and-dispose-of-ion-exchange-resins/ (accessed on 22 January 2021).
49. Zong, L.; Liu, F.; Chen, D.; Zhang, X.; Ling, C.; Li, A. A novel pyridine based polymer for highly efficient separation of nickel from high-acidity and high-concentration cobalt solutions. *Chem. Eng. J.* **2018**, *334*, 995–1005. [CrossRef]
50. Ho, Y.S.; McKay, G. Pseudo-second order model for sorption processes. *Process. Biochem.* **1999**, *34*, 451–465. [CrossRef]
51. Wang, J.; Guo, X. Adsorption kinetic models: Physical meanings, applications, and solving methods. *J. Hazard. Mater.* **2020**, *390*, 122156. [CrossRef]
52. Foo, K.; Hameed, B. Insights into the modeling of adsorption isotherm systems. *Chem. Eng. J.* **2010**, *156*, 2–10. [CrossRef]
53. Marković, D.D.; Lekić, B.; Rajaković-Ognjanović, V.N.; Onjia, A.; Rajaković, L.V. A New Approach in Regression Analysis for Modeling Adsorption Isotherms. *Sci. World J.* **2014**, *2014*. [CrossRef] [PubMed]
54. Sivarajasekar, N.; Baskar, R. Adsorption of Basic Magenta II onto H_2SO_4 activated immature Gossypium hirsutum seeds: Kinetics, isotherms, mass transfer, thermodynamics and process design. *Arab. J. Chem.* **2019**, *12*, 1322–1337. [CrossRef]
55. Uzun, I.; Guzel, F. Adsorption of some heavy metal ions from aqueous solution by activated carbon and comparison of percent adsorption results of activated carbon with those of some other adsorbents. *Turk. J. Chem.* **2000**, *24*, 291–297.
56. Satapathy, D.; Natarajan, G.S. Potassium bromate modification of the granular activated carbon and its effect on nickel adsorption. *Adsorption* **2006**, *12*, 147–154. [CrossRef]

57. Wei, Y.-Z.; Feng, Q.; Arai, T.; Kumagai, M. Adsorption and Separation Behaviour of Cobalt, Nickel, and Copper in Nitrite Medium by Anion Exchanger. *Solvent Extr. Ion Exch.* **2002**, *20*, 561–573. [CrossRef]
58. Keränen, A.; Leiviskä, T.; Salakka, A.; Tanskanen, J. Removal of nickel and vanadium from ammoniacal industrial wastewater by ion exchange and adsorption on activated carbon. *Desalin. Water Treat.* **2015**, *53*, 2645–2654. [CrossRef]
59. Xiuling, S.; HuiPu, D.; Shijun, L.; Hui, Q. Adsorption Properties of Ni(II) by D301R Anion Exchange Resin. *J. Chem.* **2014**, *2014*, 1–5. [CrossRef]
60. Deepatana, A.; Valix, M. Recovery of nickel and cobalt from organic acid complexes: Adsorption mechanisms of metal-organic complexes onto aminophosphonate chelating resin. *J. Hazard. Mater.* **2006**, *137*, 925–933. [CrossRef]
61. Kołodyńska, D.; Hubicka, H. Polyacrylate anion exchangers in sorption of heavy metal ions with non-biodegradable complexing agents. *Chem. Eng. J.* **2009**, *150*, 308–315. [CrossRef]
62. Douven, S.; Paez, C.A.; Gommes, C.J. The range of validity of sorption kinetic models. *J. Colloid Interface Sci.* **2015**, *448*, 437–450. [CrossRef]
63. Płaziński, W.; Rudziński, W. Kinetyka adsorpcji na granicy faz roztwór/ciało stałe. Znaczenie równań pseudo-first order oraz pseudo-second order. *Wiadomości Chem.* **2011**, *65*, 11–12.
64. Hu, Q.; Zhang, Z. Application of Dubinin–Radushkevich isotherm model at the solid/solution interface: A theoretical analysis. *J. Mol. Liq.* **2019**, *277*, 646–648. [CrossRef]
65. Morris, D.; Reed, G.; Short, E.; Slater, D.; Waters, D. Nickel (II) chloride complexes in aqueous solution. *J. Inorg. Nucl. Chem.* **1965**, *27*, 377–382. [CrossRef]
66. Mohanty, U.S.; Rintala, L.; Halli, P.; Avarmaa, K.; Reuter, M.A. Hydrometallurgical Approach for Leaching of Metals from Copper Rich Side Stream Originating from Base Metal Production. *Metal* **2018**, *8*, 40. [CrossRef]
67. Lee, M.S.; Nam, S.H. Chemical Equilibria of Nickel Chloride in HCl Solution at 25 °C. *Bull. Korean Chem. Soc.* **2009**, *30*, 2203–2207. [CrossRef]
68. Kołodyńska, D.; Hałas, P.; Michalski, R. Development of ion exchangers for the removal of health-hazardous perchlorate ions from aqueous systems. *Appl. Geochem.* **2019**, *101*, 75–87. [CrossRef]
69. Zagorodni, A.A.; Kotova, D.L.; Selemenev, V.F. Infrared spectroscopy of ion exchange resins: Chemical deterioration of the resins. *React. Funct. Polym.* **2002**, *53*, 157–171. [CrossRef]
70. Lazar, L.; Bulgariu, L.; Bandrabur, B.; Tataru-Farmus, R.-E.; Drobotă, M.; Gutt, G. FTIR Analisys of ion Exchange Resins with Application in Permanent Hard Water Softening. *Environ. Eng. Manag. J.* **2014**, *13*, 2145–2152. [CrossRef]
71. Ghosh, S.; Dhole, K.; Tripathy, M.K.; Kumar, R.; Sharma, R.S. FTIR spectroscopy in the characterization of the mixture of nuclear grade cation and anion exchange resins. *J. Radioanal. Nucl. Chem.* **2015**, *304*, 917–923. [CrossRef]
72. Traboulsi, A.; Dupuy, N.; Rebufa, C.; Sergent, M.; Labed, V. Investigation of gamma radiation effect on the anion exchange resin Amberlite IRA-400 in hydroxide form by Fourier transformed infrared and 13C nuclear magnetic resonance spectroscopies. *Anal. Chim. Acta* **2012**, *717*, 110–121. [CrossRef]

processes

Article

The Comparison of the Efficacy of Natural and Synthetic Aluminosilicates, Including Zeolites, in Concurrent Elimination of Lead and Copper from Multi-Component Aqueous Solutions

Bożena Kozera-Sucharda [1], Barbara Gworek [2], Igor Kondzielski [2,*] and Józef Chojnicki [1]

1 Department of Soil Science, Warsaw University of Life Sciences, Nowoursynowska Str. 159, 02-776 Warsaw, Poland; bozena.sucharda@sumiagro.pl (B.K.-S.); jozef_chojnicki@sggw.edu.pl (J.C.)
2 Institute of Environmental Protection—National Research Institute, Krucza 5/11D Str., 00-548 Warsaw, Poland; barbara.gworek@ios.edu.pl
* Correspondence: igor.kondzielski@ios.edu.pl; Tel.: +48-22-375-05-78; Fax: +48-22-375-05-01

Abstract: The unique and outstanding physical and chemical properties of aluminosilicate minerals, including zeolites, make them extremely useful in remediation processes. That is due to their demonstrated high efficiency, inexpensiveness, and environmental friendliness in processes aimed on the elimination of heavy metals from water. The paper reports the results of the examination of selectivity of the tested clay minerals and zeolites toward different heavy metals in light of the postulated sorption mechanisms. It was stated that while the most efficient at concurrent removal of lead and copper from aqueous solutions were synthetic zeolites 3A and 10A, smectite was the best in dealing with prolonged pollution with Pb^{2+} and Cu^{2+}. Determined as one of the parameters in DKR isotherm energy of the process for each combination of sorbate and sorbent, it showed that the dominant mechanism of adsorption on the tested mineral sorbents was physisorption. The exception was kaolinite, for which that energy implied ion exchange as the dominant mechanism of the process.

Keywords: water; metals; smectite; kaolinite; zeolites

Citation: Kozera-Sucharda, B.; Gworek, B.; Kondzielski, I.; Chojnicki, J. The Comparison of the Efficacy of Natural and Synthetic Aluminosilicates, Including Zeolites, in Concurrent Elimination of Lead and Copper from Multi-Component Aqueous Solutions. *Processes* **2021**, *9*, 812. https://doi.org/10.3390/pr9050812

Academic Editor: Monika Wawrzkiewicz

Received: 24 November 2020
Accepted: 2 May 2021
Published: 6 May 2021

Highlights

- Synthetic zeolites remove Pb^{2+} and Cu^{2+} ions from multi-component aqueous solutions faster and more efficiently than natural minerals;
- The natural aluminosilicate smectite has a higher sorption capacity than synthetic aluminosilicates over a wider range of concentrations;
- Lead displays about 30% higher affinity towards the tested aluminosilicates than copper;
- Natural and synthetic aluminosilicates are able to reduce the mobility of Pb^{2+} and Cu^{2+} in aquatic environments.

1. Introduction

One of the results of technological development is an increasing level of environmental pollution with various compounds, including heavy metals. That contributes to the decrease in freshwater resources and lowering of their quality and is related to the impacted recovery and self-purification abilities of natural freshwater ecosystems [1–6].

Heavy metals are among the most important environmental pollutants, because of their high potential for accumulation in various components of the environment. Mining and processing of non-ferrous metals, the activities which generate a broad spectrum of solid and liquid wastes often bearing high amounts of those elements, are enumerated among the most important sources of pollution of environment with heavy metals. Pollution from these sources impacts the environment and poses a significant threat to the human health due to the inclusion of heavy metals into the food chain.

The experimental hypothesis of this work is based on the earlier experience of the authors on reducing the incorporation of heavy metals in the trophic chain by minimising their uptake from soils by plants and the results of other researchers examining the elimination of these elements from polluted waters [7–20]. At present several various methods of elimination of heavy metals from water, based on the sorption phenomena, are characterised and used in practice. Among those newly developed and most promising are physisorption on nanomaterials and ultrafiltration with nanomembranes [21–24].

In this work it was assumed that the materials used to eliminate heavy metals from water should display following three features: high efficiency, inexpensiveness, and environmental friendliness. These conditions are met by the natural and synthetic aluminosilicate minerals.

The main task of the study was to determine the sorption capacity and selectivity of layered aluminosilicate minerals (clays) and porous aluminosilicate minerals containing networks of pores and chambers (zeolites) toward lead and copper present in a multi-solute aqueous solution. An additional goal was to identify the would-be mechanisms of Pb^{2+} and Cu^{2+} ions onto the tested sorbents. That led to the assessment of the possibility of using the tested minerals for the decontamination of water, in particular in case of accidental emissions. From a practical point of view, the obtained results should give the practitioners a clear indication on the selection of the adequate sorbent to cope with accidental high-concentration releases of lead and copper into an aquatic environment, hitherto seldom considered in scientific publications.

For that reason, the concentrations of metals used in the experiment were correlated with the Cation Exchange Capacity (CEC) of the tested minerals. That was done to clearly and unambiguously determine the efficiency of the tested minerals in removing the pollutants from the purified matrix by means of sorption.

To meet the above aims of the study, the sorption from aqueous solution onto aluminosilicate minerals of Pb^{2+} and Cu^{2+} dissolved alongside Cd^{2+}, and Zn^{2+} was examined.

The reason for that was two-fold. First, these two metals are on the list of the most common metallic environmental pollutants.

It was stated that the affinity to clay minerals is much stronger in the case of Pb^{2+}, than that for Cu^{2+}. Due to these similarities in behaviour in the environment, it was decided to examine and compare the sorption of those two elements on the selected aluminosilicates, in order to attempt to determine the similarities and differences of their behaviour.

It is important to remove those compounds from the surface water compartment, in order to limit their inclusion in the food chain [3,6,24–26].

2. Materials and Methods

In the study, two kinds of aluminosilicate minerals—natural clays, having a layered structure, and zeolites—porous minerals also named "molecular sieves" were used.

Two natural clay minerals were selected—kaolinite and smectite. The layers of those minerals are composed of sheets containing silicate tetrahedra arranged in hexagons linked with sheets formed of octahedra containing Al atoms. The Si atoms in tetrahedra may be substituted by Al atoms, while Al atoms in octahedra by Mg and Fe. Depending on the arrangement of tetrahedron and octahedron sheets in a layer, the aluminosilicates are divided into two-layer (1:1) and three-layer (2:1) structures. The main representative of two-layer aluminosilicates is kaolinite, while that of three-layer minerals is smectite.

That results in different sorption properties. In the case of two-layer aluminosilicates, having layers linked by hydrogen bonds, interlayer sorption is impossible, and the sorption occurs only on the grain surface (as in kaolinite). In the three-layer minerals, such as smectite, the layers are linked by weak intermolecular electrostatic forces, such as van der Waals forces. As a result, sorption onto these minerals occurs on the surface of the mineral, as well as in the internal interlayer space.

The kaolinite used in the study originated from deposits located in southern Poland and smectite from deposits located in Milwaukee, WI, USA.

Alongside them, four zeolites were tested—A natural Cliniptilolite and three synthetic zeolites: 3A, 10a, and 13X. Clinoptilolite came from Caucasian deposits—A zeolite-bearing rock containing 90% clinoptilolite from Sokyrnytsa mine, Zakarpatye region, Ukraine. All three synthetic zeolites, which are commercially available chemicals, were purchased from a manufacturer (IZC "Soda-Mątwy", Inowrocław, Poland) who also provided their SDS (Safety Data Sheet) cards.

Zeolites are crystalline hydrated aluminosilicates with highly variable internal structures. They consist of silicate tetrahedra linked by the oxygen bridges, in which central Si atoms may be heterovalently substituted by Al atoms or the elements belonging to the groups Ia and IIa of the periodic table (Li^+, Na^+, K^+, Mg^{2+}, Ca^{2+}, Sr^{2+}, and Ba^{2+}); preferably Mg^{2+}. Their characteristic feature are empty spaces packed with ions and molecules of water displaying a high degree of freedom of movement. Their alternate name—"molecular sieves" is due to their ability to selectively sorb the chemical molecules smaller than their pores.

Clinoptilolite, the selected natural zeolite, has an experimentally determined Si/Al ratio of 2:5, equal to 1:2.5, K^+ and Ca^{2+} as dominant exchangeable cations, and pore diameter of 0.44–0.55 nm. Synthetic zeolite 3A is a sodium-and-potassium zeolite, while 10A and 13x are both sodium zeolites. Their pore diameter is 0.38 nm for 3A, 0.9–1.0 nm for 10A and 0.9–1.0 nm for 13X. For all three zeolites the Si/Al ratio reported by the manufacturer was 2:4 (equal to 1:2).

In Table 1 below, the key properties of each tested aluminosilicate mineral are provided.

Table 1. The key properties of minerals used in the experiment.

Property		Minerals					
		Natural Aluminosilicates			Synthetic Zeolites		
		Kaolinite	Smectite	Natural Zeolite	3A	10A	13X
Grain size [mm]		<0.2	<0.2	0.05–1.0	1.0	1.0	1.0
pH	in H_2O	6.6	9.5	6.1	10.2	9.4	10.1
	in 1M KCl	5.0	7.9	4.7	8.8	7.9	8.9
Content of exchange-able cations [cmol/kg]	Total—CEC	4.9	120.7	93.9	354.1	377.9	235.2
	Na^+	0.3	68.2	1.7	146.5	364.2	223.7
	K^+	0.2	1.7	46.1	200.3	8.4	3.8
	Mg^{2+}	0.7	10.1	1.5	0.5	2.4	1.5
	Ca^{2+}	3.7	40.7	44.6	6.8	2.9	6.2
	Dominant cations	Ca^{2+}	Na^+ and Ca^{2+}	K^+ and Ca^{2+}	K^+ and Na^+	Na^+	Na^+

CEC–Cation Exchange Capacity.

The adsorption of lead and copper, as Pb^{2+} and Cu^{2+}, onto selected six minerals was carried out using aqueous solutions also containing Zn^{2+} and Cd^{2+}, prepared using the serial dilution method from respective stock solutions. Their initial concentrations were set to 2% CEC, 10% CEC, 20% CEC, 30% CEC, 50% CEC, 75% CEC, and 100% CEC (Cation Exchange Capacity) of the given mineral sorbent.

Adequate amounts of individual, analytical grade solid nitrate (V) salts (purchased from Merck™) were dissolved in deionized water to prepare stock solutions. The obtained mixtures had a total concentration of test ions equal to 100% CEC of the given mineral sorbent.

The experiment was performed in line with the provisions of the OECD Guideline 106 [27] and it consisted of two stages.

The initial stage was aimed at the determination of the adequate incubation temperature and equilibration time, and in general, it was performed to confirm the previous findings of the authors (for that reason it was shortened to the absolute minimum). The aim of the definitive test was to determine the sorption capacity of each tested mineral towards each of the test metal ions.

The whole experiment was performed in the batch mode using the pre-defined sorbent:solution ratio of 1:25.

In the definitive test, seven polymetallic solutions were used. Their concentrations are presented above. The samples were placed on a on a horizontal shaker (type EIMI WS) in a water bath having a constant temperature of T = 20 ± 1 °C and equilibrated. After that each sample was centrifuged. Clarified supernatants were collected and analyzed for the content of heavy metals using the AAS technique with flame atomization (equipment: Carl Zeiss Jenoptic). The method of analysis, and in particular, the atomization technique, was selected to fit to the analyzed concentrations, which covered the broad range from 0.1 ppm to ~6000 ppm while simultaneously limiting the number of necessary dilutions. For the same reason, depending on the range of expected concentrations in analyzed solutions, different spectral lines corresponding to the different level of sensitivity and pre-defined calibration curves covering the different ranges of concentrations were selected. For each set of analyses, a single calibration sample was used as a means of control. That analysis provided the equilibrium concentrations in solution—C_e.

From the difference between the initial and the equilibrium concentrations in test solutions, the corresponding equilibrium concentrations of each metal adsorbed onto sorbent were calculated—x/m values (indirect method of determining the sorption isotherms).

The obtained equilibrium concentrations C_e and x/m were used to determine the following parameters of the process:

- Percentage of sorption;
- Distribution coefficient Kd;
- Freundlich sorption isotherm and its parameters—K_f and 1/n;
- Langmuir's sorption isotherm and its parameters—K_L and N_S;
- DKR sorption isotherm and its parameters;

The detailed characterisation of the data-processing procedure is characterized in our previous work [28].

3. Results and Discussion

In the preliminary experiments it was found that the equilibrium state was attained after 1 h, while the optimum sorption was observed at T = 20 °C, the temperature considered as representative of average experimental conditions. For that reason, in the definitive test, samples were equilibrated for 1 h at the constant temperature T = 20 °C.

Two parameters were calculated for each tested combination M^{2+}—mineral sorbent in relation to the tested concentrations—the percentage of sorption and the distribution coefficient K_d. The results are presented below in two tables. Table 2 provides the results for sorption onto natural minerals calculated. The results are presented, in numerical form, in the two tables below: Table 2 for the sorption of Pb^{2+} and Cu^{2+} ions onto natural minerals and Table 3 for the sorption of both elements onto synthetic zeolites.

Table 2. The results of the examination of sorption of Pb^{2+} and Cu^{2+} onto tested natural mineral sorbents—% sorption and K_d values.

Sorbent: Kaolinite							
Sorbed Element: Pb^{2+}				**Sorbed Element: Cu^{2+}**			
Initial concentration of Pb^{2+} in solution expressed:		% sorption	K_d (L/kg)	Initial concentration of Cu^{2+} in solution expressed:		% sorption	K_d (L/kg)
in (cmol/L)	as % CEC			in (cmol/L)	as % CEC		
0.004	2	100.00	n. c.	0.004	2	100	n. c.
0.018	10	100.00	n. c.	0.021	10	95.24	500.00

Table 2. *Cont.*

Sorbent: Kaolinite							
Sorbed Element: Pb^{2+}				Sorbed Element: Cu^{2+}			
Initial concentration of Pb^{2+} in solution expressed:		% sorption	K_d (L/kg)	Initial concentration of Cu^{2+} in solution expressed:		% sorption	K_d (L/kg)
in (cmol/L)	as % CEC			in (cmol/L)	as % CEC		
0.037	20	88.89	437.50	0.043	20	81.39	109.38
0.055	30	89.09	204.17	0.064	30	71.88	63.89
0.074	50	82.42	117.31	0.085	50	63.53	43.55
0.111	75	74.77	74.11	0.128	75	55.47	31.14
0.185	100	61.11	39.24	0.213	100	44.60	20.13
Sorbent: Smectite							
Sorbed element: Pb^{2+}				Sorbed element: Cu^{2+}			
Initial concentration of Pb^{2+} in solution expressed:		% sorption	K_d (L/kg)	Initial concentration of Cu^{2+} in solution expressed:		% sorption	K_d (L/kg)
in (cmol/lL)	as % CEC			in (cmol/L)	as % CEC		
0.01	2	90.00	225.00	0.01	2	80.00	100.00
0.07	10	97.14	850.00	0.07	10	97.14	850.00
0.14	20	97.86	1141.67	0.15	20	98.67	1850.00
0.21	30	96.67	725.00	0.22	30	85.91	152.42
0.35	50	97.43	947.22	0.36	50	70.55	59.91
0.49	75	83.88	130.06	0.51	75	60.20	37.81
0.70	100	64.57	45.56	0.72	100	49.58	24.59
Sorbent: Natural Zeolite							
Sorbed element: Pb^{2+}				Sorbed element: Cu^{2+}			
Initial concentration of Pb^{2+} in solution expressed:		% sorption	K_d (L/kg)	Initial concentration of Cu^{2+} in solution expressed:		% sorption	K_d (L/kg)
in (cmol/L)	as % CEC			in (cmol/L)	as % CEC		
0.02	2	100.00	n. c.	0.02	2	85.00	143.33
0.10	10	100.00	n. c.	0.10	10	70.00	58.33
0.20	20	85.00	141.67	0.20	20	55.00	30.56
0.30	30	76.67	82.14	0.31	30	41.93	18.06
0.50	50	60.00	37.50	0.51	50	33.33	12.50
0.70	75	51.43	26.47	0.71	75	29.58	10.50
1.00	100	46.00	21.30	1.02	100	30.39	10.91

Table 3. The results of the examination of sorption of Pb^{2+} and Cu^{2+} onto tested synthetic zeolites—% sorption and K_d values.

Sorbent: Zeolite 3A							
Sorbed Element: Pb^{2+}				Sorbed Element: Cu^{2+}			
Initial concentration of Pb^{2+} in solution expressed:		% sorption	K_d (L/kg)	Initial concentration of Cu^{2+} in solution expressed:		% sorption	K_d (L/kg)
in (cmol/L)	as % CEC			in (cmol/L)	as % CEC		
0.06	2	100.00	n. c.	0.07	2	85.71	150.00
0.32	10	81.25	108.33	0.33	10	45.45	20.83

Table 3. *Cont.*

Sorbent: Zeolite 3A							
Sorbed Element: Pb^{2+}				Sorbed Element: Cu^{2+}			
Initial concentration of Pb^{2+} in solution expressed:		% sorption	K_d (L/kg)	Initial concentration of Cu^{2+} in solution expressed:		% sorption	K_d (L/kg)
in (cmol/L)	as % CEC			in (cmol/L)	as % CEC		
0.64	20	57.81	34.26	0.65	20	32.31	11.93
0.96	30	75.00	75.00	0.98	30	57.14	33.33
1.61	50	73.91	70.83	1.64	50	56.10	29.14
2.25	75	60.44	38.20	2.29	75	39.30	16.19
3.32	100	59.04	36.03	3.27	100	37.76	16.08
Sorbent: Zeolite 10A							
Sorbed element: Pb^{2+}				Sorbed element: Cu^{2+}			
Initial concentration of Pb^{2+} in solution expressed:		% sorption	K_d (L/kg)	Initial concentration of Cu^{2+} in solution expressed:		% sorption	K_d (L/kg)
in (cmol/L)	as % CEC			in (cmol/L)	as % CEC		
0.06	2	100.00	n. c.	0.07	2	40.00	10.00
0.32	10	96.88	775	0.33	10	57.88	33.93
0.64	20	87.50	175	0.65	20	49.23	24.24
0.96	30	88.54	193.18	0.98	30	44.90	20.37
1.61	50	85.71	150.00	1.64	50	54.88	30.40
2.25	75	73.33	68.75	2.24	75	41.52	18.70
3.22	100	66.77	50.23	3.27	100	30.58	11.01
Sorbent: Zeolite 13X							
Sorbed element: Pb^{2+}				Sorbed element: Cu^{2+}			
Initial concentration of Pb^{2+} in solution expressed:		% sorption	K_d (L/kg)	Initial concentration of Cu^{2+} in solution expressed:		% sorption	K_d (L/kg)
in (cmol/L)	as % CEC			in (cmol/L)	as % CEC		
0.05	2	100.00	n. c.	0.05	2	100.00	n. c.
0.25	10	96.00	600.00	0.26	10	88.46	191.67
0.50	20	86.00	153.57	0.54	20	75.93	78.85
0.75	30	78.67	92.19	0.79	30	58.23	34.85
1.25	50	62.40	41.49	1.32	50	50.00	25.00
1.75	75	51.43	26.47	1.85	75	42.16	18.22
2.50	100	32.40	11.98	2.64	100	35.23	13.60

The comparison of the above results with aim to determine behaviour patterns showed that neither for lead nor copper was it possible to find such a single pattern for all tested minerals. That may be attributed to the structural properties of the tested sorbents. At the same time, the sorption of lead was higher than that of copper, which may be explained by the commonly observed higher affinity of lead towards aluminosilicate mineral sorbents in general and clay minerals in particular [13].

In case of sorption of both Pb^{2+} and Cu^{2+} onto kaolinite, the gradual decrease in sorption capacity of the sorbent was observed. According to the literature, this may be attributed to the decreasing negativity of the potential of the surface of that mineral with the increase in the ionic strength of the solution [29–32].

As the explanation of the generally high efficiency of smectite in the sorption of both Pb^{2+} and Cu^{2+}, provided by the scientific literature on the subject, for that mineral, the main identified mechanism of sorption the was ion exchange. For Pb^{2+}, this was also coupled with the intersphere complex formation, while for Cu^{2+}, the additional mechanism was surface complex formation [31,33–35]. It should be noted that, for this mineral, two mechanisms of sorption were observed: physisorption on the outer and inner surfaces of the mineral, as well as ion exchange. In the case of the second postulated mechanism, Na^{+} (ion radius 0.118 nm) and Ca^{+2} (ion radius 0.112 nm) ions present in the mineral's lattice were most probably substituted by lead (ion radius 0.132 nm), as these elements have a similar ionic radius. That explains the lower amount of sorbed copper, which has a smaller ionic radius (0.087 nm) than sodium and calcium [36–39].

The sorption of Cu^{2+} onto natural zeolite was, except for the lowest concentration tested, significantly lower than that of Pb^{2+}—by 40–60% at lower concentrations (i.e., 10% CEC to 50% CEC) and by 10–20% for the two highest concentrations. It was also noticed that the decrease in adsorption of Cu^{2+} with increasing concentration was initially sharp, but then became less steep than that of Pb^{2+}.

When the data for all three natural aluminosilicate minerals used in the experiment were compared, two general observations were made:

- The level of sorption of Pb^{2+} was generally higher than that of Cu^{2+}, which is due to the high affinity of lead to the oxygen in the functional groups of the tested minerals and the lower solvation energy—for Pb^{2+} it is −1481 [kJ/mol], while Cu^{2+} ions have a solvation energy of −2100 [kJ/mol]. That, in turn, results in a higher affinity of copper ions to the free water molecules in the test system and their higher presence in solution [36,37]; and
- The decrease in the amount sorbed with concentration displayed a higher continuity for copper than for lead.

The comparative analysis of the data for synthetic zeolites led to the following two general conclusions:

- The level of sorption of Pb^{2+} was generally higher than that of Cu^{2+}, for similar reasons as indicated above for natural minerals; and
- Unlike that in natural minerals, the decrease in the amount sorbed with increasing initial concentration displayed a high degree of continuity for Pb^{2+}; while for Cu^{2+} that was observed only for sorption onto zeolite 13X.

All the above was reflected, for both Pb^{2+} and Cu^{2+}, by the K_d values, where those could be calculated (for the samples where the level of sorption was 100%, it was not possible to calculate the K_d values).

Similar general conclusions were drawn by other researchers [18], on the basis of the performed statistical analysis.

The above analysis shows that by determining of the percentage of sorption and K_d values, it was not possible to clearly identify among the seven tested minerals that, which may be considered the most efficient in concurrent elimination of Pb^{2+} and Cu^{2+} ions from multi-component aqueous solutions, which was the main goal of the study. It was, therefore, decided to apply three sorption isotherms in further analysis—Freundlich, Langmuir, and DKR. The suitability of these three isotherms in examination of sorption in multi-solute systems is well documented [40].

They were used in the following way:

- From the Freundlich isotherm, the information on the sorption strength and extent as well as on the nature of the process was derived;
- Langmuir isotherm returned the maximum sorption capacity;
- DKR isotherm enabled possible mechanisms of sorption and the capacity of sorption in micropores to be identified.

Below the numerical and graphical results of the determination of Freundlich and Langmuir isotherms are presented. The parameters of each isotherm are provided in

Table 4 for the Freundlich model and Table 5 for the Langmuir model. The plotted Freundlich isotherms are presented in Figures 1 and 2 and plots of Langmuir isotherms in Figures 3 and 4. The isotherms, represented by red solid lines with blue dots for the experimental points, are plotted with their Confidence Bands, determined at two levels of confidence—95% and 90%, marked using dark pink and light pink, respectively (dark grey and light grey, respectively, if reproduced in black-and-white format).

Table 4. Parameters of Freundlich isotherms and isotherms' statistical evaluation.

Sorbent	Sorbed Element	Parameters of Freundlich Isotherm				Statistical Parameters of the Isotherm		
		Adsorption Constant K_f (L/kg)		$1/n$		SD	r	R^2
		Value	SD	Value	SD			
Kaolinite	Pb^{2+}	6.8522	*1.1519*	0.3373	*0.0463*	0.1904	0.9839	0.9680
	Cu^{2+}	5.1054	*0.3219*	0.3670	*0.0204*	0.0677	0.9971	0.9942
Smectite	Pb^{2+}	18.1458	*3.8585*	0.2692	*0.0665*	2.0082	0.9102	0.8285
	Cu^{2+}	12.1560	*1.5544*	0.2914	*0.0554*	1.0265	0.9594	0.9204
Natural Zeolite	Pb^{2+}	14.1300	*1.6682*	0.3834	*0.0815*	1.0949	0.9687	0.9383
	Cu^{2+}	8.8328	*0.8141*	0.5938	*0.0902*	0.6324	0.9751	0.9508
Zeolite 3A	Pb^{2+}	39.6011	*2.7081*	0.6473	*0.1071*	4.5865	0.9703	0.9415
	Cu^{2+}	19.5745	*1.7402*	0.7023	*0.1406*	3.7342	0.9555	0.9129
Zeolite 10A	Pb^{2+}	52.6663	*2.5209*	0.4128	*0.0430*	3.1511	0.9889	0.9779
	Cu^{2+}	18.4016	*1.7441*	0.5374	*0.1241*	3.9257	0.9390	0.8816
Zeolite 13X	Pb^{2+}	20.7458	*1.1487*	0.2156	*0.0396*	2.0156	0.9771	0.9548
	Cu^{2+}	19.0258	*0.4424*	0.3520	*0.0263*	0.8688	0.9954	0.9909

Table 5. Parameters of Langmuir isotherms and isotherms' statistical evaluation.

Sorbent	Sorbed Element	Parameters of Langmuir Isotherm					Statistical Parameters of the Isotherm		
		K_L (L/kg)		K_L*N		N (cmol/kg)	SD	r	R^2
		Value	SD	Value	SD				
Kaolinite	Pb^{2+}	107.7272	*42.3762*	321.0384	*95.6054*	2.9801	0.2715	0.9669	0.9349
	Cu^{2+}	44.1964	*16.4879*	116.8291	*30.8984*	2.6434	0.2102	0.9714	0.9437
Smectite	Pb^{2+}	140.6089	*36.5785*	1639.6604	*359.1450*	11.6611	1.0754	0.9751	0.9508
	Cu^{2+}	91.3203	*63.0109*	725.1182	*451.5682*	7.9404	1.3491	0.9287	0.8625
Natural Zeolite	Pb^{2+}	10.1024	*5.1734*	124.5607	*47.8629*	12.4406	1.3262	0.9537	0.9095
	Cu^{2+}	2.2961	*1.3264*	25.7112	*7.8464*	11.1978	0.7770	0.9621	0.9257
Zeolite 3A	Pb^{2+}	0.8936	*0.4724*	76.9219	*18.7724*	86.0809	4.6542	0.9694	0.9398
	Cu^{2+}	0.4804	*0.3314*	30.4542	*8.6006*	63.3934	3.6394	0.9577	0.9173
Zeolite 10A	Pb^{2+}	4.6054	*1.0929*	284.6344	*48.1922*	61.8045	3.3058	0.9878	0.9757
	Cu^{2+}	1.1222	*0.5154*	42.5203	*11.5569*	37.8901	3.0334	0.9640	0.9294
Zeolite 13X	Pb^{2+}	15.9406	*4.7676*	351.7683	*91.5613*	22.0674	1.6941	0.9839	0.9681
	Cu^{2+}	4.1023	*1.5764*	99.9011	*29.5274*	24.3525	2.0823	0.9735	0.9477

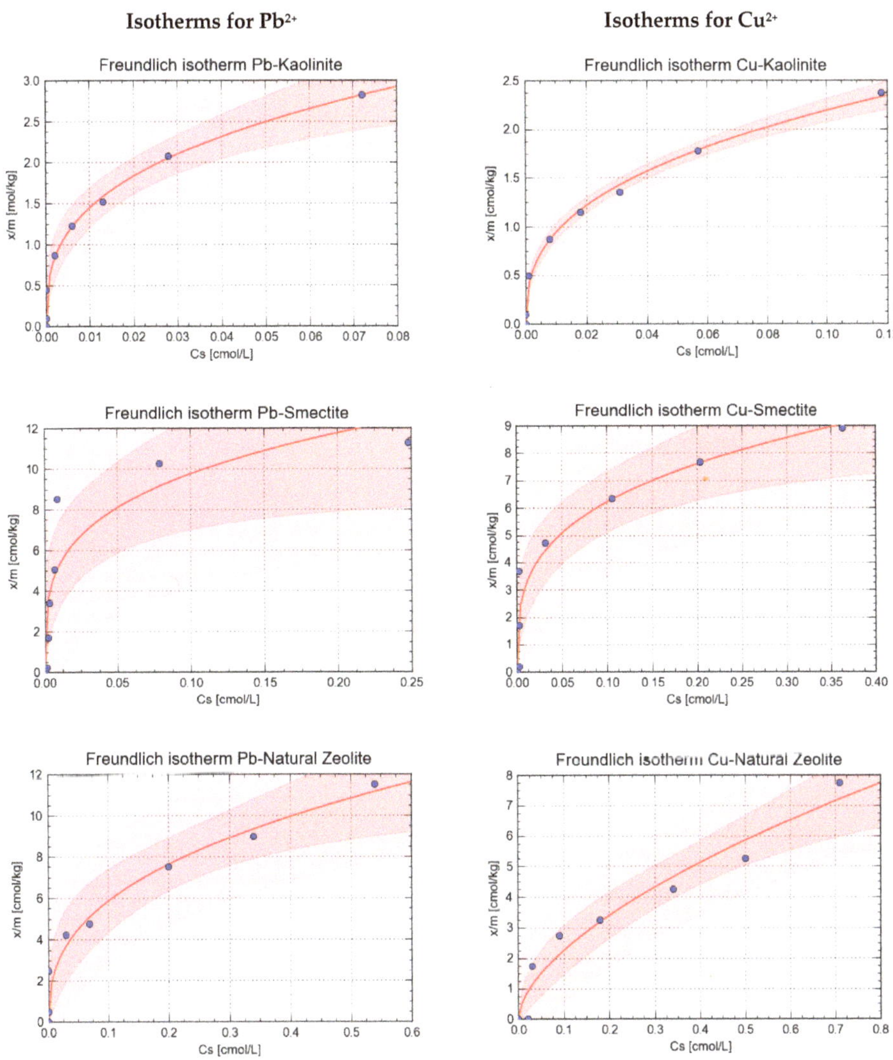

Figure 1. The Freundlich isotherms obtained in the study for sorption of Pb^{2+} and Cu^{2+} onto tested natural mineral sorbents.

The plotted isotherms were analyzed for their goodness of fit by means of visual inspection and examination of the values of correlation coefficient r and determination coefficient R^2. The visual inspection showed good compliance of the estimated curves with the input data. On the basis of the coefficients r and R^2, it may be stated that the adsorption of Pb^{2+} and Cu^{2+} onto the tested minerals was better characterised by the Freundlich model. As a result, the parameters of Freundlich isotherms were chosen as those characterizing the sorption strength and its extent.

Analysis of the strength of sorption based on the determined K_f values showed that Pb^{2+} ions were sorbed more strongly than Cu^{2+} ions, which indicated a higher affinity of lead towards the tested aluminosilicates. Additionally, the tested sorbents which were the strongest with regard to both Pb^{2+} and Cu^{2+} were synthetic zeolites, while the weakest one was kaolinite. Smectite was the strongest sorbent among the tested natural minerals.

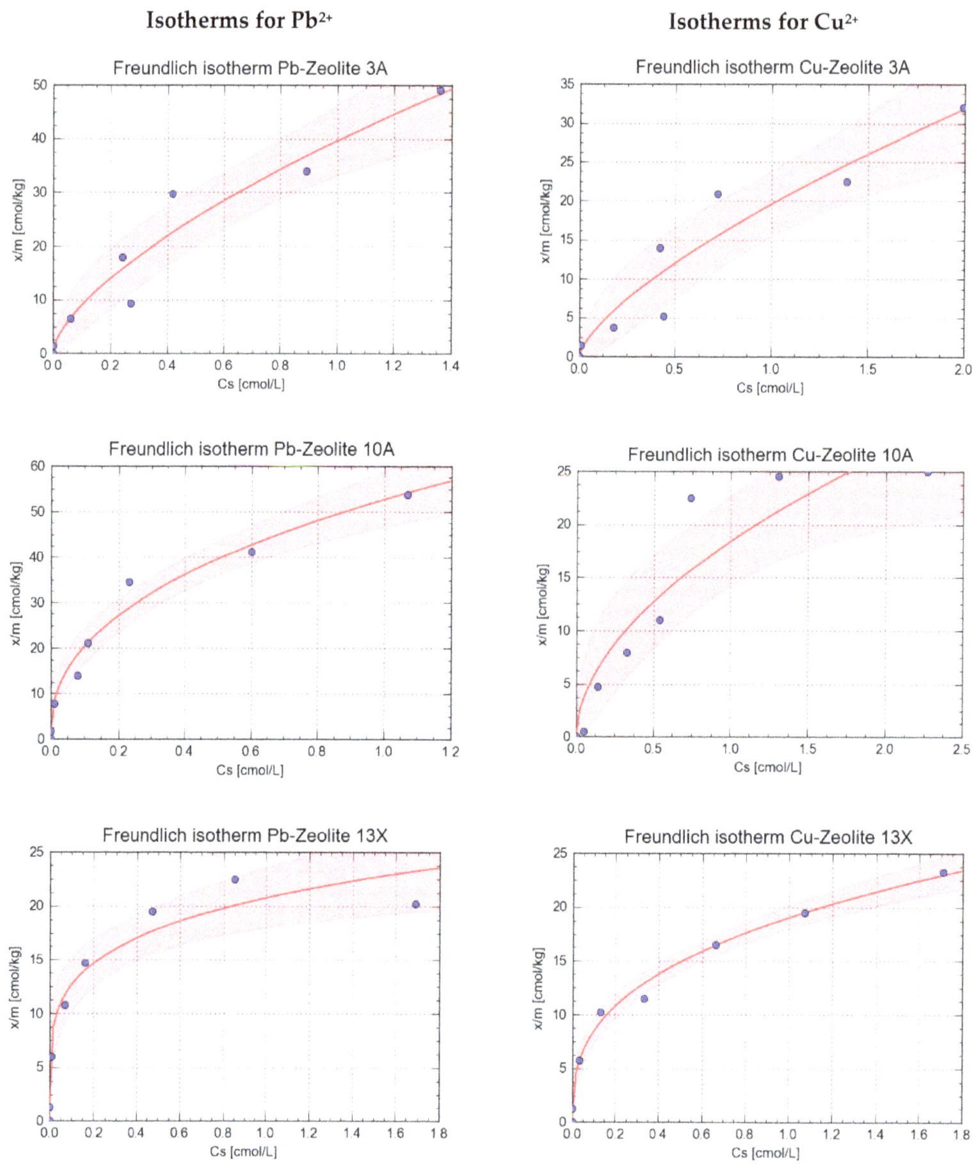

Figure 2. The Freundlich isotherms obtained in the study for sorption of Pb^{2+} and Cu^{2+} onto tested synthetic zeolites.

The arrangement of the tested minerals for their strength and extent of sorption, from strongest to the weakest, returned the following order:
for Pb^{2+}:

Zeolite 10A > Zeolite 3A > Zeolite 13X > Smectite > Natural Zeolite > Kaolinite;

0 and, for Cu^{2+}:

Zeolite 3A $\geq$ Zeolite 13X $\geq$ Zeolite 10A > Smectite > Natural Zeolite > Kaolinite.

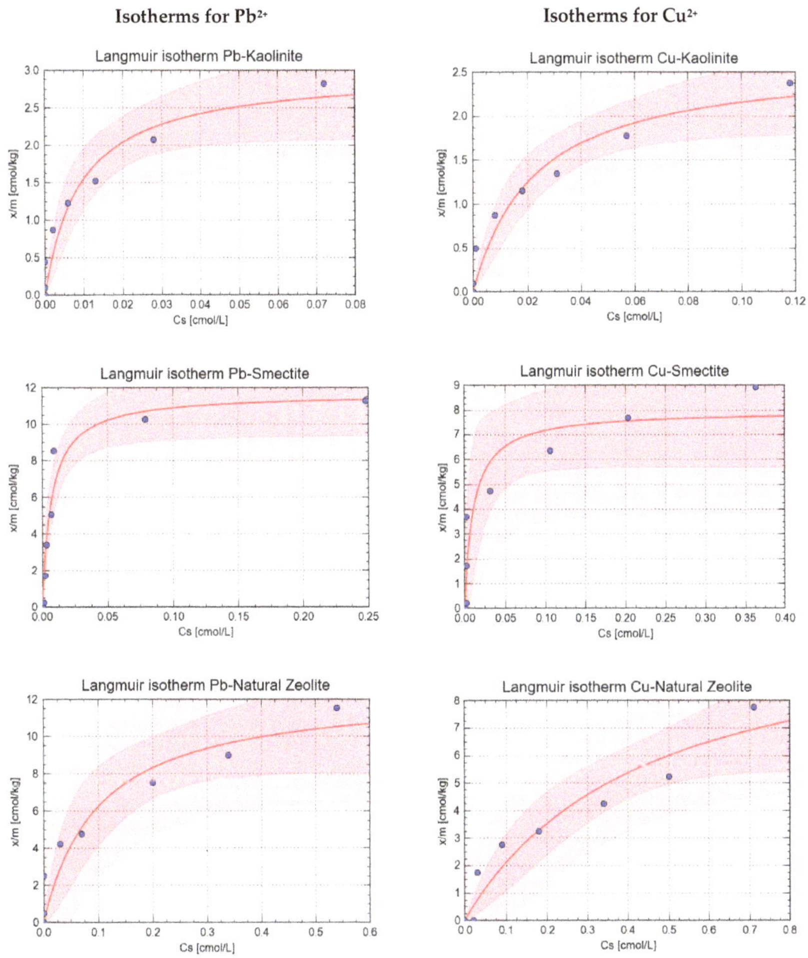

Figure 3. The Langmuir isotherms obtained in the study for sorption of Pb^{2+} and Cu^{2+} onto tested natural mineral sorbents.

It has to be noted that differences in the sorption strength of Pb^{2+} onto the three synthetic zeolites tested were clearly visible; while, in case of Cu^{2+}, those values were comparable, indicating that copper may display a similar affinity to those three aluminosilicates.

The values of the parameter $1/n$, which informs about the nature of the process, were always below 0.8. That indicated the preferential character of sorption of both Pb^{2+} and Cu^{2+} on all tested minerals. It was either favourable or pseudo-linear [41]. The order in which those values may be arranged, from highest to the lowest, thus reflecting the decrease in the linearity, is following:

for Pb^{2+}:

Zeolite 3A > Zeolite 10A > Natural Zeolite > Kaolinite > Smectite > Zeolite 13X;

and, for Cu^{2+}:

Zeolite 3A > Natural Zeolite > Zeolite 10A > Kaolinite > Zeolite 13X > Smectite.

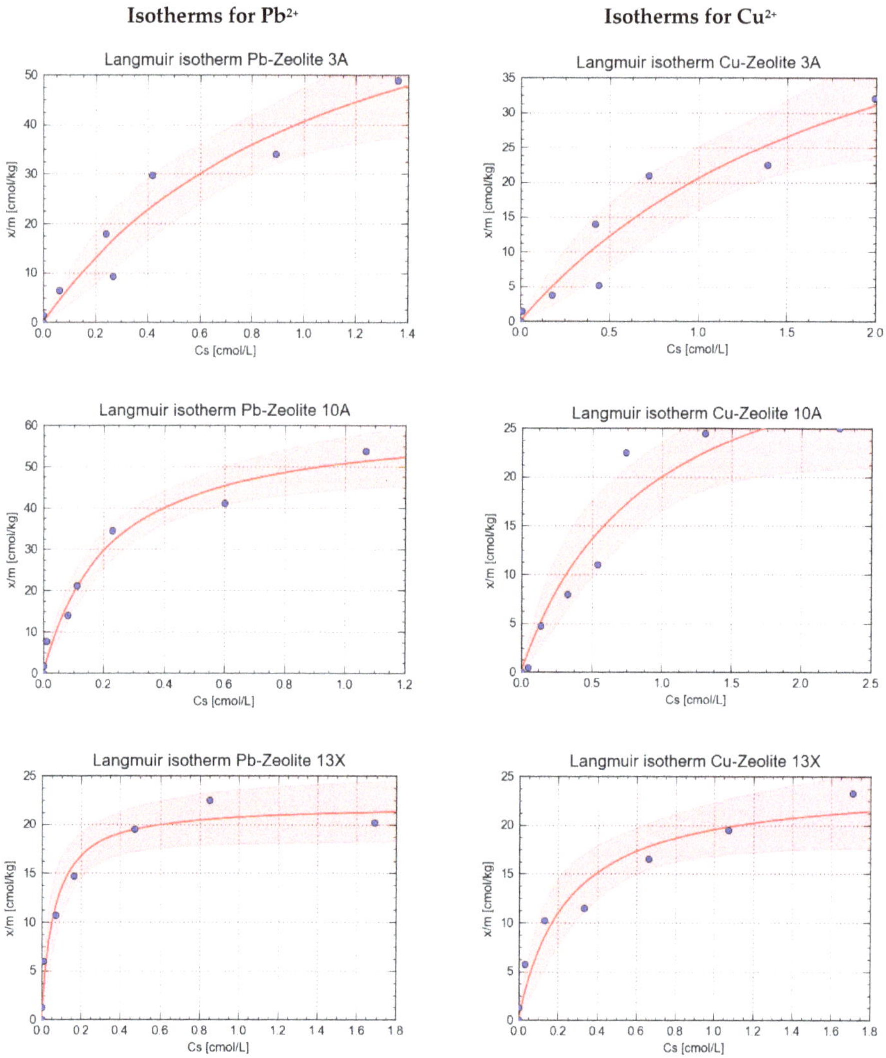

Figure 4. The Langmuir isotherms obtained in the study for sorption of Pb^{2+} and Cu^{2+} onto tested synthetic zeolites.

This may indicate that for either the broader range of concentrations or prolonged exposure to the polluted matrix, natural zeolite and the synthetic zeolites 3A and 5A will be more efficient in the elimination of pollutants than the remaining tested minerals. Additionally, for 1/n interpreted as a potential availability of different sorption sites on the sorbent's surface to the sorbed compound, it may be stated that smectite, kaolinite, and zeolite 13X will become saturated faster with the metal ions of concern than the remaining four zeolites tested.

A good correlation was observed between the Freundlich adsorption constant, characterizing the strength of sorption, and the maximum sorption capacity—the parameter N of the Langmuir sorption isotherm. It can be stated that the maximum sorption capacity of both Pb^{2+} and Cu^{2+} was higher for the synthetic zeolites than for the natural minerals tested. Once again, the lowest maximum sorption capacity was determined for kaolinite.

When arranged from highest to lowest, the maximum sorption capacity for both Pb^{2+} and Cu^{2+}, followed the order:

Zeolite 3A > Zeolite 10A > Zeolite 13X > Natural Zeolite > Smectite > Kaolinite.

It was also noticed that the maximum sorption capacity was generally higher for Pb^{2+} than for Cu^{2+}, although the differences were bigger in the synthetic zeolites. For natural minerals, they tended to be smaller, with very little difference in values observed in kaolinite and natural zeolite. This may indicate that those two aluminosilicates displayed much lower relative selectivity to lead and copper.

The numerical parameters of the DKR isotherm are presented in Table 6.

Table 6. Parameters of the DKR isotherms and isotherms' statistical evaluation.

Sorbent	Sorbed Element	Parameters of the DKR Isotherm						Statistical Parameters of the Isotherm		
		$ln\ X_m$		β (mol²/kJ²)		X_m (cmol/kg)	E (kJ/mol)	SD	r	R^2
		Value	SD	Value	SD					
Kaolinite	Pb^{2+}	1.2166	*0.0880*	−0.00618	*0.00063*	3.3757	8.9948	0.0913	0.9847	0.9697
	Cu^{2+}	0.8184	*0.1137*	−0.00573	*0.000810*	2.2669	9.3413	0.1668	0.9622	0.9258
Smectite	Pb^{2+}	2.7303	*0.2506*	−0.00797	*0.00168*	15.3375	7.9206	0.3195	0.9216	0.8494
	Cu^{2+}	2.4701	*0.1366*	−0.01773	*0.01251*	11.8236	5.3652	0.2273	0.9923	0.9847
Natural Zeolite	Pb^{2+}	2.3814	*0.08254*	−0.01325	*0.00206*	10.8200	6.1429	0.1164	0.9655	0.9322
	Cu^{2+}	1.7831	*0.1383*	−0.01802	*0.00391*	5.9483	5.2675	0.2315	0.9174	0.8416
Zeolite 3A	Pb^{2+}	3.5669	*0.2655*	−0.03864	*0.01197*	35.4067	3.5972	0.4640	0.8501	0.7227
	Cu^{2+}	3.3246	*0.2786*	−0.1036	*0.02756*	27.7879	2.1969	0.4523	0.8829	0.7795
Zeolite 10A	Pb^{2+}	3.7057	*0.1757*	−0.01452	*0.00313*	40.6785	5.8681	0.3223	0.9185	0.8436
	Cu^{2+}	3.1869	*0.1276*	−0.07012	*0.00540*	24.2132	2.6703	0.25692	0.9854	09712
Zeolite 13X	Pb^{2+}	3.0095	*0.0673*	−0.01020	*0.00121*	20.2773	7.0014	0.1296	0.9729	0.9467
	Cu^{2+}	2.9171	*0.1086*	−0.01680	*0.00331*	18.4876	5.4554	0.2082	0.9307	0.8662

The correlation and determination coefficients—r and R^2, considered as the indicators of the goodness of fit of each isotherm showed that for each combination sorbate sorbent, the fit was at least acceptable. On that basis, it can be stated that the DKR isotherm adequately characterized sorption in the test systems, confirming the appropriateness of selection of that model. This statement is similar to the analogic conclusions drawn for Freundlich and Langmuir isotherms.

The constant, X_m, which characterizes the maximum sorption capacity, was higher for the sorption of Pb^{2+} (3.38–40.68 (cmol/kg)) than that for Cu^{2+} (2.27–27.79 (cmol/kg)), which confirms the conclusions drawn using the results of the two previously presented isotherms.

The decrease in that parameter observed for the tested minerals followed the order: for Pb^{2+}:

Zeolite 10A > Zeolite 3A > Zeolite 13X > Smectite > Natural Zeolite > Kaolinite;

and, for Cu^{2+}:

Zeolite 3A > Zeolite 10A > Zeolite 13X > Smectite > Natural Zeolite > Kaolinite.

Comparison with the analysis for the analogical parameter of Langmuir isotherm—N, showed that for Cu^{2+}, the trend was identical, while for Pb^{2+} it was very similar. The observed differences in values may be observed by the conceptual differences of the two models. The DKR isotherm was oriented in the examination of sorption in the micropores, while Langmuir's model was more general.

The further comparison of the two parameters consisted of the calculation of the $N{:}X_m$ ratio, presented, alongside differently expressed X_m and N values, as seen in the Table 7. In the case of smectite, for both Pb^{2+} and Cu^{2+}, that ratio showed that N was lower than X_m, indicating a potentially high significance of interlattice sorption as the sorption mechanism. In the case of kaolinite, the difference between the two parameters was not significant. In the case of all tested zeolites, the value of X_m, was always a fraction of that of N.

Table 7. The values of maximum sorption capacity in Langmuir's (N) and DKR (X_m) isotherms and their ratios.

Sorbent	Sorbed Element	Maximum Sorption Capacity N—Langmuir's Isotherm, Expressed in:			Maximum Sorption Capacity X_m—DKR Isotherm, Expressed in:			Ratio $N{:}X_m$
		(cmol/kg)	(mmol/g)	(mg/g)	(cmol/kg)	(mmol/kg)	(mg/g)	
Kaolinite	Pb^{2+}	2.9801	0.0298	6.175	3.3757	0.0338	6.994	1:1.13
	Cu^{2+}	2.6434	0.0264	1.680	2.2669	0.0227	1.440	1.17:1
Smectite	Pb^{2+}	11.6611	0.1166	24.162	15.3375	0.1534	31.779	1:1.31
	Cu^{2+}	7.9404	0.0790	5.046	11.8236	0.1182	7.513	1:1.49
Natural Zeolite	Pb^{2+}	12.4406	0.1244	25.777	10.8200	0.1082	22.419	1.15:1
	Cu^{2+}	11.1978	0.1120	7.116	5.9483	0.0595	3.780	1.88:1
Zeolite 3A	Pb^{2+}	86.0809	0.8608	178.360	35.4067	0.3541	73.363	2.43:1
	Cu^{2+}	63.3934	0.6339	40.284	27.7879	0.2779	17.658	2.28:1
Zeolite 10A	Pb^{2+}	61.8045	0.6180	128.060	40.6785	0.4068	84.286	1.52:1
	Cu^{2+}	37.8901	0.3789	24.078	24.2132	0.2421	15.386	1.56:1
Zeolite 13X	Pb^{2+}	22.0674	0.2207	45.724	20.2773	0.2028	42.015	1.09:1
	Cu^{2+}	24.3525	0.2435	15.475	18.4876	0.1849	11.748	1.32:1

The second parameter of sorption, determined indirectly from DKR isotherm, was the energy of sorption. It provides the information on the possible mechanism of the process [30,42,43], based on the following classification:

- Physisorption is postulated as the dominant mechanism of sorption when $E < 8$ kJ/mol;
- For E in the range of 8 to 16 kJ/mol, ion exchange is indicated as the dominant mechanism of sorption;
- Finally, when $E > 16$ kJ/mol, sorption occurs mainly as chemisorption, which is the strongest.

In the experiment, the range of the energy of sorption was 3.60–8.99 [kJ/mol] for Pb^{2+} and 2.20–9.34 [kJ/mol] for Cu^{2+}.

Only for kaolinite was ion exchange the dominant mechanism of sorption for both elements. For the remaining minerals tested it was physical sorption.

In the case of Pb^{2+}, the decrease in the sorption energy E may be arranged as follows:

Kaolinite > Smectite $\geq$ Zeolite 13X > Natural Zeolite > Zeolite 10A > Zeolite 3A;

A similar arrangement for Cu^{2+} follows:

Kaolinite > Zeolite 13X $\geq$ Smectite > Natural Zeolite > Zeolite 10A > Zeolite 3A.

It should be indicated that the difference in E for the sorption of Cu^{2+} onto smectite and zeolite 13X was small. A similar observation was made in the case of Pb^{2+}; therefore, in reality, the order of the determined sorption energies for both elements may be the same.

Table 7 contains the values of the maximum sorption capacity of each tested mineral towards either Pb^{2+} and Cu^{2+} ions, determined using the Langmuir's and DKR isotherms. Alongside the values derived directly from isotherms—in (cmol/kg), the values in (mmol/g) and (mg/g) are provided, as they are more commonly encountered in the scientific literature on the subject. To convert (cmol/kg) to (mg/g), the relevant molar weights: 207.2 g/mol for Pb and 63.546 g/mol for Cu, were used. The conversion was performed to compare the results with those presented in other scientific papers on the same subject [10,16,17,20,29,44–46]. That comparison demonstrated that the tested minerals displayed similar or greater sorption capacities than the similar sorbents and other novel materials tested to eliminate heavy metals from wastewaters. As a result, the tested minerals were shown to meet the criteria of high efficiency, inexpensiveness, and environmental friendliness in purification of water polluted with heavy metal ions.

4. Conclusions

On the basis of the obtained results, it may be stated that:

(1) The efficiency of the tested synthetic zeolites—3A, 10A and 13X, in the concurrent elimination of Pb^{2+} and Cu^{2+} ions from aqueous solutions was greater than that of the tested natural minerals—kaolinite, smectite, and natural zeolite, which may be explained by the higher mineralogical homogeneity of the formers as well as, in the case of zeolites 3A and 10A, by their structural properties, in particular, a high share of mesopores (having the diameter in the range of 1.5 to 200 nm), constituting 68% of the total porosity of those two sorbents;

(2) The analysis of the parameters of sorption isotherm models showed that of the six tested mineral sorbents, the most efficient in the simultaneous removal of Pb^{2+} and Cu^{2+} ions from aqueous solutions were zeolite 3A and zeolite 10A. Therefore, those two sorbents should be recommended for the rapid reduction in the level of pollution with those two elements and their spread in the environment;

(3) It was demonstrated that smectite displayed a relatively high and constant sorption capacity over a broader range of concentrations, which indicates that it will be efficient in coping with a prolonged low- and medium-level lead and copper pollution in aquatic environments;

(4) The determined adsorption energies indicated that for the five tested minerals—smectite, natural zeolite, and synthetic zeolites 3A, 10A, 13X, physisorption may be postulated as the predominant mechanism of sorption. For kaolinite, the weakest sorbent tested, the mechanism was ion exchange. Additionally, the lowest sorption energies were determined for zeolite 3A and zeolite 10A, further confirming their highest sorption capacity.

(5) For all tested minerals, the sorption of lead was about 30% higher than that of copper, which may indicate the selectivity of the process with preference for Pb^{2+} ions.

Author Contributions: B.G. Conceptualization, Methodology, Investigation, Supervision, Writing—Review and Editing; B.K.-S. Conceptualization, Methodology, Investigation, Validation; I.K. Formal analysis, Visualization, Writing—Original Draft, Writing—Review and Editing. J.C. Conceptualization, Methodology, Investigation. All authors have read and agreed to the published version of the manuscript.

Funding: This research received no external funding.

Institutional Review Board Statement: Not applicable.

Informed Consent Statement: Not applicable.

Acknowledgments: This research was supported by statutory activity of the University of Life Sciences and the Institute of Environmental Protection, National Research Institute. The authors would like to thank Zbigniew Zagórski for identification of zeolites.

Conflicts of Interest: The authors declare no conflict of interest.

References

1. Karlaviciene, V.; Svendine, S.; Marciulioniene, D.E.; Randerson, P.; Rimeika, M.; Hogland, W. The impact of storm water runoff on a small urban stream. *J. Soils Sediments* **2009**, *9*, 6–12. [CrossRef]
2. Lynch, S.F.L.; Batty, L.C.; Byrne, P. Environmental Risk of Metal Mining Contaminated River Bank Sediment at Redox-Transitional Zones. *Minerals* **2014**, *4*, 52–73. [CrossRef]
3. Newman, M.C. *Fundamentals of Ecotoxicology. The Science of Pollution*, 4th ed.; CRC Press, Taylor & Francis Group: Boca Raton, FL, USA, 2015.
4. Hudson-Edwards, K.A.; Dold, E. Mine Waste Characterization, Management and Remediation. *Minerals* **2015**, *5*, 82–85. [CrossRef]
5. Ramesh, K.; Elango, L. Impact of Groundwater Quality from Industrial East Coastal Town, Southern India. *Int. J. Eng. Res. Appl.* **2014**, *4*, 346–354.
6. Pierzynski, G.M.; Sims, T.J.; Vance, G.F. *Soils and Environmental Quality*, 2nd ed.; CRC Press, Taylor & Francis Group: Boca Raton, FL, USA, 2000.
7. Gworek, B. Inactivation of cadmium in contaminated Soils using synthetic zeolites. *Environ. Pollut.* **1992**, *75*, 269–271. [CrossRef]
8. Gworek, B. Inactivation of lead anthropogenic soils by synthetic zeolites and plants growth. *Plant Soil* **1992**, *143*, 71–74. [CrossRef]
9. Gworek, B. The Effect of Zeolites on Cooper Uptake by plants growing in Contaminated Soils. *J. Incl. Phenom. Mol. Recognit. Chem.* **1993**, *15*, 1–7. [CrossRef]
10. El-Azim, H.A.; Mourad, F.A. Removal of Heavy Metals Cd (II), Fe (III) and Ni (II), from Aqueous Solutions by Natural (Clinoptilolite) Zeolites and Application to Industrial Wastewater. *Asian J. Environ. Ecol.* **2018**, *7*, 1–13. [CrossRef]
11. Oste, L.A.; Lexmond, T.M.; Van Riemsdijk, W.H. Metal Immobilization in Soils Using Synthetic Zeolites. *J. Environ. Qual.* **2002**, *31*, 813–821. [CrossRef]
12. Karatas, M. Removal of Pb(II) from water by natural zeolitic tuff: Kinetics and thermodynamics. *J. Hazard. Mater.* **2012**, *199–200*, 383–389. [CrossRef]
13. Ibrahim, H.S.; Jamil, T.; Hegazy, E.Z. Application of Zeolite Prepared from Egyptian Kaolin for the Removal of Heavy Metals: II. Isotherm Models. *J. Hazard. Mater.* **2010**, *182*, 842–847. [CrossRef]
14. Blanchard, G.; Maunaye, M.; Martin, G. Removal of heavy metals from waters by means of natural zeolites. *Water Res.* **1984**, *18*, 1501–1507. [CrossRef]
15. Wingenfelder, U.; Hansen, C.; Furrer Schulin, R. Removal of Heavy Metals from Mine Waters by Natural Zeolites. *Environ. Sci. Technol.* **2005**, *39*, 4606–4613. [CrossRef]
16. Fanta, F.T.; Dubale, A.A.; Bebizuh, D.F.; Atlabachew, M. Copper doped zeolite composite for antimicrobial activity and heavy metal removal from waste water. *BMC Chem.* **2019**, *13*, 44. [CrossRef] [PubMed]
17. Bhattacharyya, K.G.; Gupta, S.S. Adsorptive accumulation of Cd(II), Co(II), Cu(II), Pb(II), and Ni(II) from water on montmorillonite: Influence of acid activation. *J. Colloid Interface Sci.* **2007**, *310*, 411–424. [CrossRef] [PubMed]
18. Sprynskyy, M.; Buszewski, B.; Terzyk, A.P.; Namieśnik, J. Study of the selection mechanism of heavy metal (Pb^{2+}, Cu^{2+}, Ni^{2+}, and Cd^{2+}) adsorption on clinoptilolite. *J. Colloid Interface Sci.* **2006**, *304*, 21–28. [CrossRef] [PubMed]
19. Usman, A.; Yakov, R.A.; Kuzyakov, Y.; Lorenz, K.; Stahr, K. Remediation of a soil contaminated with heavy metals by immobilizing compounds. *J. Plant Nutr. Soil Sci.* **2006**, *169*, 205–212. [CrossRef]
20. Zhao, Y. Review of the Natural, Modified, and Synthetic Zeolites for Heavy Metals Removal from Wastewater. *Environ. Eng. Sci.* **2016**, *33*, 443–454. [CrossRef]
21. Wu, J.; Xue, S.; Bridges, D.; Yu, Y.; Zhang, L.; Pooram, J.; Hill, C.; Wu, J.; Hu, A. Fe-based ceramic nanocomposite membranes fabricated via e-spinning and vacuum filtration for Cd^{2+} ions removal. *Chemosphere* **2019**, *230*, 527–535. [CrossRef]
22. Mahmoud, M.E.; Osman, M.M.; Yakout, A.A.; Abdelfattah, A.M. Green nanosilica@folic Acid (VB9) nanocomposite for engineered adsorptive water remediation of bivalent lead, cadmium and copper. *Powder Technol.* **2019**, *344*, 719–729. [CrossRef]
23. Castro-Muñoz, R.; González-Melgoza, L.L.; Garciá-Depraect, O. Ongoing progress on novel nanocomposite membranes for the separation of heavy metals from contaminated water. *Chemosphere* **2021**, *270*, 129421. [CrossRef] [PubMed]
24. Mittal, H.; Maity, A.; Ray, S.S. The adsorption of Pb^{2+} and Cu^{2+} onto gum ghatti-grafted poly(acrylamide-co-acrylonitrile) biodegradable hydrogel: Isotherms and kinetic models. *J. Phys. Chem. B* **2015**, *119*, 2026–2089. [CrossRef] [PubMed]
25. Agency for Toxic Substances and disease Registry (ATSDR). *Toxicological Profile for Copper;* US Department of Health and Human Services, Centers for Diseases Control: Atlanta, GA, USA, 2004.
26. Agency for Toxic Substances and disease Registry (ATSDR). *Toxicological Profile for Lead;* US Department of Health and Human Services, Centers for Diseases Control: Atlanta, GA, USA, 2007.
27. OECD. OECD Guideline for the Testing of Chemicals No. 106 "Adsorption-Desorption Using a Batch Equilibrium Method". 2000. Adopted 21st January 2000. Available online: https://www.oecd-ilibrary.org/environment/test-no-106-adsorption-desorption-using-a-batch-equilibrium-method_9789264069602-en (accessed on 20 November 2020).
28. Kozera-Sucharda, B.; Gworek, B.; Kondzielski, I. The simultaneous Removal of Zinc and Cadmium from Multicomponent Aqueous Solutions by Their Sorption onto Selected Natural and Synthetic Zeolites. *Minerals* **2020**, *10*, 343. [CrossRef]
29. Kralik, M. Adsorption, chemisorption, and catalysis. *Chem. Pap.* **2014**, *68*, 1625–1638. [CrossRef]
30. Apiratikul, R.; Pavasant, P. Sorption of Cu^{2+}, Cd^{2+} and Pb^{2+} using modified zeolite from coal fly ash. *Chem. Eng. J.* **2008**, *144*, 245–258. [CrossRef]

31. Uddin, M.K. A review on the adsorption of heavy metals by clay minerals, with special focus on the past decade. *Chem. Eng. J.* **2017**, *308*, 438–462. [CrossRef]
32. Coles, C.A.; Yong, R.N. Aspects of kaolinite characterization and retention of Pb and Cd. *Appl. Clay Sci.* **2002**, *22*, 39–45. [CrossRef]
33. Chyaari, I.; Fakhfakh, E.; Charkoun, S.; Bouzid, J.; Boujelben, N.; Feki, M.; Rocha, F.; Jamoussi, F. Lead removal from aqueous solutions by a Tunisian smectitic clay. *J. Hazard. Mater.* **2008**, *156*, 545–551. [CrossRef]
34. Eren, E.; Afsin, B. An investigation of Cu(II) adsorption by raw and acid activated bentonite: A combined potentiometric, thermodynamic, XRD, IR, DTA study. *J. Hazard. Mater.* **2008**, *151*, 682–691. [CrossRef] [PubMed]
35. Eren, E. Removal of copper ions by modified Unye clay, Turkey. *J. Hazard. Mater.* **2008**, *159*, 235–244. [CrossRef]
36. Baker, H.M.; Massadeh, A.M.; Younes, H.A. Natural Jordanian zeolite: Removal of heavy metal ions from water samples using column and batch methods. *Environ. Monit. Assess.* **2009**, *157*, 319–330. [CrossRef]
37. Bohli, T.; Villaescusa, I.; Quederini, A. Comparative Study of Bivalent Cationic Metals Adsorption Pb(II0, Cd(II), Ni(II) and Cu(II) on Olive Stones Chemically Activated Carbon. *J. Chem. Eng. Process Technol.* **2013**, *4*, 1000158. [CrossRef]
38. Sepehrian, H.; Ahmadi, S.J.; Waqif-Husain, S.; Faghihian, H.; Alighanbari, H. Adsorption Studies of Heavy Metal Ions on mesoporous Aluminosilicate, Novel Cation Exchanger. *J. Hazard. Mater.* **2010**, *176*, 252–256. [CrossRef] [PubMed]
39. Eloussaief, M.; Hamza, W.; Kallel, N.; Benzin, M. Wastewaters decontamination: Mechanisms of Pb(II), Zn(II), and Cd(II) competitive adsorption on tunisian smectite in single and multi-solute systems. *Environ. Prog. Sustain. Energy* **2013**, *32*, 223–238. [CrossRef]
40. Allen, S.J.; Mckay, G.; Porter, J.F. Adsorption isotherm models for basic dye adsorption by peat in single and binary component system. *Colloid Interface Sci.* **2004**, *280*, 322–333. [CrossRef]
41. Tseng, R.-L.; Wu, F.-C. Inferring the favorable adsorption level and the concurrent multi-stage process with the Freundlich constant. *J. Hazard. Mater.* **2008**, *155*, 277–287. [CrossRef] [PubMed]
42. Mobacherpour, I.; Salahi, E.; Pazouki, M. Comparative of the removal of Pb^{2+}, Cd^{2+} and Ni^{2+} by nano crystallite hydroxyapatite from aqueous solutions: Adsorption isotherms study. *Arab. J. Chem.* **2012**, *5*, 439–446. [CrossRef]
43. Erdem, E.; Karapinar, N.; Donat, R. The removal of heavy metals by natural zeolites. *J. Colloid Interface Sci.* **2004**, *280*, 309–314. [CrossRef]
44. De Gisi, S.; Giusy, L.; Grassi, M.; Notarnicola, M. Characteristics and adsorption capacities of low-cost sorbents for wastewater treatment: A review. *Sustain. Mater. Technol.* **2016**, *9*, 10–40. [CrossRef]
45. Du, T.; Zhou, L.-F.; Zhang, Q.; Liu, L.-Y.; Li, G.; Luo, W.-B.; Liu, H.-K. Mesoporous structured aluminosilicate with excellent adsorption performances for water purification. *Sustain. Mater. Technol.* **2018**, *17*, e00080. [CrossRef]
46. Kumari, P.; Alam, M.; Siddiqi, W.A. Usage of nanoparticles as adsorbents for waste water treatment: An emerging trend. *Sustain. Mater. Technol.* **2019**, *22*, e00128. [CrossRef]

processes

MDPI

Article

Magnetic Multiwall Carbon Nanotube Decorated with Novel Functionalities: Synthesis and Application as Adsorbents for Lead Removal from Aqueous Medium

Ghadir Hanbali [1], Shehdeh Jodeh [1,*], Othman Hamed [1,*], Roland Bol [2], Bayan Khalaf [1], Asma Qdemat [3], Subhi Samhan [4] and Omar Dagdag [5]

1 Department of Chemistry, Faculty of Science, An-Najah National University, P.O. Box 7, Nablus 00001, Palestine; g.hanbali@najah.edu (G.H.); bayan.kh107@hotmail.com (B.K.)
2 Institute of Bio and Geosciences, Agrosphere (IBG-3), Forschungszentrum Jülich GmbH, 52425 Jülich, Germany; r.bol@fz-juelich.de
3 Jülich Center for Neutron Science, Peter Grunberg Institute, Forschungszentrum Jülich GmbH, 52425 Jülich, Germany; qdemat@fz-juelich.de
4 Palestine Water Authority, Ramallah 00001, Palestine; subhisamhan@yahoo.com
5 Laboratory of Agroresources, Polymers and Process Engineering (LAPPE), Department of Chemistry, Faculty of Science, Ibn Tofail University, BP 133, 14000 Kenitra, Morocco; omar.dagdag@uit.ca.ma
* Correspondence: sjodeh@hotmail.com (S.J.); ohamed@najah.edu (O.H.); Tel.: +970-599-590-498 (S.J.); +970-594-466-271 (O.H.)

Received: 19 July 2020; Accepted: 10 August 2020; Published: 14 August 2020

Abstract: Water pollution is one of the major challenges facing modern society because of industrial development and urban growth. This study is directed towards assessing the use of multiwall carbon nanotube, after derivatization and magnetization, as a new and renewable absorbent, to remove toxic metal ions from waste streams. The adsorbents were prepared by, first oxidation of multiwall carbon nanotube, then derivatizing the oxidized product with hydroxyl amine, hydrazine and amino acid. The adsorbents were characterized by various techniques. The adsorption efficiency of the multiwall carbon nanotube adsorbents toward Pb(II) was investigated. The effect of adsorbent's dose, temperature, pH, and time on the adsorption efficiency were studied and the adsorption parameters that gave the highest efficiency were determined. The derivatives have unique coordination sites that included amine, hydroxyl, and carboxyl groups, which are excellent chelating agents for metal ions. The thermodynamic and kinetic results analysis results indicated spontaneous adsorption of Pb(II) by the multiwall carbon nanotube adsorbents at room temperature. The adsorption process followed pseudo-second-order and Langmuir isotherm model. The adsorbents were regenerated using 0.1 N HCl.

Keywords: magnetic multiwall carbon nanotube; adsorption; kinetics; isotherm; thermodynamic; lead

1. Introduction

Metal water pollution is a major environmental concern as some are highly toxic elements, even at low concentrations [1]. The toxicity of heavy metals is an inevitable consequence as they accumulate in the soft tissues of the human body [2,3]. These metals might enter the human body through different ways, among these is drinking water, and unlike organic pollutants, heavy metal ions are nonbiodegradable.

Heavy metal ions exists in waste streams of various industries, including leather tanning, metal plating, radiator manufacturing, batteries manufacturing, mining operations, smelting, alloy industries, and fertilizer industries [4].

Lead is considered one of the most toxic metals, it has access to the human body via ingestion, inhalation or skin assimilation [5]. Once inside the body, lead is absorbed and stored in bones, blood, and tissues [6]. Lead can cause severe damage to the brain, kidneys and, ultimately, it could lead to death. By simulating calcium, lead can cross the blood-brain barrier and destroys myelin sheaths of neurons, decreases their numbers, inhibits with neurotransmission routes and reduces the growth of neurons [7].

In nature, lead can compete with other minerals in and on plants surfaces and may inhibit photosynthesis, and negatively effect on plant growth and survival at sufficiently high concentrations [8].

The bioaccumulation of lead in food sources and its toxicity to biological systems attributable to increased concentrations over time has called for a significant pressure for removal. Several technologies have successfully removed lead from polluted water; among these are adsorption, membrane filtration, ion exchange bioremediation, precipitation and coagulation, and heterogeneous photocatalyst [9–11]. Among the most effective technologies, adsorption has been proven the highest removal efficiency at a reasonable cost. In addition, adsorbents can be regenerated and reused [12,13]. An unlimited number of effective adsorbents were developed and reported in the literature. Among the most effective metal adsorbents is multiwall carbon nanotube (MWCNT), it has shown significant success and has been employed in many of commercial applications [14]. Several published studies reported that carbon nanotubes are among the most effective in water treatment and adsorption of heavy metals such as copper, lead, cadmium. It also showed high efficiency in polar and nonpolar organic molecules. This was related to the various physical and chemical interactions MWCNT can afford in the from π–π interactions, covalent bonding, hydrogen bonding [15], and electrostatic interactions [16].

Chemical modification or doping is believed to optimize MWCNT properties for a certain specific application. A common method is to enhance the dispersion, and optimize the use of multi-wall carbon nanotubes through chemical functionality. This enables the formation of chemical interconnection between MWCNT and targeted materials. Both non-covalent and covalent structures have been employed to improve solubility and physical properties of MWCNT [17].

MWCNT functionalization with P, O, and N containing groups on the surface of MWCNT allow the researchers to control the surface area, surface charge, hydrophobicity/hydrophilicity, dispersion. and improve adsorption capacity and selectivity towards the heavy metals [18,19]. For instance, Li et al. [20] showed that acid-refluxed carbon nanotubes (CNTs) have enhanced ability to adsorb Pb(II) ions from water. Wang et al. [21] showed that adding oxygen functional groups on acidic MWCNT improved the adsorption capacity of lead ion. Tofighy and Mohamadi [22] reported that the adsorption on oxidized CNTs sheets followed the order: $Pb^{2+} > Cd^{2+} > Co^{2+} > Zn^{2+} > Cu^{2+}$. Wang et al. [23] showed that oxidized MWCNT caused a rise in the zeta potential value resulting in the formation of negative surface charge because of the creation and ionization of functional groups (–COOH &–OH). It has been observed that the adsorption of Pb(II) onto acidic MWCNT was not uniform and mainly collected on the cap and defective sites of the MWCNT as adsorbent.

During recent years, the focus has been on magnetic adsorbents to avoid problems associated with adsorption regeneration. The main advantages of magnetic composites include: high strength, strong adsorption rate and improved adsorption ability [24]. The composite adsorbent could effectively achieve the separation of solid–liquid under an active magnetic field without filtration or centrifugation. This results from the properties of the magnetism, which simplifies post-processing. In addition, magnetic recycling can help prevent nano-adsorbents from occurring in the natural environment, and mitigate future hazards [25].

This work focused on a MWCNT grafted with various functionalities as a potential new adsorbent for toxic metal ions. The MWCNT was first oxidized then converted to acid chloride. MWCNT functionalized with acid chloride was reacted separately with hydroxylamine, cystine, and hydrazine.

The prepared grafted MWCNT were magnetized and then evaluated as an adsorbent for Pb(II) removal from water. The effect of various factors on the adsorption efficiencies of grafted MWCNT were evaluated.

The novelty of this work can be summarized as the grafted MWCNTs presented in this work represent the first example in the oxidized MWCNT literature modified with such functionalities and applied for removal of Pb(II). In addition, previous studies on lead removal using a single functional group provided lower removal than the one presented in this study.

2. Materials and Methods

Chemicals and MaterialsAll chemicals used in this work were purchased from Sigma-Aldrich (Jerusalem, Israel) and used as received.

2.1. Instrument and Characterization

Infrared spectra were recorded using (FTIR-SHIMADZU, Shimadzu Corporation, Kyoto, Japan, Model: FTIR-8700). The following parameters were used: resolution 4 cm^{-1}, spectral range 600–4000 cm^{-1}, number of scans 128. Flame atomic absorption spectroscopy (iCE 3300, Thermo Fisher Scientific, Cambridge, UK) was used to determine lead ion concentration. Raman spectra were recorded on Bruker RFS 100 FT-Raman spectrometer, scanning electron microscopy ((Hitachi SX-650 machine (Tokyo, Japan)) was used to study the surface morphology or the adsorbents, Brunauer–Emmett–Teller surface area analysis (Micromeritics, Norcross, GA, USA), Vibrating Sample Magnetometer (VSM-LAKESHORE 7404, Boston, MA, USA). All results of the characterization can be found in our previous study [26].

2.2. Magnetization of Modified Multiwall Carbon Nanotube

The procedure includes four steps, at the beginning, MWCNT was oxidized then the developed carboxylic acid functionality was converted to acid chloride by reacting it with oxalyl chloride. Then the acid chloride was converted to hydroxamate, hydrazine and amino acids by reacting it with hydroxyl amine, hydrazine and cystine, respectively (1).

In the first step, MWCNT (0.1 g) was treated with 100 mL of 69% HNO_3 in a flask of 500 mL. The flask was vibrated in an ultrasonic bath for 30 min at 25 °C. Next, the mixture was diluted with deionized water to reach 400 mL and then filtered through a polycarbonate membrane (0.22 μm). The same procedure was repeated exactly with H_2O_2 (30% v/v) instead of HNO_3. Hydrogen peroxide was used to complete the oxidation process, but mildly. The pH of the filtrate was roughly 7.0 by washing the solid with deionized water, then the product was dried under a 24 h vacuum to produce the carboxylic acid-functionalized MWCNT (MWCNT-COOH) [27].

Oxidized MWCNT (MWCNT-COOH) (0.1 g), were stirred in 2 mL of oxalyle chloride in the presence of 2–3 drops of dimethylformamide (DMF), and 2 mL of triethyl amine (TEA) at 70 °C for 24 h under N_2. After cooling to room temperature, the excess TEA was washed repeatedly with anhydrous tetrahydrofuran (THF) and then dried at 70 °C under vacuum in order to remove any traces of adsorbed TEA on the surface of acylated MWCNT. This samples are labelled as MWCNT-COCl [28].

Three separates of solutions of hydroxylamine (0.2 g), cystine (0.5 g) and of hydrazine (200 uL) in 1 mL THF and 0.5 mL pyridine was prepared. To each solution, w 0.05 g of MWCNT-COCl was added. The mixtures were stirred for 30 min at room temperature, then refluxed at 100 °C for 96 h. The residual hydroxylamine, cystine, and hydrazine were removed by rinsing with ethanol and sonication for 15 min. This rinsing process was repeated three times, until a clear ethanol was produced. The remaining solid was suspended in dichloromethane, sonicated and centrifuged. The collected black solid was dried under vacuum and labelled as magnetized multiwall carbon nanotube functionalized by hydroxyl amine (MWCNT-HA), magnetized multiwall carbon nanotube functionalized by cysteine (MWCNT-CYS), and magnetized multiwall carbon nanotube functionalized by hydrazine (MWCNT-HYD) [29].

A mixed solution of 0.1 M ferric chloride hexahydrate ($FeCl_3 \cdot 6H_2O$) and 0.05 M ferrous chloride tetrahydrate ($FeCl_2 \cdot 4H_2O$) was prepared with one to two molar ratios and then mixed with functionalized MWCNT and suspended for 2 h.

To precipitate iron oxides, 5.0 M NH_4OH solution was added dropwise until pH adjusted at 10 at 70 °C, and then kept under continuous stirring for 1 h. The solid was allowed to cool and magnet separate, and then rinsed with distilled water and ethanol. The composite obtained was dried for 2 h in a furnace at 100 °C. This samples are labelled as (m-MWCNT-HA, m-MWCNT-CYS, m-MWCNT-HYD) [30].

2.3. Adsorption Study

All adsorption runs were carried out in plastic vials (50-mL) that were placed in a water bath equipped with a thermostat and a shaker. The effect of various variables such as metal ion concertation, pH values, adsorbent dosage, adsorption time, and temperature on adsorption efficiency were studied. The adsorption experiments were performed on lead (II) ion. After every adsorption run, a clear supernatant sample from the adsorption mixture was collected via magnet and analyzed by a flame atomic adsorption analysis at 193.7 nm to determine the residual lead (II) concentration. All adsorption experiments were carried out in triplicate, and the mean was reported. The adsorbent efficiency and the amount of lead (II) ions adsorbed that was adsorbed (q_e, mg/g) was calculated by the Equation (1):

$$q_e = \frac{V(Co - Ce)}{W} \tag{1}$$

where V is the volume of the solution (L), C_o is the initial lead concentration (mg·L^{-1}), C_e is lead concentration at equilibrium (mg·L^{-1}), and W is the adsorbent's mass (g).

2.3.1. Adsorption Isotherms Models

Adsorption isotherms models are important for the description of the interaction mechanism between adsorbate and adsorbent. Several models are available for this purpose, among these, the most common are Langmuir and Freundlich. These models take into account essential comparative guidelines with small distinction in their approaches [31].

The Langmuir model adopts the formation of a monolayer of adsorbate on a homogeneous surface of an adsorbent and expressed as:

$$\frac{C_e}{q_e} = \frac{1}{q_m}C_e + \frac{1}{q_m K_L} \tag{2}$$

where C_e is the equilibrium concentration of adsorbate (mmol·L^{-1}), q_e is the amount of adsorbate adsorbed per unit weight of adsorbent (mmol·g^{-1}), q_m is the adsorption capacity (mmol·g^{-1}), or monolayer capacity, and K_L is a constant (L·mmol^{-1}).

Langmuir isotherm can be identified by constant dimensions separation factor (R_L) as shown by the following equation [32]

$$R_L = \frac{1}{(1 + K_L C_o)} \tag{3}$$

where C_o is the highest initial concentration of adsorbate (mg·L^{-1}), K_L (L·mg^{-1}) is Langmuir constant.

The value of the R_L refers to the form of isotherm to be either unfavorable ($R_L > 1$), linear ($R_L = 1$), favorable ($0 < R_L < 1$) or irreversible if ($R_L = 0$).

The Freundlich isotherm describes the adsorption between the adsorbates and the adsorbents with a heterogeneous surface. The rate of adsorption/absorption varies according to the degree of energy at the adsorptive sites. Freundlich's equation is expressed as shown in Equation (3):

$$\ln Q_e = i\frac{1}{n} ln\ iC_e + lnK_F \tag{4}$$

where K_F (mmol/g) and 1/n are the constant characteristics of the system [32].

K_F is an indicator of adsorption capacity of the adsorbent and 1/n is an indicator of favorability of adsorption process. If (10 > n > 0) this donates a favorable adsorption process.

Langmuir and Freundlich's isotherms could be used to define the relationship between the amounts of Pb(II) adsorbed on m-MWCNT-HA, m-MWCNT-CYS, m-MWCNT-HYD adsorbents and its equilibrium concentration in solutions.

2.3.2. Adsorption Kinetics

Numerous adsorption kinetic models have been set to define kinetics and rate-determining steps. These models give evidence about the performance of the adsorption system and the rate at which a specific component is removed using a specific adsorbent. Besides, it determines whether the adsorption process is physical or chemical in nature and which step is the rate-determining step. Examples of the adsorption kinetic models include pseudo-first-order, intra-particle diffusion kinetic model, pseudo-second-order models, intra-particle diffusion kinetic model, first-order reversible reaction model, Elovich's model, etc. [33].

Pseudo-first-order kinetics are developed for describing adsorption kinetics, and are considered as the earliest model. The equation for this model can be written as follows:

$$\log(q_e - q_t) = \log q_e - \left(\frac{K_1}{2.303}\right)t \tag{5}$$

where q_e and q_t are the masses of the adsorbate in equilibrium or at time t per unit mass of adsorbent (mg g^{-1}). k_1 is the rate constant of the first-order pseudo-adsorption model (mg·g^{-1}·min^{-1}).

The graph of log ($q_e - q_t$) as a function of t gives a straight line for first-order pseudo adsorption with log q_e as Y intercept [34].

The pseudo-second-order models are primarily based on the idea that the step to determine may be chemical adsorption, which involves exchanging or sharing electrons between adsorbate and adsorbent.

The net equation for pseudo-second-order can be addressed as an equation:

$$\frac{t}{q_t} = \frac{1}{q_e}t + \frac{1}{K_2 q_e^2}i \tag{6}$$

where K_2 is the equilibrium rate constant of the adsorption pseudo-second order (g·mg^{-1} min^{-1}).

The graph of t/q_t versus t should give a linear relationship that allows the calculation of a second order rate constants, K_2 from the Y intersect and q_e from the slope [30,35,36].

The metal ions adsorption is mainly achieved through three steps, including migration of metal ion from liquid to surface of the adsorbents (film diffusion process), diffusion of metal ions within the porous structure (intraparticle diffusion process, and metal ion adsorption on the surface to adsorbent). In general, the third step isn't time-consuming and is not considered as a rate-controlling step.

Intra-particle diffusion kinetic model suggested by Weber and Morris. The net equation of this kinetic model is:

$$q_t = K_{id}t^{0,5} + C \tag{7}$$

where K_{id} is the constant diffusion rate (mg·$g^{-1}$$min^{-1/2}$). C is a constant representing the thickness of the boundary layer (mg g^{-1}). A graph of q_t with respect to $t^{0.5}$ will show a linear relationship with the constant C as Y intersect. When the curve moves through the origin, adsorption is controlled by the process of inter-particle diffusion. Otherwise, it is dominated by film diffusion [37].

Activation energy can also be measured by an Arrhenius equation:

$$\text{Ln } K_2 = \text{Ln A} - \frac{\text{Ea}}{\text{RT}} \tag{8}$$

In general, if E_a between 5 kj/mol to 40 kj/mol the adsorption is physisorption whereas value (40 kj·mol^{-1} to 800 kj·mol^{-1}) refers to chemisorption.

2.3.3. Adsorption Thermodynamics

The thermodynamic study is done by ascertaining enthalpy, free energy, entropy.

Thermodynamic parameters are required to determine whether the process is spontaneous or not. Gibbs free energy change, ΔG^o, is an indication of the spontaneity of a chemical reaction and therefore is an important criterion for spontaneity. Both enthalpy (ΔH^o) and entropy (ΔS^o) factors must be considered to determine Gibb's free energy of the process. Reactions occur automatically at a certain temperature if ΔG^o is a negative quantity.

The following equation is the general equation that Connect between thermodynamics parameters

$$\Delta G^o = \Delta H^o - T\Delta S^o \quad (9)$$

where T is the absolute temperature (K).

The change in Gibbs energy can be expressed by the following equation:

$$\Delta G^o = -RT \ln K_D \quad (10)$$

where, K_D is the constant of thermodynamic equilibrium equal to (q_e/c_e) with a unit of (L·g^{-1}). R is the gas constant, 8.314 J·mol^{-1}K^{-1}.

The net equation of the last two equations can be expressed as the follows:

$$\ln K_d = -\frac{\Delta H^o}{RT} + \frac{\Delta S^o}{R} \quad (11)$$

The plot of InK_d against (1/T) gives a line with (ΔS^o /R) as the Y-intersect and ($-\Delta H^o$/R) as the slope. The resulting graph is identified as a Van't Hoff diagram [38].

2.3.4. Regeneration of Adsorbents

The adsorbent was separated by magnet, washed with 0.1 N HCl (10 mL) solution to release adsorbed metal ions, and then washed with distilled water. It was then left to dry at room temperature for 24 h. The three adsorbents were treated in the same manner.

2.4. Real Water Sample

The actual sample was used to study adsorbents efficacy for removal heavy metals. A real water sample was collected from Jericho city. The initial and final concentrations of the metals were measured using ICP/MS (Elan 9000 model, PerkinElmer, Inc., Waltham, MA, USA). A 0.02 g of adsorbent was mixed in 50 mL plastic vials with 10 mL of water sample under maximum condition; the amount adsorbed was calculated by comparing the concentrations of the metals before and after the adsorption.

3. Results and Discussion

3.1. Adsorbents Analysis

MWCNT was oxidized following a published procedure [26]. The developed carboxylic acid functionality was converted to acid chloride by reacting it with oxalyl chloride as shown in Scheme 1. Then the acid chloride was converted to hydroxamate, hydrazine and amino acids by reacting it with hydroxyl amine (HA), hydrazine (HYD) and cystine (CYS), respectively. The presence of these functionalities was confirmed by FTIR, which shows that when MWCNT-COOH was modified by HA, HYD, and CYS, several new peaks appeared in the spectra. Based on the literature: the peak at 1640 cm^{-1} is due to C=O stretching vibration. C–N stretching vibration and N–H bending vibration appear at 1550 cm^{-1} (Figure S1). Raman spectroscopy shows the spectra of MWCNT, oxidized MWCNT, and devised as MWCNT. It could be clearly noticed that the D-band of MWCNT-COOH shifted from 1360 to 1349 cm^{-1} and G-band from 1605 to 1600 cm^{-1} and this shows that a large defection was created after oxidation. Additionally, the ratio of the intensity of the D-band peak to that of the G-band peak (ID/IG) of the pure MWCNT was 1.33, while that for the MWCNT–functionalized with HA, CYS, HYD was 1.71 (Figure S2). The images of scanning electron microscopy of all samples show disordered and highly porous morphologies. The unmodified MWCNT showed less agglomeration than others (Figure S3). In addition, chemical modification leads to removal of the ends of MWCNT, which increases surface area as confirmed by Brunauer–Emmett–Teller (BET) (Table 1). The magnetization of prepared MWCNT confirmed by VSM, which shows that the magnetization curve for MWCNT sample has a semi-linear representation, which indicates a paramagnetic state while magnetization curve for m-MWCNT-HA, m-MWCNT-CYS, and m-MWCNT-HYD, showed a sigmoidal response with no hysteresis, which is an indication of the presence of a saturated superparamagnetic component in the samples (Figure S4). All characteristic techniques confirmed that HA, CYS and HUD were successfully drafted on MWCNT, and also created functionals, which are expected to have excellent activities toward metal ions [39].

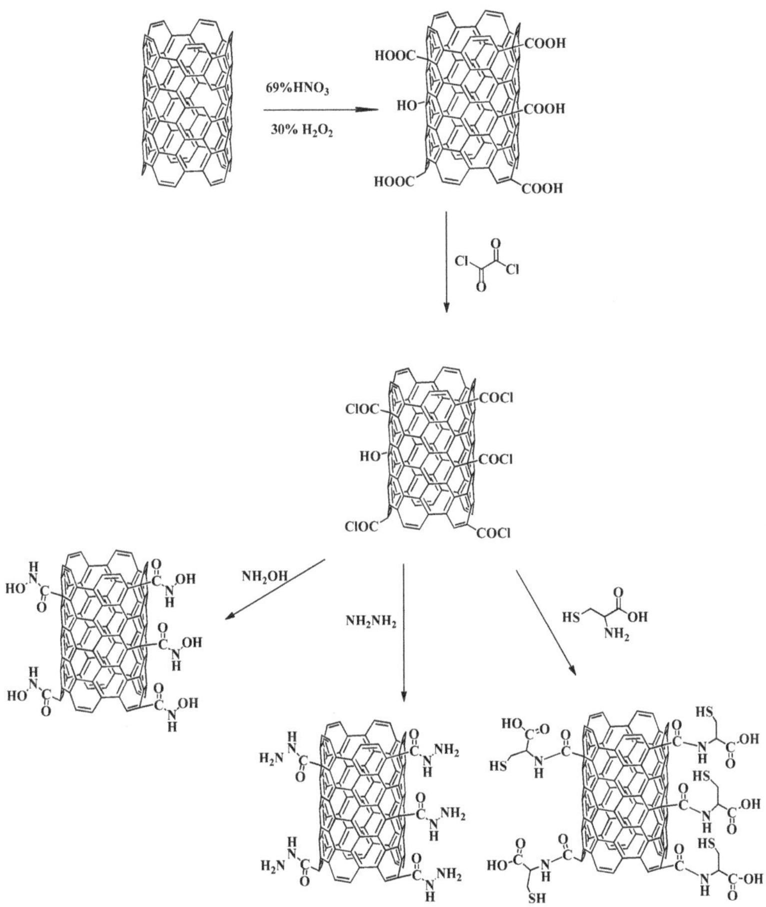

Scheme 1. Preparation of magnetic functionalized multiwall carbon nanotube.

With reference to previous studies, there are several studies related to oxidation and surface modification of MWCNT. The novelty of this work can be summarized as the grafted MWCNTs presented in this work comprise the first example in the literature of oxidized MWCNT modified with such functionalities and applied for lead removal. Additionally, previous studies on the removal of lead using a single functional group provided lower removal than that presented in this study (Table 1). Other studies with higher removals than that presented in this study were based on using photodegradation or membranes, which are extremely expensive methods compared with the method presented in our study.

Table 1. Previous studies on lead removal from water.

Adsorbents	Maximum *Adsorption* Capacity (*qm*) or % Removal	Conditions	Surface Area (m^2/g)	References
MWCNTs	1	pH: 5, CNTs dosage: 0.05 g, Temp:280–321 K	134	[40]
MWCNTs AC	4 18	pH: 5, Temp:298–323 K, CNTs dosage:1 g, Contact time: 20–120 min	162.16 1124.8	[41]
Oxidized CNTs	104.0	pH: 7, Temp: 298 K, C_0: 100–1200	66	[42]
CNTs	62.5%	pH: 5, CNTs dosage: 0.05 g, Temp: 298 K, C_0: 5–60	98.6	[43]
Oxidized MWCNTs	97.08%	pH: 6–11, CNTs dosage: 0.05–0.3 g	75.4	[44]
m-MWCNT-HA m-MWCNT-CYS m-MWCNT-HYD	99.8% 97.4% 97.5%	pH: 8, Temp: 298 K, C_0: 10 ppm	151 154.5 187	Our research

3.2. *Lead Ion Adsorption*

The adsorption was carried out using a batch process. During the prior process, a known weight of adsorbent (0.02 g) was suspended in an aqueous solution of lead ion, separated by a magnet, and analyzed. The analysis was performed on the filtrate to determine the amount of the residual lead ions. The effects of several variables such as contact time, adsorbent dosage, temperature, and pH were evaluated to determine the best conditions for the highest adsorption efficiency. The adsorption study was performed on Pb(II) ions using (m-MWCNT-HA), (m-MWCNT-CTS), (m-MWCNT-HYD) adsorbents. Each experiment was repeated three times, and the average was used when we analyzed the data.

3.2.1. Effect of Contact Time

This experiment was performed to determine the optimal adsorption time. A 10 mL solution of Pb(II) with 10 ppm concentration was placed in a vial and shaken with 0.02 g of an adsorbent for a varied period of time; ranging from 1 to 120 min.

Adsorption curves are presented in Figure 1a. The figure shows that in the beginning, the development of monolayer was very rapid and the maximum removal was attained after approximately 30 min for all three adsorbents. After that, the adsorption leveled off and became nearly constant.

The high percentage of removal of Pb(II) in the beginning was due to the presence of the vacant sites which decrease steadily with time passing. In general, excellent adsorption efficiency for Pb(II) was exhibited by the three adsorbents.

The same type of interaction appears when using m-MWCNT-HA and m-MWCNT-CYS.

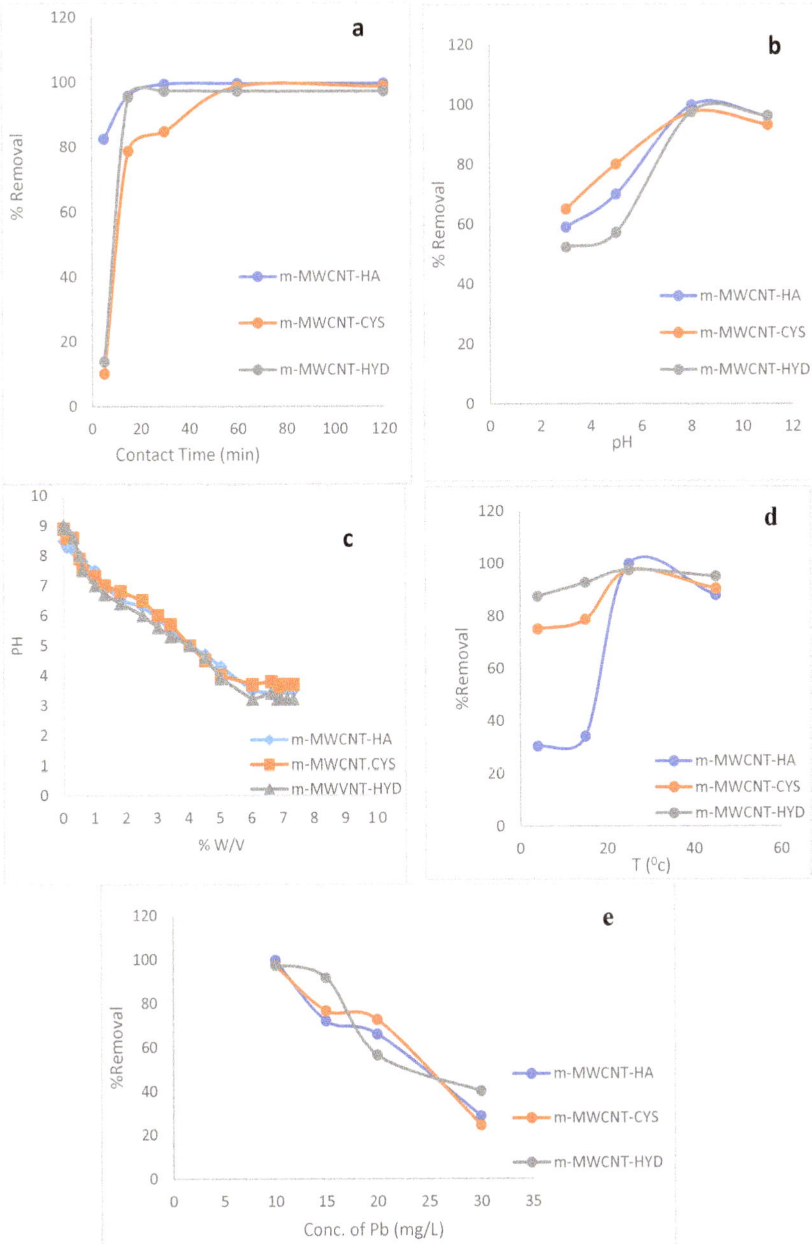

Figure 1. Effect of (**a**) contact time, (**b**) pH, (**c**) experimental mass titration curves, (**d**) temperature, (**e**) Pb(II) initial concentration on magnetized multiwall carbon nanotube functionalized by hydroxyl amine (m-MWCNT-HA), magnetized multiwall carbon nanotube functionalized by cysteine (m-MWCNT-CYS), and magnetized multiwall carbon nanotube functionalized by hydrazine (m-MWCNT-HYD).

3.2.2. Effect of pH

The effect of pH on the adsorption efficiency is shown in Figure 1b. The pH effect on adsorption is an important factor as it affects the surface charge of the adsorbents. At pH values of 3.0 or less, the amine functionality becomes ammonium whereas, at high pH value over 5.5, the amine is not protonated and the carboxylic acid functionality becomes carboxylate. The highest adsorption efficiency was observed at a pH 8.0 then it started to decline. At pH value of about 8.0, the functional groups present in the adsorbents (amine and carboxyl) carry a lone of pair of electrons causing them to behave as a strong chelating agent.

The pH effect can also be explained in terms of pH_{pzc} of adsorbents. In pH_{pzc} the positive and negative groups are equal and thus sum surface charges become zero.

Mass titration technique (MT) was used to determine zero-point charge (pzc) [45]. When observing Figure 1c, it can be investigated that pzc equals 3.5, 3.7, 3.2 for m-MWCNT-HA, m-MWCNT-CYS, m-MWCNT-HYD, respectively, with the highest adsorption value appeared at $pH > pH_{pzc}$. Consequently, electrostatic attractions among negatively charged surface of adsorbent and metal-ions could occur and influence to sorption sorbent mass (Figure 2).

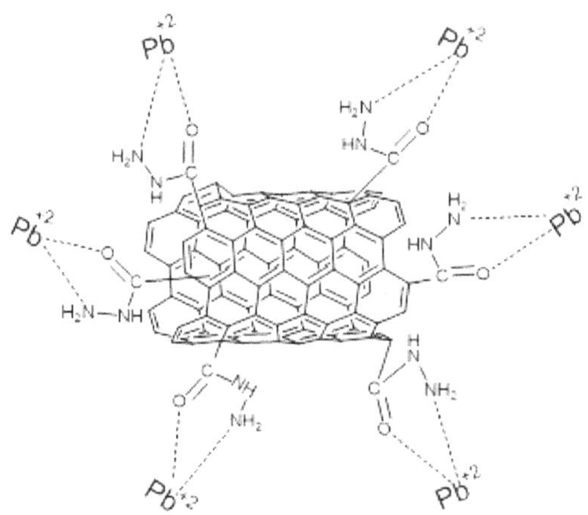

Figure 2. Mechanism of lead adsorption on m-MWCNT-HYD.

3.2.3. Effect of Temperature on Lead Adsorption

Adsorption equilibrium is reached when the chemical potential of the adsorbed solute and the chemical potential the solute in aqueous solution are equal. This equilibrium is temperature-dependence since the chemical potential is a function of temperature.

In this study, the dependence of adsorption equilibrium of Pb(II) on m-MWCNT-HA, m-MWCNT-CYS, and m-MWCNT-HYD was evaluated using different temperatures (5–50 °C) at a pH value of 8. The effect of temperature on the percentage of lead adsorption is shown in Figure 1d, as shown in the figure the maximum adsorption was at room temperature for the three adsorbents.

3.2.4. Effect of Lead Initial Concentration

The results of Pb(II) initial concertation on adsorption efficiency is shown Figure 1e. The highest removal percentage of Pb(II) for the three adsorbents at 10.0 ppm concentration.

The results demonstrate that at lower concentrations, the ratio of the number of metal ions to the existing sorption sites is low and thus the adsorption under these conditions is independent of the initial concentration. However, as the concentration of Pb(II) increases, the sites available for adsorption become less, and therefore, the removal of metals becomes highly dependent on the initial concentrations. The results show that the rate of Pb(II) ion removal decreases as the initial concentration of metal ions increases.

3.3. Equilibrium Isotherm Models for Lead Adsorption

Langmuir and Freundlich's isotherms were applied to describe the relationship between amounts Pb(II) adsorbed on m-MWCNT-HA, m-MWCNT-CYS, and m-MWCNT-HYD adsorbents and its equilibrium concentration in solutions, results are presented in Figure 3a,b.

Freundlich and Langmuir adsorption isotherms parameters were calculated. The values of R^2 using Langmuir adsorption isotherm are approximately one. The value of R_L was found to be. 0.002, 0.012, 0.007 for m-MWCNT-HA, m-MWCNT-CYS, m-MWCNT-HYD, respectively, and this indicates that Langmuir adsorption isotherm is favorable.

From Freundlich isotherm data shown in Table 2, the value of R^2 is much smaller than 1 (0.15, 0.02, 0.26) for m-MWCNT-HA, m-MWCNT-CYS, m-MWCNT-HYD, respectively), this indicates that the adsorption is not suited with Freundlich isotherm.

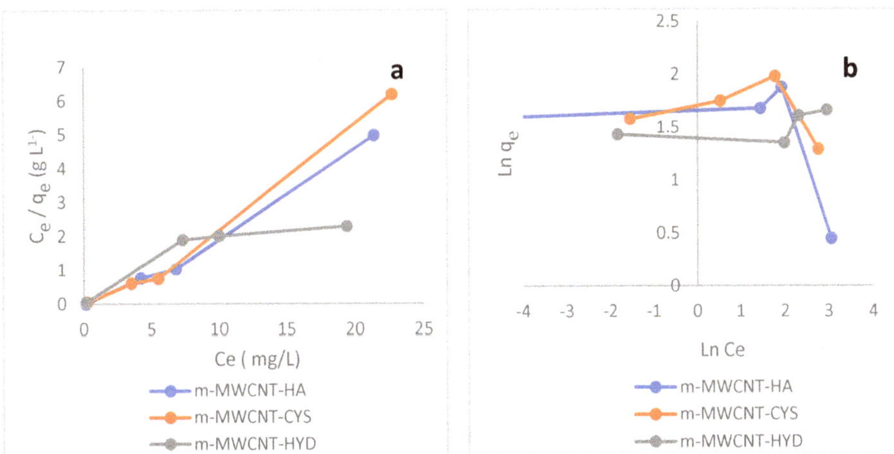

Figure 3. (**a**) Langmuir isotherm plot, and (**b**) Freundlich isotherm plot for Pb(II) adsorption.

Table 2. The Langmuir and Freundlich parameters for Pb(II) adsorption on m-MWCNT-HA, m-MWCNT-CYS, and m-MWCNT-HYD.

	Langmuir Isotherm				**Freundlich Isotherm**		
	q_m (mg·g^{-1})	K_L (g·L^{-1})	R_L	R^2	K_F (mg·g^{-1})	n (L·mg^{-1})	R^2
m-MWCNT-HA	1.15	39.13	0.002	0.99	4.22	−12.80	0.15
m-MWCNT-CYS	3.31	0.76	0.012	0.98	5.31	−41.60	0.02
m-MWCNT-HYD	9.09	0.20	0.007	0.72	4.30	28.50	0.26

3.4. Adsorption Kinetic Model

To study the mechanism of the adsorption process, pseudo first-order, pseudo second-order and intra-particle diffusion kinetic models were evaluated.

Adsorption kinetic models are represented in Figure 4. The kinetic study revealed that, the adsorption of Pb(II) on m-MWCNT-HA, m-MWCNT-CYS, m-MWCNT-HYD follows the pseudo-second-order kinetic mechanism. The correlation coefficient R^2 and q_e experimental and theoretical which in this kinetic model are approximately one (Table 3).

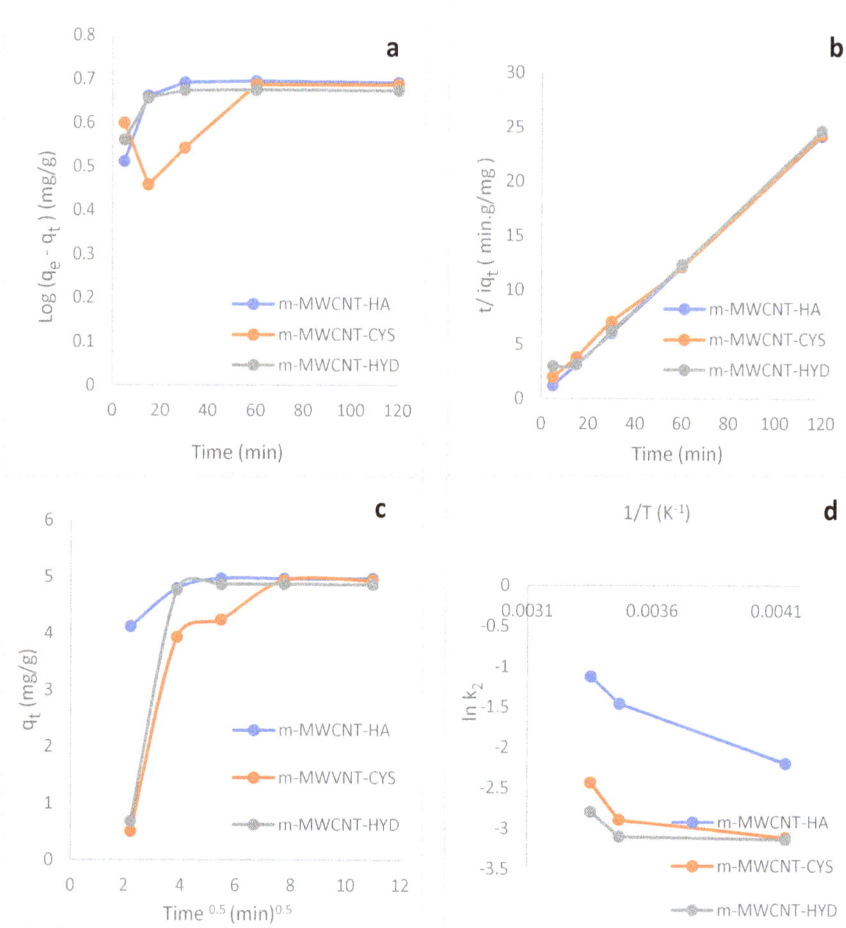

Figure 4. (**a**) Pseudo first-order plot. (**b**) Pseudo second-order plot. (**c**) Intra-particle diffusion plot. (**d**) Arrhenius plot for the adsorption of Pb(II) on m-MWCNT-HA, m-MWCNT-CYS, and m-MWCNT-HYD.

Table 3. Parameters for Pb(II) adsorption of kinetic models of pseudo-first order, pseudo-second and intraparticle diffusion on m-MWCNT-HA, m-MWCNT-CYS and m-MWCNT-HYD.

	Pseudo First-Order Kinetics				**Pseudo Second-Order**			
	q_e exp	q_e ($mg\ g^{-1}$) Calc	K_1 ($mg\ g^{-1}\ min^{-1}$)	R^2	q_e ($mg\ g^{-1}\ min^{-1}$)	K_2 ($mg\ g^{-1}\ min^{-1}$)	E_a (kj)	R^2
m-MWCNT -HA	4.96	4.03	−0.0023	0.33	4.99	0.32	10.50	0.99
m-MWCNT-CYS	4.93	3.35	−0.0034	0.51	5.15	0.03	5.65	0.99
m-MWCNT-HYD	4.85	4.16	−0.0013	0.32	5.39	0.04	2.50	0.91

Intra-Particle Diffusion		
C ($mg\ g^{-1}$)	**K_{id} ($mg\ g^{-1} min^{-0.5}$)**	**R^2**
4.29	0.07	0.51
1.172	0.41	0.60
1.90	0.34	0.40

In the intraparticle diffusion plot shown in Figure 3, the curve is not passing through the origin, so the adsorption is controlled by the external diffusion, and the interparticle diffusion mechanism.

The E_a value was calculated form Arrhenius equation by plotting ln K_2 versus 1/T (Figure 4d). The obtained value of Ea indicates that the adsorption processes of Pb(II) ions by the three adsorbents m-MWCNT-HA, m-MWCNT-HYD and m-MWCNT-CYS are physical.

3.5. Adsorption Thermodynamic

The Van't Hoff plot was used to calculate the common thermodynamics parameters: ΔS^o, ΔH^o, and ΔG^o for Pb(II) adsorption on m-MWCNT-HA, m-MWCNT-CYS, m-MWCNT-HYD as shown in Figure 5.

As shown in Table 4, the values of free energy were negative. This sign is a clear indication that the adsorption is spontaneous and favorable while the values of enthalpy were positive. This is a clear indication of the endothermic process. The positive ΔS^o values reveal an increase in randomness in the solid and liquid surface, indicating the accumulation of Pb(II).

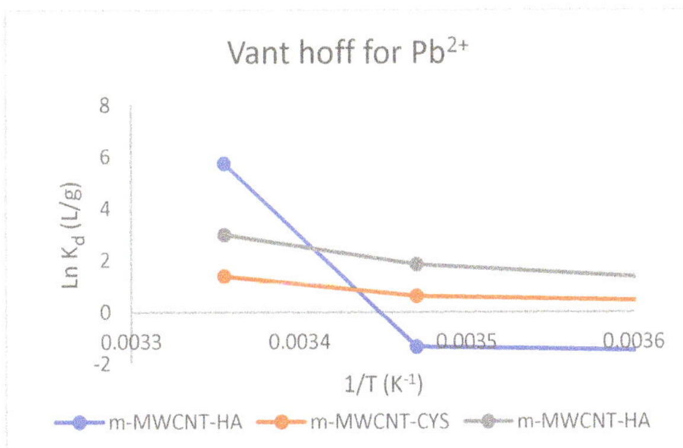

Figure 5. Van't Hoff plot for Pb(II) adsorption on m-MWCNT-HA, m-MWCNT-CYS, and m-MWCNT-HYD.

Table 4. Thermodynamic parameters for the adsorption of Pb(II) on m-MWCNT-HA, m-MWCNT-CYS, and m-MWCNT-HYD.

	ΔH^o (kj)	ΔS^o (J K^{-1})	ΔG^o (25 °C)
m-MWCNT-HA	205.60	0.72	−10.15
m-MWCNT-CYS	17.26	0.06	−2.11
m-MWCNT-HYD	50.60	0.19	−6.90

As shown in Table 4, the adsorption of Pb(II) on m-MWCNT-HA, m-MWCNT-CYS, m-MWCNT-HYD endothermic and spontaneous process ($\Delta H^o > 0$) and spontaneous ($\Delta G^o < 0$).

3.6. Adsorbent Regeneration

The adsorbents regeneration and reuse as absorbents for Pb(II) was evaluated. The three adsorbents were tested in three cycles, m-MWCNT-HA maintained its efficiency toward Pb(II), while the other two adsorbent m-MWCNT-CYS and m-MWCNT-HYD showed some reduction in the efficiency in the third cycles (Figure 6). To remediate the MWCNTs after regeneration, future studies will be needed.

Biodegradation of CNTs by enzymes was studied by Allen et al. (2008). They reported that oxidized single-wall CNTs (SWCNTs) could be degraded by enzymatic oxidation with horseradish peroxidase (HRP), a plant enzyme. They also reported that degradation took place within 10 days using hydrogen peroxides. This degradation will be leading to production of oxidized polyaromatic hydrocarbons and ultimately CO_2 [46]. Serra et al., studied the circular zero-residue process using microalgae for efficient water decontamination, biofuel production, and carbon dioxide fixation [47].

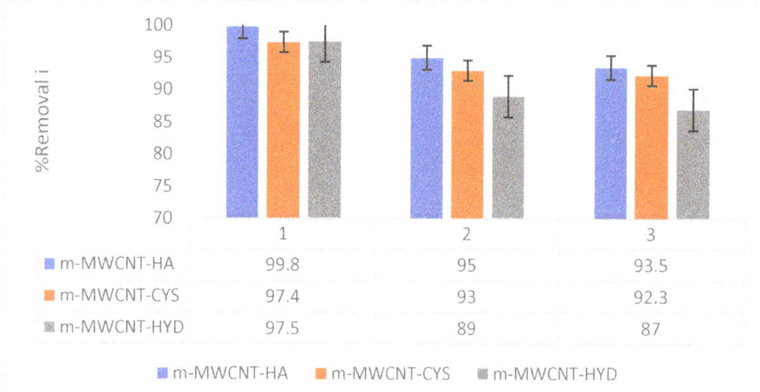

Figure 6. Three trials of adsorption—desorption of MWCNT using different functional groups towards removal efficiency of Pb(II).

3.7. *Real Water Samples*

The removal of metal by prepared adsorbents from real environmental water sample was studied to investigate the applicability of the proposed solid adsorbent. The results were presented in Table 5. It is clear from the table that m-MWCNT-HA, m-MWCNT-CYS, m-MWCNT-HYD has excellent efficiency towards most of the metal present in real water sample. The results obtained showed that the efficiency for the removal of lead were decreased when using real sample due to presence of dissolved organic matter and other metals. The removal efficiency decreased to about 64%, 73.1% and 67.1% when using m-MWCNT-HA, m-MWCNT-HYD and m-MWCNT-CYS respectively. This confirm the applicability of adsorbents for environmental.

Table 5. Effect of adsorption of metals form real water sample.

Analyte	ppb Real Water	ppb After Treatment with m-MW CNT-HA	% Removal after Treatment with m-MW CNT-HA	ppb After Treatment with m-MW CNT-CYS	% Removal after Treatment with m-MW CNT-CYS	ppb after Treatment with m-MW CNT-HYD	% Removal after Treatment with m-MW CNT-HYD
Al	9.2	1.7	81.5	2.9	68.4	0.9	90.2
As	1.7	0.7	58.8	0.66	61.1	0.23	86.4
Ba	29.6	0.1	99.6	0.5	98.3	11.3	61.8
Be	0.1	0.1	0	0.05	50	0.1	0
B	120.8	88	27.1	89.8	25.6	77.6	35.7
Cd	19	13.5	28.9	1.7	91	1.7	91
Cr	22	0.5	97.2	13.7	37.7	11	50
Co	15	5.1	66	3.1	79.3	2.2	85.3
Cu	2.1	0.99	52.8	0.8	61.9	0.11	94.7
Fe	520.5	170	67.5	77	85.2	140.5	73
Pb	169	60.7	64	45.3	73.1	55.6	67.1
Mn	290	160	44.8	35.6	87.7	40.6	86
Ni	40.8	31.9	21.8	20.1	50.7	24.5	39.9
Se	80.7	76.2	5.6	80.7	0	70.2	13
Tl	0.7	0.4	42.8	0.4	42.8	0.2	71.4
V	13.7	6.2	54.7	13.4	2.1	2.3	83.2
Zn	90	40	55.5	55.1	38.7	36	60

4. Conclusions

In this study, a magnetic multi-wall carbon nanotube with hydroxylamine, cystine and hydrazine was synthesis and tested for the removal of Pb(II) from water as the first example in the literature of oxidized MWCNT modified with such functionalities. The efficiency of the prepared derivatives toward

Pb(II) was studied as a function of adsorbent dose, pH, metal ion initial concertation, temperature and time. The three adsorbents showed excellent efficiency toward Pb(II) and % of removal was quantitative. The highest efficiency was determined to be at room temperature and a pH of 8.0. The kinetic study revealed that the Pb(II) adsorption by the three adsorbents followed pseudo-second-order and Langmuir isotherm model. The thermodynamic analysis showed a negative free energy, indicating a spontaneous adsorption process at room temperatures. The adsorbents were regenerated by treatment with 0.1 N HCl. After third regeneration and reuse cycles, the efficiency of the regenerated adsorbents has shown a small reduction in efficiency. The presence of dissolved organic matter and other metals in real water sample has a significant effect on lead removal efficiency.

Supplementary Materials: The following are available online at http://www.mdpi.com/2227-9717/8/8/986/s1, Figure S1: FT-IR spectrum for MWCNTS, m-MWCNT-COOH, m-MWCNT-HA, m-MWCNT-Cys, m-MW CNT-HYD, Figure S2: Raman spectrum for MWCNTS, m-MWCNT-COOH, m-MWCNT-HA, m-MWCNT-Cys, m-MWCNT-HYD, Figure S3: SEM images of the original and the modified carbon nano tube (right) together with diameter distribution (left). (a) MWCNTs, (b) MWCNT-COOH, (c): m-MWCNT-HA, (d) m-MWCNT-CYS and (e) m-MWCNT-HYD. Scale bars: 5 µm, Figure S4: Magnetic field dependence of the magnetization measured at (a) 300K (M–H loops) of m-MWCNT-HA, m-MWCNT-CYS, m-MWCNT-HYD and MWCNT; (b) the magnification of the central area of the hysteresis loops.

Author Contributions: G.H.: Investigation, formal analysis, writing—original draft. S.J. and O.H.: investigation, formal analysis, writing—original draft, funding acquisition. R.B., O.D., and A.Q. assisted in sample analysis; B.K. and S.S. assisted in sample preparation. All authors have read and agreed to the published version of the manuscript.

Funding: This study was supported by *Middle East Desalination Research* Center (MEDRC) and Palestinian Water Authority (PWA). Part of research was funded by the German Federal Ministry of Education and Research (BMBF).

Acknowledgments: The authors express their thanks to the MEDRC and PWA for their financial support during this study. Other thanks go to PADUCO 2 for some financial support. Many thanks go to the department of chemistry at An-Najah National University (ANNU) for facilitating the use of their instrumentations. This work was in part supported by the research project "Palestinian German Scientific Bridge (PGSB)" carried out by the Forschungszentrum Jülich and Palestinian Academy for Science and Technology- PALAST and funded by the German Federal Ministry of Education and Research (BMBF).

Conflicts of Interest: The authors report no relationships that could be construed as a conflict of interest.

Abbreviations

CNTs	Carbon nanotube
MWCNT	Multiwall carbon nanotube
MWCNT-COOH	Oxidized Multiwall carbon nanotube
m-MWCNT-HA	Magnetized Multiwall carbon nanotube functionalized by hydroxyl amine
m-MWCNT-CYS	Magnetized Multiwall carbon nanotube functionalized by cysteine
m-MWCNT-HYD	Magnetized Multiwall carbon nanotube functionalized by hydrazine

References

1. Marcovecchio, J.E.; Botté, S.E.; Freije, R.H. Heavy metals, major metals, trace elements. *Handb. Water Anal.* **2007**, *2*, 275–311.
2. Lanphear, B.P.; Hornung, R.; Khoury, J.; Yolton, K.; Baghurst, P.; Bellinger, D.C.; Canfield, R.L.; Dietrich, K.N.; Bornschein, R.; Greene, T. Low-level environmental lead exposure and children's intellectual function: An international pooled analysis. *Environ. Health Perspect.* **2005**, *113*, 894–899. [CrossRef] [PubMed]
3. Stellmach, J.J. The commercial success of chrome tanning: A study and commemorative. *JALCA* **1990**, *85*, 407–424.
4. Rao, M.M.; Rao, G.C.; Seshaiah, K.; Choudary, N.; Wang, M. Activated carbon from Ceiba pentandra hulls, an agricultural waste, as an adsorbent in the removal of lead and zinc from aqueous solutions. *Waste Manag.* **2008**, *28*, 849–858. [CrossRef] [PubMed]
5. Golub, M.S. *Metals, Fertility, and Reproductive Toxicity*; CRC Press: Boca Raton, CA, USA, 2005; Available online: https://books.google.co.il/books?hl=iw&lr=&id=Qt8LEB7_HyQC&oi=fnd&pg=PA1&dq=5.%09Golub,+M.S.+Metals,+Fertility,+and+Reproductive+Toxicity%3B+CRC+Press:+2005&ots=sQWGAh5z9m&sig=9F6mCcF22tJdtnFtoWk78F6xG2E&redir_esc=y#v=onepage&q=5.%09Golub%2C%20M.S.%20Metals%

2C%20Fertility%2C%20and%20Reproductive%20Toxicity%3B%20CRC%20Press%3A%202005&f=false (accessed on 14 May 2020).

6. Gulson, B.; Mahaffey, K.; Mizon, K.; Korsch, M.; Cameron, M.; Vimpani, G. Contribution of tissue lead to blood lead in adult female subjects based on stable lead isotope methods. *J. Lab. Clin. Med.* **1995**, *125*, 703–712.
7. Hengstler, J.G.; Bolm-Audorff, U.; Faldum, A.; Janssen, K.; Reifenrath, M.; Götte, W.; Jung, D.; Mayer-Popken, O.; Fuchs, J.; Gebhard, S. Occupational exposure to heavy metals: DNA damage induction and DNA repair inhibition prove co-exposures to cadmium, cobalt and lead as more dangerous than hitherto expected. *Carcinogenesis* **2003**, *24*, 63–73. [CrossRef]
8. Assi, M.A.; Hezmee, M.N.M.; Haron, A.W.; Sabri, M.Y.M.; Rajion, M.A. The detrimental effects of lead on human and animal health. *Vet. World* **2016**, *9*, 660. [CrossRef]
9. Roy, A.; Bhattacharya, J. *Nanotechnology In Industrial Wastewater Treatment*; IWA Publishing: Howrah, West Bengal, India, 2015.
10. Mensah, S.O. Dietary Risk Assessment of Green Leafy Vegetables (GLV) Due to Heavy Metals from Selected Mining Areas. Ph.D. Thesis, Kwame Nkrumah University of Science & Technology, Kumasi, Ghana, 2018.
11. Malassa, H.; Al-Qutob, M.; Al-Khatib, M.; Al-Rimawi, F. Determination of different trace heavy metals in ground water of South West Bank/Palestine by ICP/MS. *J. Environ. Prot.* **2013**, *4*, 818. [CrossRef]
12. Ghanem, M.; Samhan, N. Groundwater Pollution Assessment in Tulkarem Area, Palestine. *J. Earth Sci. Geotech. Eng.* **2012**, *2*, 1–16.
13. El-Nahhal, Y. Contamination of Groundwater with Heavy Metals in Gaza Strip. In Proceedings of the Tenth International Water Technology Conference IWTC10, Alexandria, Egypt, 25 March 2006.
14. Jiang, J.; Sandler, S.I.; Schenk, M.; Smit, B. Adsorption and separation of linear and branched alkanes on carbon nanotube bundles from configurational-bias Monte Carlo simulation. *Phys. Rev. B* **2005**, *72*, 045447. [CrossRef]
15. Du, R.; Wu, J.; Chen, L.; Huang, H.; Zhang, X.; Zhang, J. Hierarchical Hydrogen Bonds Directed Multi-Functional Carbon Nanotube-Based Supramolecular Hydrogels. *Small* **2014**, *10*, 1387–1393. [CrossRef]
16. Chen, L.; Xin, H.; Fang, Y.; Zhang, C.; Zhang, F.; Cao, X.; Zhang, C.; Li, X. Application of metal oxide heterostructures in arsenic removal from contaminated water. *J. Nanomater.* **2014**, *2014*, 793610. [CrossRef]
17. Lin, T.; Bajpai, V.; Ji, T.; Dai, L. Chemistry of carbon nanotubes. *Aust. J. Chem.* **2003**, *56*, 635–651. [CrossRef]
18. Abbas, A.; Al-Amer, A.M.; Laoui, T.; Al-Marri, M.J.; Nasser, M.S.; Khraisheh, M.; Atieh, M.A. Heavy Metal Removal from Aqueous Solution by Advanced Carbon Nanotubes: Critical Review of Adsorption Applications. *Sep. Purif. Technol.* **2016**, *157*, 141–161.
19. Kosa, S.A.; Al-Zhrani, G.; Salam, M.A. Removal of heavy metals from aqueous solutions by multi-walled carbon nanotubes modified with 8-hydroxyquinoline. *Chem. Eng. J.* **2012**, *181*, 159–168. [CrossRef]
20. Li, Y.-H.; Wang, S.; Wei, J.; Zhang, X.; Xu, C.; Luan, Z.; Wu, D.; Wei, B. Lead adsorption on carbon nanotubes. *Chem. Phys. Lett.* **2002**, *357*, 263–266. [CrossRef]
21. Wang, H.; Zhou, A.; Peng, F.; Yu, H.; Chen, L. Adsorption characteristic of acidified carbon nanotubes for heavy metal Pb (II) in aqueous solution. *Mater. Sci. Eng. A* **2007**, *466*, 201–206. [CrossRef]
22. Tofighy, M.A.; Mohammadi, T. Adsorption of divalent heavy metal ions from water using carbon nanotube sheets. *J. Hazard. Mater.* **2011**, *185*, 140–147. [CrossRef]
23. Wang, H.; Zhou, A.; Peng, F.; Yu, H.; Yang, J. Mechanism study on adsorption of acidified multiwalled carbon nanotubes to Pb (II). *J. Colloid Interface Sci.* **2007**, *316*, 277–283. [CrossRef]
24. Zhang, G.; Qu, J.; Liu, H.; Cooper, A.T.; Wu, R. CuFe2O4/activated carbon composite: A novel magnetic adsorbent for the removal of acid orange II and catalytic regeneration. *Chemosphere* **2007**, *68*, 1058–1066. [CrossRef]
25. Reshadi, M.A.M.; Bazargan, A.; McKay, G. A review of the application of adsorbents for landfill leachate treatment: Focus on magnetic adsorption. *Sci. Total Environ.* **2020**, *731*, 138863. [CrossRef] [PubMed]
26. Hanbali, G.; Jodeh, S.; Hamed, O.; Bol, R.; Khalaf, B.; Qdemat, A.; Samhan, S. Enhanced Ibuprofen Adsorption and Desorption on Synthesized Functionalized Magnetic Multiwall Carbon Nanotubes from Aqueous Solution. *Materials* **2020**, *13*, 3329. [CrossRef] [PubMed]
27. Kassaee, M.; Zandi, H.; Akbari, J.; Motamedi, E. An efficient and mild carboxylation of multiwall carbon nanotubes using H2O2 in the presence of heteropolyacid. *Chin. Chem. Lett.* **2012**, *23*, 470–473. [CrossRef]

28. Zhang, Q.; Wu, J.; Gao, L.; Liu, T.; Zhong, W.; Sui, G.; Zheng, G.; Fang, W.; Yang, X. Dispersion stability of functionalized MWCNT in the epoxy–amine system and its effects on mechanical and interfacial properties of carbon fiber composites. *Mater. Des.* **2016**, *94*, 392–402. [CrossRef]
29. Shukla, K.S. Synthesis and Characterization of Aromatic-Heterocyclic Derivatives of Sulphonamides as Potential Antimicrobial Agents. *J. Enzyme Inhib. Med. Chem.* **2009**, *24*, 986–992.
30. Gupta, V.; Agarwal, S.; Saleh, T.A. Chromium removal by combining the magnetic properties of iron oxide with adsorption properties of carbon nanotubes. *Water Res.* **2011**, *45*, 2207–2212.
31. Al-Khateeb, L.A.; Hakami, W.; Salam, M.A. Removal of non-steroidal anti-inflammatory drugs from water using high surface area nanographene: Kinetic and thermodynamic studies. *J. Mol. Liq.* **2017**, *241*, 733–741. [CrossRef]
32. Jodeh, S.; Warad, I. Synthesis of 1-(Pyrrol-2-yl) Imine Modified Silica as a New Sorbent for the Removal of Hexavalent Chromium from Water. Maste's Thesis, An-Najah National University, Nablus, Palestine, 2016.
33. Padilla-Ortega, E.; Leyva-Ramos, R.; Mendoza-Barron, J.; Guerrero-Coronado, R.; Jacobo-Azuara, A.; Aragon-Piña, A. Adsorption of heavy metal ions from aqueous solution onto sepiolite. *Adsorpt. Sci. Technol.* **2011**, *29*, 569–584. [CrossRef]
34. Nethaji, S.; Sivasamy, A.; Mandal, A. Adsorption isotherms, kinetics and mechanism for the adsorption of cationic and anionic dyes onto carbonaceous particles prepared from Juglans regia shell biomass. *Int. J. Environ. Sci. Technol.* **2013**, *10*, 231–242. [CrossRef]
35. Coelho, G.F.; Gonçalves, A.C.; Schwantes, D.; Rodríguez, E.Á.; Tarley, C.R.T.; Dragunski, D.; Junior, É.C. Removal of Cd (II), Pb (II) and Cr (III) from water using modified residues of *Anacardium occidentale* L. *Appl. Water Sci.* **2018**, *8*, 96. [CrossRef]
36. Rees, F.; Simonnot, M.-O.; Morel, J.-L. Short-term effects of biochar on soil heavy metal mobility are controlled by intra-particle diffusion and soil pH increase. *Eur. J. Soil Sci.* **2014**, *65*, 149–161. [CrossRef]
37. Rashed, M.N. Adsorption technique for the removal of organic pollutants from water and wastewater. In *Organic Pollutants-Monitoring, Risk and Treatment*; IntechOpen: Aswan, Egypt, 2013.
38. Cheung, W.; Szeto, Y.; McKay, G. Intraparticle diffusion processes during acid dye adsorption onto chitosan. *Bioresour. Technol.* **2007**, *98*, 2897–2904. [CrossRef] [PubMed]
39. Lima, E.C.; Hosseini-Bandegharaei, A.; Moreno-Piraján, J.C.; Anastopoulos, I. A critical review of the estimation of the thermodynamic parameters on adsorption equilibria. Wrong use of equilibrium constant in the Van't Hoof equation for calculation of thermodynamic parameters of adsorption. *J. Mol. Liq.* **2019**, *273*, 425–434. [CrossRef]
40. Li, Y.-H.; Wang, S.; Luan, Z.; Ding, J.; Xu, C.; Wu, D. Adsorption of cadmium (II) from aqueous solution by surface oxidized carbon nanotubes. *Carbon* **2003**, *41*, 1057–1062. [CrossRef]
41. Atieh, M.A.; Bakather, O.Y.; Al-Tawbini, B.; Bukhari, A.A.; Abuilaiwi, F.A.; Fettouhi, M.B. Effect of carboxylic functional group functionalized on carbon nanotubes surface on the removal of lead from water. *Bioinorg. Chem. Appl.* **2010**, *2010*, 603978. [CrossRef] [PubMed]
42. Vuković, G.D.; Marinković, A.D.; Čolić, M.; Ristić, M.Đ.; Aleksić, R.; Perić-Grujić, A.A.; Uskoković, P.S. Removal of cadmium from aqueous solutions by oxidized and ethylenediamine-functionalized multi-walled carbon nanotubes. *Chem. Eng. J.* **2010**, *157*, 238–248. [CrossRef]
43. Hsieh, S.-H.; Horng, J.-J.; Tsai, C.-K. Growth of carbon nanotube on micro-sized Al 2 O 3 particle and its application to adsorption of metal ions. *J. Mater. Res.* **2006**, *21*, 1269–1273. [CrossRef]
44. Xu, D.; Tan, X.; Chen, C.; Wang, X. Removal of Pb (II) from aqueous solution by oxidized multiwalled carbon nanotubes. *J. Hazard. Mater.* **2008**, *154*, 407–416. [CrossRef]
45. Fiol, N.; Villaescusa, I. Determination of sorbent point zero charge: Usefulness in sorption studies. *Environ. Chem. Lett.* **2009**, *7*, 79–84. [CrossRef]

46. Allen, B.L.; Kotchey, G.P.; Chen, Y.; Yanamala, N.V.; Klein-Seetharaman, J.; Kagan, V.E.; Star, A. Mechanistic investigations of horseradish peroxidase-catalyzed degradation of single-walled carbon nanotubes. *J. Am. Chem. Soc.* **2009**, *131*, 17194–17205. [CrossRef]
47. Serrà, A.; Artal, R.; García-Amorós, J.; Gómez, E.; Philippe, L. Circular zero-residue process using microalgae for efficient water decontamination, biofuel production, and carbon dioxide fixation. *Chem. Eng. J.* **2020**, *388*, 124278. [CrossRef]

processes

MDPI

Article

Evaluation of the Adsorption Efficiency on the Removal of Lead(II) Ions from Aqueous Solutions Using *Azadirachta indica* Leaves as an Adsorbent

Abubakr Elkhaleefa [1,2], Ismat H. Ali [3,*], Eid I. Brima [3,4], Ihab Shigidi [1], Ahmed. B. Elhag [5,6] and Babiker Karama [7]

1 Department of Chemical Engineering, College of Engineering, King Khalid University, Abha 61413, Saudi Arabia; amelkhalee@kku.edu.sa (A.E.); etaha@kku.edu.sa (I.S.)
2 Department of Chemical Engineering and Chemical Technology, University of Gezira, Wadmedani 21113, Sudan
3 Department of Chemistry, College of Science, King Khalid University, Abha 61413, Saudi Arabia; ebrima@kku.edu.sa
4 School of Allied Health Science, De Montfort University, The Gateway, Leicester LE1 9BH, UK
5 Department of Civil Engineering, College of Engineering, King Khalid University, Abha 61413, Saudi Arabia; abalhaj@kku.edu.sa
6 Department of Geology, Faculty of Science, Kordofan University, Elobied 51111, Sudan
7 Department of Chemical Engineering, Karary University, Khartoum 14411, Sudan; babikerka@hotmail.com
* Correspondence: ismathassanali@gmail.com

Citation: Elkhaleefa, A.; Ali, I.H.; Brima, E.I.; Shigidi, I.; Elhag, A..B.; Karama, B. Evaluation of the Adsorption Efficiency on the Removal of Lead(II) Ions from Aqueous Solutions Using *Azadirachta indica* Leaves as an Adsorbent. *Processes* **2021**, *9*, 559. https://doi.org/10.3390/pr9030559

Academic Editors: Monika Wawrzkiewicz and Anna Wołowicz

Received: 2 February 2021
Accepted: 24 February 2021
Published: 23 March 2021

Publisher's Note: MDPI stays neutral with regard to jurisdictional claims in published maps and institutional affiliations.

Abstract: The efficiency of *Azadirachta indica* (neem leaves) on the removal of Pb(II) ions by adsorption from aqueous solution was investigated in this study. The efficiency of these leaves (without chemical or thermal treatment) for the adsorption of Pb(II) ions has not previously been reported. Batch experiments were performed to study the effect of the particle size, pH, adsorbent dose, contact time, initial Pb(II) ion concentration, and temperature. The maximum removal of 93.5% was achieved from an original Pb(II) ion solution concentration of 50 mg/L after 40 min, at pH 7, with 0.60 g of an adsorbent dose. The maximum adsorption capacity recorded was 39.7 mg/g. The adsorption process was also studied by examining Langmuir, Freundlich, Temkin isotherm, and Dubinin–Radushkevich (D-R) isotherm models. The results revealed that the adsorption system follows the pseudo-second-order model and fitted the Freundlich model. Several thermodynamic factors, namely, the standard free energy (ΔG°), enthalpy (ΔH°), and entropy (ΔS°) changes, were also calculated. The results demonstrated that the adsorption is a spontaneous, physical, and exothermic process. The surface area, pore size, and volume of adsorbent particles were measured and presented using a surface area analyzer (BET); the morphology was scanned and presented with the scanning electron microscope technique (SEM); and the functional groups were investigated using μ-FTIR.

Keywords: biosorption; isotherms; lead (II); *Azadirachta indica* leaves

1. Introduction

Industrial activities have significantly contributed to environmental pollution. Toxic metals are one of the major pollutants causing serious damage to the environment, in addition to being a documented source of many diseases [1]. The heavy metal pollutants in wastewater are highly toxic, hazardous to plants and animals, and result in a shortage of clean water [2,3]. Lead is among these toxic elements, and has acute and chronic health effects on humans [4]. Lead pollution affected nearly 18–22 million people in 2010, motivating an investigation of water purification processes [5].

With respect to metal removal, wastewater purification includes various methods, such as biosorption, adsorption, electrochemical treatments, and oxidation [6,7]. Heavy metal biosorption is one of the favorable fields of remedy due to the low cost and easy accessibility of raw materials, as well as their eco-friendly nature [8]. Other benefits of biosorption are the selectivity for particular metals; low operational time; and most importantly, the absence of chemical waste generation [9]. Different types of cheap plant materials have been investigated as prospective biosorbents for heavy metals [10].

Biosorption is known to be more effective in removing lead metal contaminants from aqueous solutions, where the efficiency rate in batch adsorption tests is higher than 90% [11]. However, the success of biosorption in the removal of lead metal from wastewater depends on several experimental parameters, such as the temperature, pH, and original solution metal concentration [12,13]. The biosorption removal of Pb(II) ions using natural biosorbent from cactus cladodes revealed that the particle size, mass of biosorbent, and contact time are additional factors influencing the operation success [14]. The optimum biosorption of Pb(II) ions increases with an increase in the biosorbent particle size and dosage [14,15]. Various natural materials have been used for lead removal. These include pumpkin seeds, bamboo, cocoa shells, coconut shell, calotropis roots, peanut shells, Moringa bark, orange bark, cashew nut shells, and activated carbon (maize tassel, periwinkle shells, oil palm, and rice husk) [16,17]. Other natural materials have been reported to be efficient sorbents for various pollutants. *Platanus orientalis* leaf powder was used as an efficient sorbent for Cu(II) ion removal from aqueous solutions. The maximum adsorption capacity for this material was determined to be 49.94 mg/g [18]. *Urtica* has been reported as an effective sorbent for the removal of Methylene Blue dye from aqueous solutions, with a maximum adsorption capacity of 101.01 mg/g [19].

The chemical screening of neem leaves showed positive results for steroids, saponins, flavonoids, alkaloids, tannins, and amino acid [20]. Most of these constituents possess electron withdrawing groups which may enhance the adsorption capacity of the neem leaves.

This study aims to evaluate *Azadirachta indica* leaves' powder (AILP) as an eco-friendly, efficient, and cheap biosorbent for Pb(II) ion removal.

2. Materials and Methods

2.1. Preparation and Adsorption Processes

Azadirachta indica (neem leaves) samples were collected from Jizan city, south of Saudi Arabia, and washed with deionized and distilled water. Then samples were dried at room temperature. The leaves were ground by a grinder into powder form, before being sieved. A stock solution of 50 mg/L was prepared using $Pb(NO_3)_2$ (Sigma-Aldrich, Saint Louis, MissourI, USA, ACS reagent $\geq$ 99%.)

2.2. Batch Adsorption Experiment

Batch experiments were performed using 250 mL stoppered bottles containing the desired mass at a pH of 5.5 with 50 mL of Pb(II) solution (50 mg/L). The mixture was shaken at 298 K (except when examining the temperature effect) by a temperature-controlled water bath shaker. At the end of the shaking time, the solid adsorbent was isolated from the solution by filtration, and the remaining concentration of Pb(II) ions was then determined using an atomic absorption spectrometer (AAS) (SpectrAA 220, Varian, UK). Parameters influencing the sorption efficiency were investigated by changing one factor and keeping the others constant. The adsorption extent was determined using Equation (1):

$$q_e = \frac{C_o - C_e}{M} \times V, \tag{1}$$

where V is the volume of the Pb(II) solution in the bottle, q_e is the metal uptake capacity (mg/g), M is the mass of AILP (g), C_o is the initial Pb(II) concentration, and C_e is the Pb(II) concentration at equilibrium. The total removal efficiency (R%) was calculated using Equation (2).

$$R\% = \frac{C_o - C_e}{C_o} \times 100 \tag{2}$$

The investigated parameters, including the particle size, shaking time, mass of adsorbent, temperature, and solution concentration, were investigated.

2.3. *Characterization of AILP*

The surface area, pore-volume, and pore size were investigated using a Quanta Chrome NOVA 4200E Surface Area Analyzer. The morphology of the powder was examined using a Jeol 6360 (Japan) scanning electron microscope at an accelerating voltage of 20 kV. The functional groups of the powder before and after adsorption were investigated by μ-FT-IR (Cary 630 FTIR from Agilent, Santa Clara, USA) in the range of 400–4000 cm^{-1} at a spectral resolution of 8 cm^{-1}.

2.4. *Reusability of AILP*

AILP was reactivated by vigorous shaking in an excess quantity of de-ionized water for 12 hrs. AILP was separated by filtration and then dried by heating it overnight at 105 °C. The optimal conditions were maintained throughout all reusability experiments.

3. Results

3.1. *Characterization of AILP*

3.1.1. Surface Area of AILP

The AILP sample exhibited a high surface area of 183 m^2/g, enhancing its adsorption capabilities. The results revealed that the average pore diameter is 10.3 nm and the total pore volume is 0.355 cm^3/g. Comparative data are given in Table 1. These data demonstrate that the AILP surface area is larger than that of the reported sorbents, indicting a high adsorption capacity.

Table 1. The surface area, total pore volume, and average pore diameter for various adsorbents.

Adsorbent	Surface Area (m^2/g)	Total Pore Volume (cm^3/g)	Average Pore Diameter (nm)	Reference
AILP	183	0.335	10.3	This study
Date seeds	124	0.469	9.8	[21]
P(AN-AA)/AMP composite	39	0.35	34.4	[22]
Red clay	35	0.57	6.5	[23]
Modified sugarcane	7	0.15	Not reported	[24]

3.1.2. IR Spectrum

The μ-FT-IR spectrum is used to estimate the contribution of functional groups on the adsorbent surface in the adsorption process. The FT-IR analysis was performed before and after the adsorption process (Figure 1). The results revealed distinguished bands of Pb-O stretching vibrations that appeared at 462 cm^{-1} and OH stretching of the hydroxide groups. These bands may confirm the adsorption of Pb(II) ions on the adsorbent surface. Other bands appeared at 1298, 1350, and 1723 cm^{-1}, and these bands are probably related to the stretching of -C-N-, -C=N-, and -C=C-, respectively, in polyheterocycles. However, the FT-IR spectra of the adsorbent before and after adsorption were very similar due to the physisorption process.

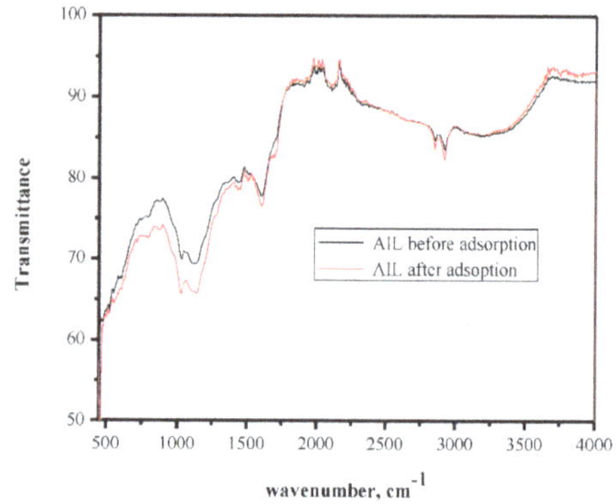

Figure 1. FT-IR spectrum of *Azadirachta indica* leaves' powder (AILP) before and after adsorption.

3.1.3. Scanning Electron Microscope (SEM)

The SEM images of AILP before and after adsorption are presented in Figure 2. The SEM images illustrate differences in the surface morphologies of the AILP surfaces before and after adsorption. The surface of the AILP before adsorption showed micro flakes with an approximate size of 0.5 μm (Figure 2a). Additionally, the analysis of the AILP surface after adsorption revealed differences in the surface structure and the adsorption process led to the formation of many floccules and spots due to the adsorption of Pb(II) ions on the AILP surface, leading to a rough appearance.

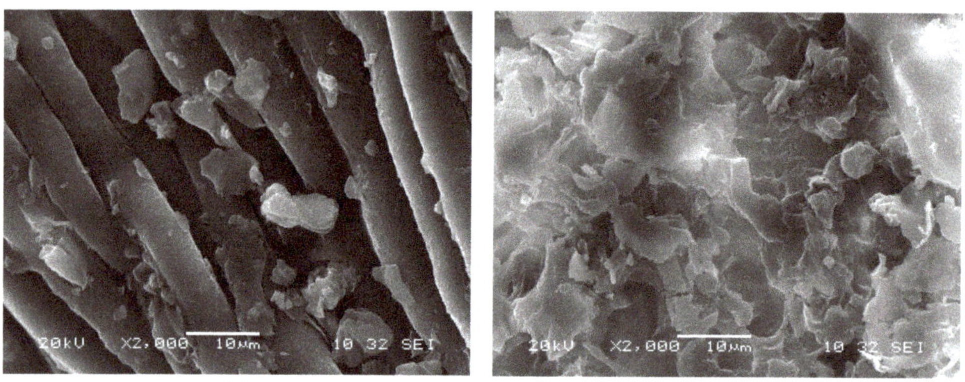

(a) (b)

Figure 2. Scanning Electron Microscope (SEM) images of the AILP surface (**a**) before and (**b**) after adsorption.

3.2. *Effect of the Adsorbent Particle Size*

The removal efficiency of Pb(II) ions from aqueous solutions by the AILP increased from 82.9% to 99.1% as the AILP size decreased from 800 to 100 μm, at 298 K, for a 50 mg/L initial Pb(II) ions concentration (Figure 3). With a decrease in AILP particle sizes, the surface area of the AILP increased, so the number of active sites accessible on the AILP surface had a better exposure to the adsorbate. Similar findings have also been reported by other researchers [12].

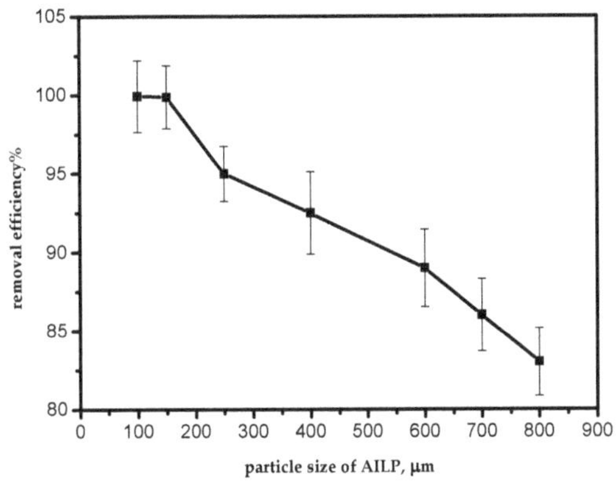

Figure 3. Effect of the adsorbent particle size.

3.3. *Effect of the pH*

The maximum adsorption was achieved at pH 5.0 (Figure 4). It is presidentially noted that with a pH higher than 5.0, the adsorption activity decreased sharply. However, at pH values lower than 5.0, the adsorption process could be hindered by the existence of a high concentration of positive hydrogen ions, and the adsorption sites became positively charged, exerting a repelling effect on the Pb(II) cations. With an increase in the pH, the negative charge density on AILP increased as a result of deprotonation of the metal-binding sites, hence increasing the metal uptake; such findings are in agreement with previous studies. It is well-known that lead ions start to precipitate above pH 5.0–6.0 in the form of $Pb(OH)_2$, suppressing the adsorption of lead ions [1].

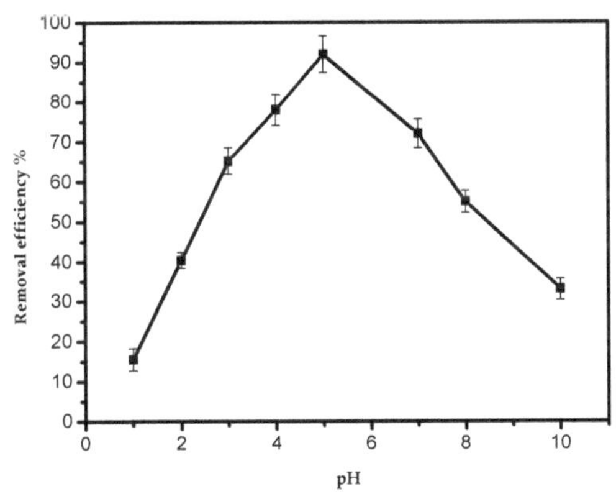

Figure 4. Effect of the pH on the removal efficiency of Pb(II) ions by AILP.

3.4. *Effect of the Adsorbent Dosage*

To investigate the effect of the sorbent mass on the efficiency of the adsorption process, a series of batch experiments were conducted with various sorbent masses of 0.05, 0.10, 0.20, 0.30, 0.40, 0.60, 0.80, and 1.0 g per 50 mL of aqueous solution of Pb(II) ions. The effect is presented in Figure 5. As the mass of the sorbent increased, the adsorption efficiency also increased. This can be attributed to an increase in the AILP surface area with its mass. The maximum adsorption efficiency for Pb(II) ions was achieved at 0.60 g/50 mL. From the results, it is clear that an increase in the adsorbent mass above 0.6 g had no effect on the adsorption efficiency of Pb(II) ions from the aqueous solution.

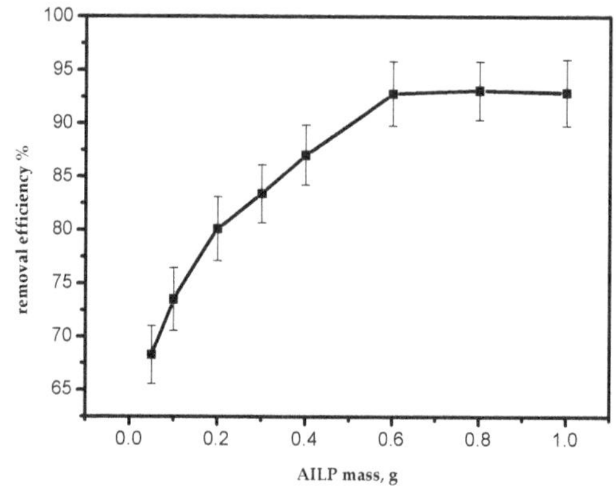

Figure 5. Effect of the adsorbent mass on the removal efficiency of Pb(II) ions by AILP.

3.5. *Effect of the Agitation Time*

The effect of the agitation time on the adsorption of Pb(II) ions using the AILP is presented in Figure 6. The results showed that the time of 40 min was sufficient for achieving the optimum adsorption, without any influence from increasing the contact time. As demonstrated in Figure 6, the adsorption process could be divided into three phases. The first phase was significantly fast, representing about 88% removal within 30 min. The second phase showed slower adsorption. The first stage was caused by the large surface area of the AILP. With the progressive occupation of these sites, the process became slower in the second stage due to less adsorption sites being available for Pb(II) ions. The third phase was the desorption of Pb(II) ions from the AILP surface. Some of the adsorbed Pb(II) ions might have leached from the AILP into the solution and this is probably a result of the long agitation time.

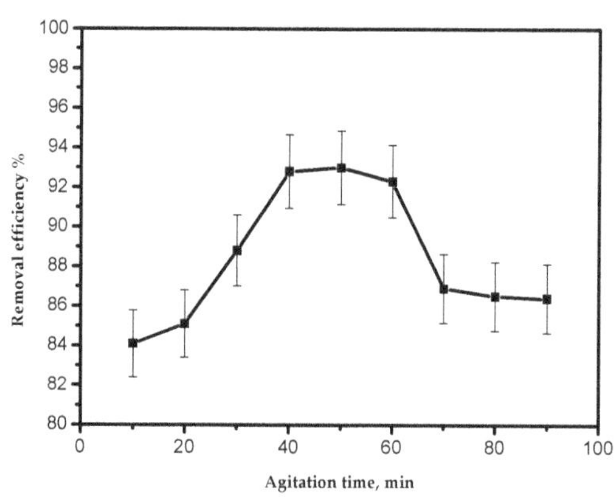

Figure 6. Effect of the agitation time on the removal efficiency of Pb(II) ions by AILP.

3.6. *Adsorption Kinetics*

The aim of investigating the adsorption kinetics is to explore the mechanism of adsorption and evaluate the efficiency of the adsorbent. The pseudo-first-order, pseudo-second-order, and intra-particle diffusion kinetic models were used to investigate the obtained data on the adsorption of Pb(II) ions on the AILP.

3.6.1. Pseudo-First-Order Kinetic Model

The equation of this model is given by Equation (3):

$$\ln(q_e - q_t) = -kt + \ln q_e, \quad (3)$$

where q_e is the amount of solute (mg/g) removed at equilibrium, q_t is the quantity of solute adsorbed at any time (t), and k is the rate constant. It was found that the data do not fit this model (R^2 = 0.0006).

3.6.2. Pseudo-Second-Order Kinetic Model

The equation of this model is

$$\frac{t}{q_t} = \frac{1}{k_2 q_e^2} + \frac{t}{q_e}. \quad (4)$$

A typical plot of the pseudo-second-order equation is shown in Figure 7. The straight line in the plot proves that the experimental data appropriately obey the pseudo-second-order model. Values of q_e and k_2 were determined from the intercept and slope of Figure 8. These parameters and the correlation coefficient are presented in Table 2. This agrees with previous results and explanations [1,25]. It can be concluded from previous studies that the adsorption kinetics model was governed by the adsorbent nature.

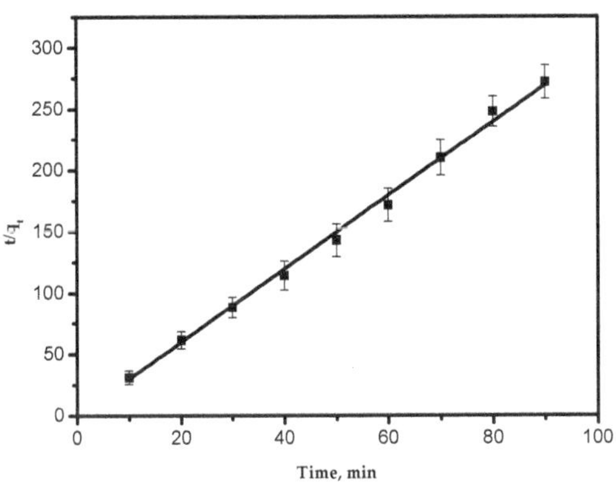

Figure 7. Pseudo-second-order adsorption kinetics.

3.6.3. Intra-Particle Diffusion Kinetic Model

This model is shown by Equation (5):

$$q_t = k_{id}\, t^{1/2} + I, \quad (5)$$

where k_{id} is the intra-particle diffusion rate constant, mg/g/min$^{1/2}$. The value of (k_{id}) was calculated from Equation (5) and is presented in Table 2. The relationship between qt and $t^{1/2}$ was not linear over the entire time range. The first curved part of the plot can be attributed to the boundary layer diffusion effects and it represents intra-particle diffusion controlled by the rate constant k_{id} (Figure 8).

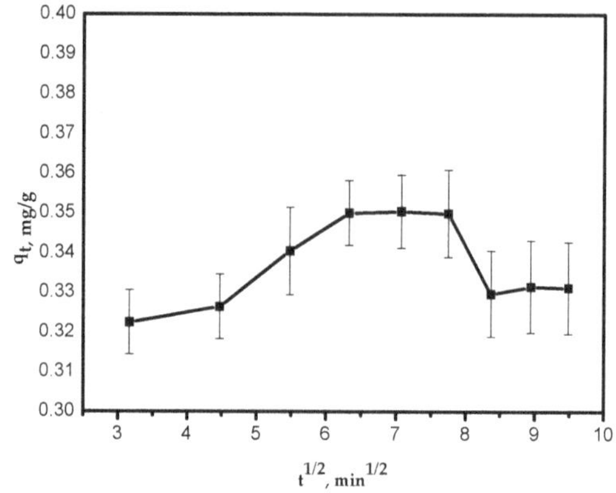

Figure 8. Intra-particle diffusion kinetic model.

Table 2. Kinetic parameters of the removal of Pb(II) ions from aqueous solutions by AILP.

Kinetic Model	Parameters	
Pseudo-second-order model	q_e (mg/g)	3.00
	k_2 (g/mg min$^{(1/2)}$)	0.21
	R^2	0.9976
Intraparticle diffusion model	k_{id} mg/g min$^{(1/2)}$	0.0014
	I	0.328
	R^2	0.07

3.7. *Effect of the Initial Pb(II) Concentration*

The effect of the initial concentration of Pb(II) ions on the adsorption efficiency is presented in Figure 9, where it is evident that the adsorption efficiency of Pb(II) ions decreased as the initial Pb(II) ion concentration increased. The decrease in adsorption efficiency can be attributed to the deficiency of sufficient adsorption sites for accumulating the large increase of metal ions available in the solution [21].

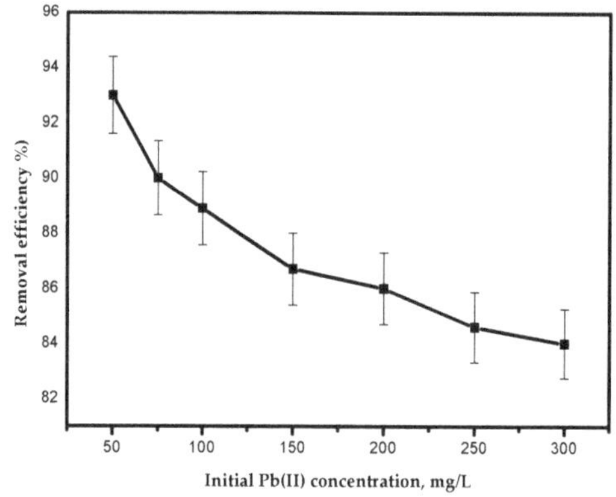

Figure 9. Effect of the initial concentration of Pb(II) ions on the removal efficiency.

3.8. *Adsorption Isotherm*

The adsorption isotherm models were investigated to explore the interaction mechanism of Pb(II) ions on the adsorbent sites [1,21]. The adsorption isotherm models were analyzed by specific parameters, whose values express the surface characteristics and affinity of the adsorbent towards heavy metal ion adsorption [21]. Out of numerous isotherm models, four were examined for this study, namely, Freundlich, Langmuir, Temkin, and Dubinin–Radushkevich (D-R) models.

3.8.1. Langmuir Isotherm

Equation (6) presents the Langmuir linear equation:

$$\frac{C_e}{q_e} = \frac{C_e}{q_m} + \frac{1}{bq_m}. \quad (6)$$

The results obtained in this study were fitted to the Langmuir isotherm model to determine the maximum adsorption capacity (q_{max}) of AILP on Pb(II) ions. The linear plot of C_e/q_e vs. C_e (Figure 10a) obtained indicates the applicability of the Langmuir model. Values of b and q_m were calculated graphically from the intercept and slope, respectively, and are presented in Table 3. From the Langmuir isotherm, the adsorption process can be classified as follows: when $R_L = 0$, a favorable process occurs when $0 < R_L < 1$, linear process occurs when $R_L = 1$, and unfavorable process occurs when $R_L > 1$. The R_L values for the initial AILP concentrations were between 0.13 and 0.49, which indicates that the adsorption of Pb(II) ions on AILP is a favorable process.

3.8.2. Freundlich Isotherm

This is an experimental expression used for adsorption on heterogeneous surfaces with the interaction between adsorbed molecules. This model suggests an exponential drop in adsorption energy upon completion of the adsorption centers of the adsorbent. The linear equation of the Freundlich model is shown by Equation (7) [21]:

$$\ln q_e = \ln K_f + \frac{1}{n} \ln C_e, \quad (7)$$

where K_f is the relative adsorption capacity of the adsorbent that is related to the bonding energy and n is the heterogeneity element stating the deviancy from the linearity of adsorption. A plot of ln q_e versus ln C_e was used to confirm the Freundlich adsorption isotherm. Values of K_f and n were calculated from the intercept and slope (Figure 10b) and are tabulated in Table 3. An explanation of n, which controls the relationship between the solution concentration and adsorption, is as follows: If $n < 1$, adsorption is a physical process; if $n = 1$, adsorption is linear; and if $n > 1$, adsorption is a chemical process. The obtained n value was found to be 0.60, indicating physical processes.

3.8.3. Temkin Isotherm

In this model, the heat of adsorption AILP molecules in the layer is supposed to decrease linearly with the coverage of the adsorbent surface because of a reduction in the adsorbate–adsorbent interactions. The model is defined by Equation (8):

$$q_e = B \ln A + B \ln C_e, \quad (8)$$

where $B = (RT)/b_t$, R is the universal gas constant, and T is the temperature in Kelvin. The constant b_t is associated with the heat of adsorption (J/mol). A is the equilibrium binding constant corresponding to the maximum binding energy.

Values of A and B were determined from the slope and intercept of Equation (6) and Figure 10c and are shown in Table 3. The b_t value was 318 J/mol, indicating physical adsorption processes [21].

3.8.4. Dubinin–Radushkevich (D-R) Isotherm

The D-R Isotherm model is commonly used to conclude the mean free energy of the adsorption process [1]. The formula of this model is described by Equation (9):

$$\ln q_e = \ln q_{max} - \beta \varepsilon^2, \quad (9)$$

where ε is defined as

$$\varepsilon = RT \ln\left(\frac{1}{1 + C_e}\right). \quad (10)$$

T denotes the temperature (K), R is the universal gas constant, and q_{max} and β are D-R isotherm constants. β and q_{max} were obtained from the slope and intercept of Equation (9) and Figure 10d and the values are presented in Table 3.

The mean free energy of adsorption process, E_f, is defined as the free energy change when one mole of ions is moved to the solid surface from infinity in solution and was determined from Equation (11).

$$E_f = \frac{1}{\sqrt{2\beta}} \quad (11)$$

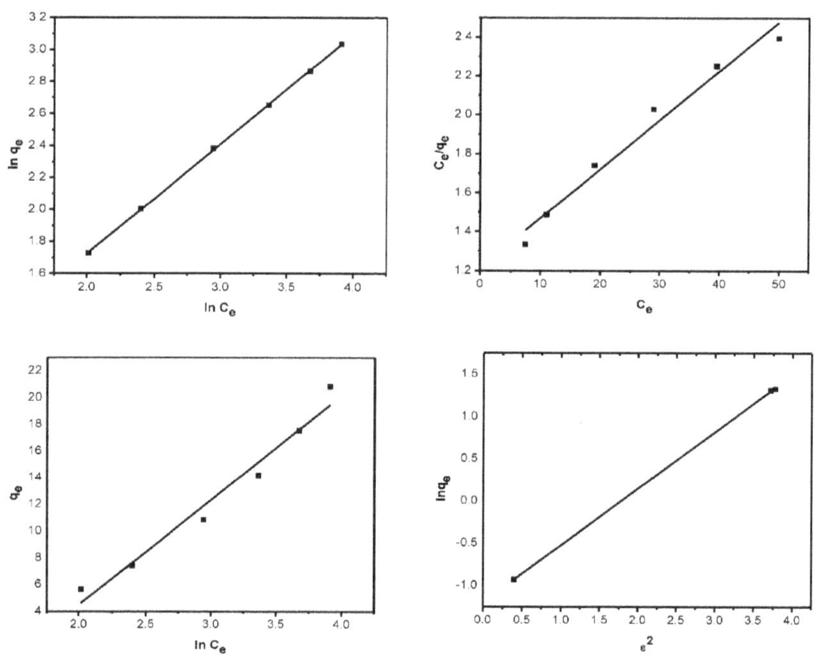

Figure 10. Adsorption isotherm models: (**a**) Langmuir; (**b**) Freundlich; (**c**) Temkin; and (**d**) D-R.

It is well-established that the E_f value is crucial for predicting the adsorption type. For $8 < E_f < 16$ kJ/mol, adsorption is considered a chemisorption process, and, when $E_f < 8$ kJ/mol, adsorption is classified as a physical process. In this study, the lower value of E_f (2.2 KJ/mol) indicates that the adsorption process was of a physical type.

Table 3. Adsorption isotherm constants obtained from the four adsorption isotherm models.

Adsorption Model	Isotherm Constants	Value
Langmuir	q_{max} (mg/g) K_L (L/g) R^2	39.7 0.021 0.9756
Freundlich	n K_f (mg/g)/(mg/L) R^2	0.6 1.0 0.9997
Temkin	A (L/g) B R^2	1.3 7.8 0.9703
D-R	B q_{max} (mg/g) R^2	1×10^{-7} 4.05 0.9997

3.9. *Effect of the Temperature*

The temperature effect on the adsorption of Pb(II) ions on AILP was examined at a range from 298 to 318 K and is presented in Figure 11. The results revealed that the adsorption efficiency decreased slightly from 95.9% to 93.4% as the temperature increased from 298 to 318 K. This can be attributed to the possible damage of adsorption sites at elevated temperatures.

3.10. *Thermodynamic Parameters*

Enthalpy ΔH°, free energy ΔG°, and entropy ΔS° changes were determined using Equations (12)–(14):

$$\Delta G^o = RT \ln K_D, \tag{12}$$

$$\Delta G^o = \Delta H^o - T\Delta S^o, \tag{13}$$

$$\ln K_D = \frac{-\Delta H^o}{RT} + \frac{\Delta S^o}{R}, \tag{14}$$

where K_D is the distribution coefficient and equivalent to the preliminary Pb(II) concentration (C_o) divided by the final Pb(II) concentration (C_e).

Figure 12 shows a linear plot of ln K_D versus 1/T. The thermodynamic parameters were determined and are presented in Table 4. The negative value of ΔH° indicates the exothermic nature of the adsorption process. The positive value of entropy (ΔS°) proposes an increased randomness at the solid–solution interface. Negative values of Gibbs free energy (ΔG°) prove that the adsorption process is spontaneous, which becomes more positive with an increase in temperature. Therefore, this indicates that less adsorption occurs at elevated temperatures. This is in good agreement with the results obtained from Figure 10c.

Table 4. Thermodynamic parameters of the adsorption of Pb(II) ions on AILP.

T, K	K_D	ΔG, kJ/mol	ΔS, J/mol K	ΔH, kJ/mol
298	24.70	−8.0		
308	19.99	−7.6	40	−18.6
318	15.40	−7.0		

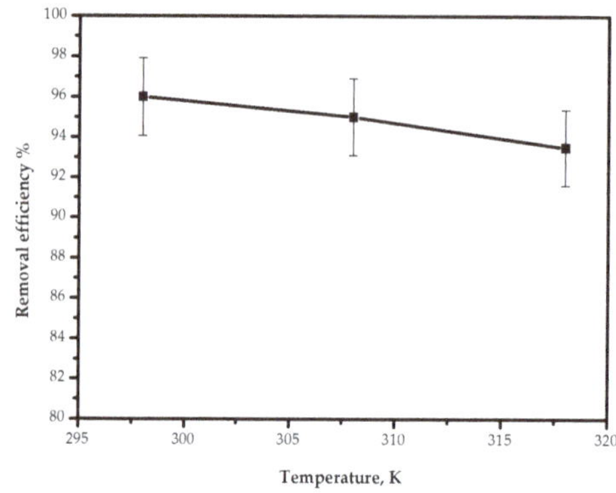

Figure 11. Effect of the temperature on the removal of Pb(II) ions by AILP.

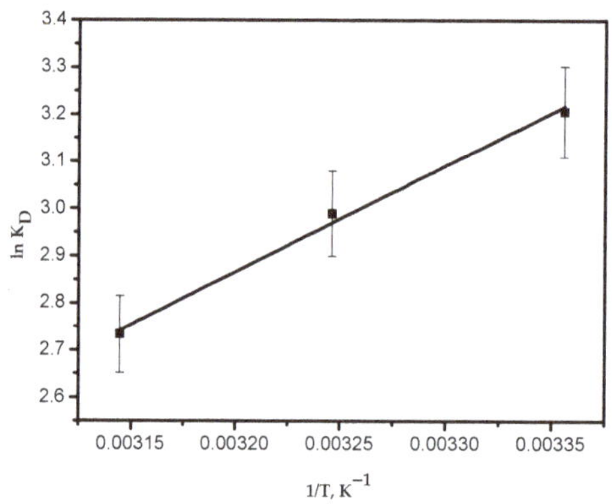

Figure 12. Plot of ln K_D versus 1/T.

3.11. Reusability of AILP

The results of reusing the AILP are presented in Figure 13. It appeared that the adsorption effectiveness of the AILP decreased gradually with the reusability. The results confirm that the AILP can be efficiently recycled 2–3 times when removing Pb(II) ions from their aqueous solutions. The adsorption efficiency of the AILP decreased to 68.5% when it was reused for a fifth time.

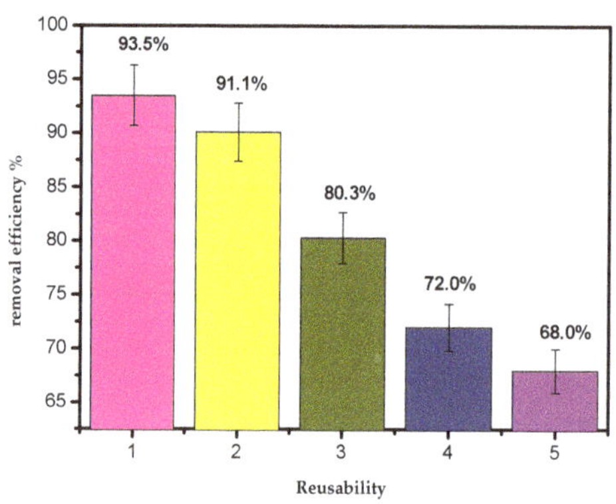

Figure 13. The reusability of AILP.

3.12. Possible Adsorption Mechanism

The adsorption of Pb(II) ions on the AILP surface possibly jointly occurs by means of chemical interactions and physisorption. The ATR-FT-IR analysis proves that a Pb-O bond is formed after the adsorption process; on the other hand, values of ΔG° and b_t suggest a physisorption process.

3.13. Comparison of AILP with Other Sorbents

An evaluation of q_{max} of the AILP according to the reported q_{max} values for other biosorbents is shown in Table 5. Variances in q_{max} can be attributed to the specific nature and properties of each adsorbent, surface area, and construction of the adsorbent. Making this comparison with other adsorbents confirms that AILP is a good and promising adsorbent.

Table 5. Comparison of AILP and other biosorbents.

Adsorbent	q_{max}	Kinetics	Isotherm	References
AILP	39.7	2nd order	Frendluich	This study
Juniperus procera	30.3	2nd order	Langmuir	[1]
Coconut shell	22.6	1st order	Langmuir	[26]
Pumpkin seed shell	14.3	2nd order	Langmuir	[27]
Cashew nut shell	17.8	2nd order	Frendluich	[28]
Polypyrrole-based AC	50.0	2nd order	Frendluich	[29]
Polygonum orientale Linn	98.4	2nd order	Langmuir	[30]

4. Conclusions

In this study, *Azadirachta indica* leaves' powder (AILP), which is a natural adsorbent material, eco-friendly, and inexpensive, was investigated. AILP was used for the removal of Pb(II) ions from aqueous solution. The removal efficiency reached above 95%. The powder of the leaves spontaneously retained considerable amounts of Pb(II) ions. The maximum adsorption capacity for neem leaves was determined as being 39.7 mg/g. Experimental investigations defined the optimum conditions for the competent removal of 50 mg/L of Pb(II) ion to be 0.60 g, 40 min of contact time, and a pH of 7. These are significant characteristics for this tested adsorbent, making it an excellent choice for cleaning polluted water due to it (1) being a natural material and (2) cost-effective, as it requires no chemical treatment; (3) its availability, with this material being abundantly available in the region; and (4) its effectiveness, with it being capable of removing Pb(II) ions with a high efficiency. These positive characteristics would make this adsorbent a suitable candidate for removing other poisonous metal ions.

Author Contributions: Conceptualization A.E. and I.H.A.; methodology, A.E.; validation, E.I.B.; formal analysis, E.I.B.; investigation A.E.; resources, A.E.; writing—original draft preparation, A.E. and I.H.A.; writing—review and editing, E.I.B. and I.S.; project administration, I.H.A.; funding acquisition, A.B.E.; supervision, B.K. All authors have read and agreed to the published version of the manuscript.

Funding: The authors extend their appreciation to the Deanship of Scientific Research at King Khalid University for funding this work through a research groups program under grant number R.G.P.1/139/40.

Institutional Review Board Statement: Not applicable.

Informed Consent Statement: Not applicable.

Data Availability Statement: Data is contained within the article.

Conflicts of Interest: The authors declare no conflict of interest.

References

1. Ali, I.H.; Al Mesfer, M.K.; Khan, M.I.; Mohd, M.; Alghamdi, M.M. Exploring Adsorption Process of Lead (II) and Chromium (VI) Ions from Aqueous Solutions on Acid Activated Carbon Prepared from *Juniperus procera* Leaves. *Processes* **2019**, *7*, 217. [CrossRef]
2. Fu, F.; Wang, Q. Removal of heavy metal ions from wastewaters: A review. *J. Environ. Manag.* **2011**, *92*, 407–418. [CrossRef]
3. Morosanu, I.; Teodosiu, C.; Paduraru, C.; Ibanescu, D.; Tofan, L. Biosorption of lead ions from aqueous effluents by rapeseed biomass. *New Biotechnol.* **2017**, *39*, 110–124. [CrossRef]
4. Gupta, V.K.; Rastogi, A. Biosorption of lead from aqueous solutions by green algae Spirogyra species: Kinetics and equilibrium studies. *J. Hazard. Mater.* **2008**, *152*, 407–414. [CrossRef]
5. Amer, M.W.; Ahmad, R.A.; Awwad, A.M. Biosorption of Cu(II), Ni(II), Zn(II) and Pb(II) ions from aqueoussolution by *Sophora japonica* pods powder. *Int. J. Ind. Chem.* **2015**, *6*, 67–75. [CrossRef]
6. Das, D.; Chakraborty, S.; Bhattacharjee, C.; Chowdhury, R. Biosorption of lead ions (Pb^{2+}) from simulated wastewater using residual biomass of microalgae. *Desalination Water Treat.* **2016**, *57*, 4576–4586. [CrossRef]
7. Papirio, S.; Frunzo, L.; Mattei, M.R.; Ferraro, A.; Race, M.; D'Acunto, B.; Esposito, G. Heavy metal removal from wastewaters by biosorption: Mechanisms and modeling. In *Sustainable Heavy Metal Remediation*, 1st ed.; Rene, E.R., Sahinkaya, E., Lewis, A., Lens, P.N.L., Eds.; Springer: Berlin/Heidelberg, Germany, 2017; Volume 8, pp. 25–63.
8. Xuan, Z. Study on the equilibrium, kinetics and isotherm of biosorption of lead ions onto pretreated chemically modified orange peel. *Biochem. Eng. J.* **2006**, *31*, 160–164. [CrossRef]
9. Mungasavalli, P.D.; Viraraghavan, T.; Jin, Y. Biosorption of chromium from aqueous solutions by pr treated Aspergillus niger: Batch and column studies. *Colloids Surf. A Physicochem. Eng. Asp.* **2007**, *301*, 214–223. [CrossRef]
10. Ucun, H.; Kemal, B.Y.; Kaya, Y.; Cakici, A.; Algur, O.F. Biosorption of lead (II) from aqueous solution by cone biomass of *Pinus sylvestris*. *Desalination* **2003**, *154*, 233–238. [CrossRef]
11. Singanan, M. Removal of lead (II) and cadmium (II) ions from wastewater using activated biocarbon. *Sci. Asia* **2011**, *37*, 115–119. [CrossRef]
12. Sulaymon, A.H.; Mohammed, A.A.; Al-Musawi, T.J. Competitive biosorption of lead, cadmium, copper, and arsenic ions using algae. *Environ. Sci. Pollut. Res.* **2013**, *20*, 3011–3023. [CrossRef] [PubMed]
13. Verma, A.; Kumar, S.; Kumar, S. Biosorption of lead ions from the aqueous solution by *Sargassum filipendula*: Equilibrium and kinetic studies. *J. Environ. Chem. Eng.* **2016**, *4*, 4587–4599. [CrossRef]
14. Barka, N.; Abdennouri, M.; El Makhfouk, M.; Qourzal, S. Biosorption characteristics of cadmium and lead onto eco-friendly dried cactus (Opuntia ficus indica) cladodes. *J. Environ. Chem. Eng.* **2013**, *1*, 144–149. [CrossRef]
15. Iram, S.; Shabbir, R.; Zafar, H.; Javaid, M. Biosorption and bioaccumulation of copper and lead by heavy metal-resistant fungal isolates. *Arab. J. Sci. Eng.* **2015**, *40*, 1867–1873. [CrossRef]
16. Okoye, A.; Ejikeme, P.; Onukwuli, O. Lead removal from wastewater using fluted pumpkin seed shell activated carbon: Adsorption modeling and kinetics. *Int. J. Environ. Sci. Technol.* **2010**, *7*, 793–800. [CrossRef]
17. Senthil, K.P. Adsorption of lead (II) ions from simulated wastewater using natural waste: A kinetic, thermodynamic and equilibrium study. *Environ. Prog. Sustain. Energy* **2014**, *33*, 55–64. [CrossRef]
18. Abadian, S.; Rahbar-Kelishami, A.; Norouzbeigi, R.; Peydayesh, M. Cu(II) adsorption onto Platanus orientalis leaf powder: Kinetic, isotherm, and thermodynamic studies. *Res. Chem. Intermed.* **2015**, *41*, 7669–7681. [CrossRef]
19. Peydayesh, M.; Isanejad, M.; Mohammadi, T.; Jafari, S.M.R.S. Assessment of Urtica as a low-cost adsorbent for methylene blue removal: Kinetic, equilibrium, and thermodynamic studies. *Chem. Pap.* **2015**, *69*, 930–937. [CrossRef]
20. Al-Hashemi, Z.S.S.; Hossain, M.A. Biological activities of different neem leaf crude extracts used locally in Ayurvedic medicine. *Pac. Sci. Rev. A Nat. Sci. Eng.* **2016**, *18*, 128–131. [CrossRef]
21. Elkhaleefa, A.; Ali, I.H.; Brima, E.I.; Elhag, A.B.; Karama, B. Efficient Removal of Ni(II) from Aqueous Solution by Date Seeds Powder Biosorbent: Adsorption Kinetics, Isotherm and Thermodynamics. *Processes* **2020**, *8*, 1001. [CrossRef]
22. El-Zahhar, A.A.; Idris, A.M. Mercury(II) decontamination using a newly synthesized poly(acrylonitrile-acrylic acid)/ammonium molybdophosphate composite exchanger. *Toxin Rev.* **2020**, *39*, 1–13. [CrossRef]
23. Khan, M.I.; Al Mesfer, M.K.; Danish, M.; Ali, I.H.; Shoukry, H.; Patel, R.; Gardy, J.; Rehan, M. Potential of Saudi natural clay as an effective adsorbent in heavy metals removal from wastewater. *Desalination Water Treat.* **2019**, *15*, 140–151. [CrossRef]
24. Eleryan, A.; El Nemr, A.; Abubakr, M.; Idris, A.M.; Alghamdi, M.M.; El-Zahhar, A.A.; Tarek, O.; Said, T.O.; Sahlabji, T. Feasible and eco-friendly removal of hexavalent chromium toxicant from aqueous solutions using chemically modified sugarcane bagasse cellulose. *Toxin Rev.* **2020**, *39*, 1–12. [CrossRef]
25. Kavitha, G.; Sridevi, V.; Venkateswarlu, P.; Babu, N.C. Biosorption of chromium from aqueous solution by *Gracilaria corticata* (Red Algae) and its statistical analysis using response surface methodology. *Open Access Libr. J.* **2016**, *3*, 1–32. [CrossRef]
26. Taha, M.F.; Kiat, C.F.; Shaharun, M.S.; Ramli, A. Removal of Ni(II), Zn(II) and Pb(II) ions from single metal aqueous solution using activated carbon prepared from rice husk. *Int. J. Environ. Ecol. Eng.* **2011**, *5*, 855–860.
27. Xiong, C.; Yao, C. Synthesis, characterization and application of triethylenetetraamine modified polystyrene resin in removal of mercury, cadmium and lead from aqueous solutions. *Chem. Eng. J.* **2009**, *155*, 844–850. [CrossRef]
28. Xiong, C.; Yao, C. Preparation and application of acrylic acid grafted polytetraflouroethylene fiber as a weak acid cation exchanger for adsorption of Er(III). *J. Hazard. Mater.* **2009**, *170*, 1125–1132. [CrossRef]

29. Abdulaziz Ali Alghamdi, A.A.; Al-Odayni, A.; Saeed, W.S.; Al-Kahtani, A.; Alharthi, F.A.; Aouak, T. Efficient Adsorption of Lead (II) from Aqueous Phase Solutions Using Polypyrrole-Based Activated Carbon. *Materials* **2019**, *12*, 1–16.
30. Wang, L.; Zhang, J.; Zhao, R.; Li, Y.; Li, C.; Zhang, C. Adsorption of Pb (II) on activated carbon prepared from Polygonum orientale Linn.: Kinetics, isotherms, pH, and ionic strength studies. *Bioresour. Technol.* **2010**, *101*, 5808–5814. [CrossRef] [PubMed]

processes

MDPI

Article

Efficient Removal of Ni(II) from Aqueous Solution by Date Seeds Powder Biosorbent: Adsorption Kinetics, Isotherm and Thermodynamics

Abubakr Elkhaleefa [1,2], Ismat H. Ali [3,*], Eid I. Brima [3,4], A. B. Elhag [5,6] and Babiker Karama [7]

1 Department of Chemical Engineering, College of Engineering, King Khalid University, Abha 61413, Saudi Arabia; amelkhalee@kku.edu.sa
2 Department of Chemical Engineering and Chemical Technology, University of Gezira, Wadmedani 21113, Sudan
3 Department of Chemistry, College of Science, King Khalid University, Abha 61413, Saudi Arabia; ebrima@kku.edu.sa
4 School of Allied Health Science, De Montfort University, The Gateway, Leicester LE1 9BH, UK
5 Department of Civil Engineering, College of Engineering, King Khalid University, Abha 61413, Saudi Arabia; abalhaj@kku.edu.sa
6 Department of Geology, Faculty of Science, Kordofan University, Elobied 51111, Sudan
7 Department of Chemical Engineering, Karary University, Khartoum 14411, Sudan; babikerka@hotmail.com
* Correspondence: ihali@kku.edu.sa

Received: 20 July 2020; Accepted: 14 August 2020; Published: 17 August 2020

Abstract: Adsorption investigations in batch approaches were performed to explore the biosorption of Ni(II) ions from aqueous solutions on date seeds powder. The effects of pH, particle size, initial concentration of Ni(II) ions, adsorbent mass, temperature, and contact on the adsorption efficacy were studied. The maximum removal obtained was 90% for an original Ni(II) ion solution concentration of 50 ppm was attained at pH 7 after 30 min and with 0.30 g of an added adsorbent. The four adsorption models, namely Freundlich, Langmuir, Dubinin–Radushkevich (D–R), and Temkin were examined to fit the experimental findings. The adsorption system obeys the Freundlich model. The system was found to follow the pseudo-second order kinetic model. Thermodynamic factors; entropy (ΔS°), enthalpy (ΔH°), and Gibbs free energy (ΔG°) changes were also assessed. Results proved that adsorption of Ni(II) ions is exothermic and spontaneous. Sticking probability value was found to be less than unity, concluding that the process is dominated by physical adsorption.

Keywords: adsorption; date seeds; kinetics; isotherm; thermodynamics

1. Introduction

The presence of heavy metals in water streams is among one of the most dangerous environmental problems arising from the disposal of untreated industrial effluents [1–4]. Many industries comprise of final treatment processes where discharged metal compounds may lead to pollution in the effluent water [2,5,6]. Most of these heavy metals are non-biodegradable or with long biological half-life leading to potential accumulation and human exposure through food or water [1].

Ni(II) ions exist naturally in water as nitrates, sulfides, and oxides. Nickel intake above the permissible limit causes skin dermatitis, fibrosis, vomiting, pulmonary, nausea, and many other diseases [3,6–9]. Most of the methods used for Ni(II) removal from artificial wastewater-like cation exchange and precipitation are costly and produce toxic sludge [2,9]. Recently, some economical, renewable, and effective agricultural and natural materials have been studied as alternative biosorbents [2].

Date (*Phoenix dactylifera*) seeds, which are composed of lignin, hemicellulose, and cellulose are effective biosorbents; thus, eliminates various contaminants from wastewater [4,10]. The efficiency of this inexpensive biosorbent is due to oxygenated functional groups in the lignocellulosic materials such as cellulose and phenolic compounds [10]. The removal of Ni(II) ions onto bentonite/grapheme oxide was previously studied [11] and the results revealed that it follows Langmuir isotherm with high uptake capacity. Mehrmad et al. [12] have reported that the removal of Ni(II) ions by functionalized henna powder depends on the experimental circumstances, mainly the Ni(II) concentration, the biosorbent mass, and the pH of the medium, the Freundlich and Langmuir isotherm models were used to define the process. The kinetic data were fitted with the pseudo-second-order reaction. The adsorption of Ni(II) ions from wastewater by natural clay had previously been investigated [13] and it was reported that the process is controlled by the pH value of the medium; the sorption process was rapid whereas the maximum adsorption capacity had been attained within 120 min and it was found that the system follows a pseudo-second-order reaction.

The adsorbent used in this study was prepared by a direct, facile, and economic technique. The process does not require any chemicals addition nor high-temperature calcination. It is believed that only few adsorbents were tested without any modification in the removal of heavy metals from aqueous solutions while being powerfully recyclable. *Phoenix dactylifera* is one of the most plentiful plants in Saudi Arabia and the region. Using date seed powder as an adsorbent for Ni(II) ions is considered the novelty of this research.

This work aims to assess the potential of date seed powder to act as an inexpensive and environment-friendly material for the removal of Ni(II) from artificial wastewater. This study was designed to assess, compare, and characterize the adsorption of Ni(II) ions by date seeds powder (DSP) without the addition of any chemicals or thermal treatment. DSP was initially characterized using Brunauer–Emmett–Teller (BET) surface area, scanning electron microscope technique (SEM), and attenuated total reflection-Fourier transform infrared spectrometer (ATR-FTIR). The factors that influence the adsorption efficiency such as mass, concentration of Ni(II) ion, pH, temperature, and contact time, were investigated in this work.

2. Materials and Methods

2.1. Collection and Treatment of Adsorbent

Dates seeds were collected locally from Abha city, Saudi Arabia. Seeds were rinsed with tap water and then by deionized water, dried at room temperature and were then grounded to powder size using a ball mill before being sieved. The efficiency of the room-temperature dried DSP was compared with small amount of oven-dried DSP and no differences were noticed on the results. Thus, to minimize the cost, drying of DSP by oven was not used in this study.

2.2. Reagents

Deionized water (>18 Ω/cm, Milli-Q) was used throughout this work for the solutions and DSP preparations. A stock solution of 1000 ppm Ni(II) ions was prepared using $NiNO_3$ (LOBA Chemie, Laboratory Reagents and fine Chemicals, Mumbai, India).

2.3. Batch Adsorption

Adsorption batch experiments were conducted at different operating conditions (pH, time, adsorbent dosage, adsorbent particles size, and temperature) by adding the desired amount of DSP to 50 mL of Ni(II) ion solution under each particular condition. NaOH (0.25 M) and/or HCl (0.25 M) were used to control the pH value. Mechanical thermostated shaker (WSB, Witeg, Belrose, Germany) was used throughout all the experiments. The solutions were filtered and then analyzed using Atomic

Absorption Spectroscopy (AAS) (SpectrAA 220, Varian, Australia) to measure the remaining Ni(II) ions concentrations. The removal efficiencies (R%) were calculated using Equation (1).

$$R\% = \frac{C_o - C_e}{C_o} \times 100 \quad (1)$$

where C_0 and C_e are the initial and equilibrium Ni(II) ion concentrations, respectively.

2.4. Characterization of Biosorbent

The Brunauer–Emmett–Teller (BET) surface area, pore size and pore volume after and before the adsorption process were investigated using Quanta Chrome NOVA 4200E Surface Area Analyzer. The morphology of the DSP was investigated using a scanning electron microscope technique (SEM) JEOL 6360 (Japan). Accelerating voltage of 20 kV was used. The functional groups of DSP before and after the adsorption process were investigated by ATR-FTIR (Cary 630 FTIR from Agilent) in the range of 4000–400 cm^{-1} at a spectral resolution of 8 cm^{-1}. DSP samples were analyzed without any pretreatment.

3. Results and Discussion

3.1. Characterization of the DSP

Table 1 displays the results of the analysis of DSP by surface area analyzer. Results prove that DSP poses a mesoporous arrangement. Mesopores are detected over the entire sample surface forming a highly uniform and interpenetrating permeable media. Moreover, results also confirm that DSP has a large surface area compared to some other adsorbents used to adsorb Ni(II) ions [13].

Table 1. Properties of the DSP surface.

Surface Area (m^2/g)	Total Pore Volume (cm^3/g)	Average Pore Diameter (A°)
124	46.9×10^{-2}	98

3.2. ATR—FTIR Spectrum

Investigations of the ATR-FTIR spectra from DSP after and before the adsorption process (Figure 1) proves the presence of functional groups, which are among the major characteristic to DSP. The presence of developed aliphatic groups was identified by the absorption band ascribed to the stretching vibrations of carbon-hydrogen bonds in the range between (2980–2840 cm^{-1}). OH functional groups were recognized by the broad band ascribed to stretching vibrations of oxygen-hydrogen bonds in the range from 3600 to 3100 cm^{-1}, whereas, ether structures were identified by the band ascribed to stretching vibrations of carbon-oxygen in the range (1100–1000 cm^{-1}). Moreover, CO groups were also detected by the band ascribed to stretching vibrations of C=O bonds in the range from 1750 to 1500 cm^{-1}. On the other hand, the bands of aromatic compounds bond groups overlay with those obtained by the bonds of other structures. Generally, the FTIR spectrum proves the multiplicity of the structure of DSP. The presence of these aforementioned functional groups on the DSP surface indicates its potential ability to act as a promising adsorbent [14].

3.3. Scanning Electron Microscope Technique (SEM)

The SEM technique investigations provided an insight into the diverse morphology of the DSP where some larger constituents show asymmetrical form, some other constituents have an extended rod-like construction whereas other smaller constituents display rectangular form. Overall, most of the constituents have a reedy and coarse structure with irregular ends. Particle size of the DSP ranges from 5–15 μm. It can be noted from Figure 2a the availability of numerous available cavities and holes enabling Ni(II) ions to be adsorbed, while Figure 2b shows that these holes are occupied by Ni(II) ions indicating good adsorption capacity for DSP.

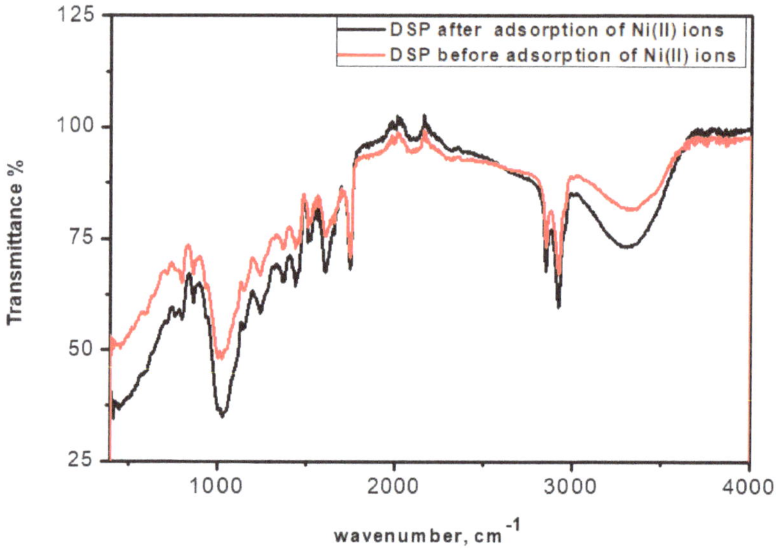

Figure 1. Attenuated total reflection-Fourier transform infrared spectrometer (ATR-FTIR) spectra for the date seeds powder (DSP) before and after Ni(II) ions adsorption process.

Figure 2. SEM technique images of the DSP surface (**a**) before and (**b**) after adsorption.

3.4. *Effect of pH Values*

A pH range of 1–11 was assessed to define the optimum value for the removal of Ni(II) ions by adsorption. Results are presented in Figure 3. It has been noted that the adsorption efficiency increases with increasing pH value up to 7, after which no alteration was noted with further increase in the pH value. These results attributed to the competition between Ni(II) ions and H^+ ions for adsorption spots on the DSP surface at low pH values [15]. As the pH value increases, less H^+ ions are present; hence, more adsorption sites are available for Ni(II) ions. The optimal pH value was defined as 7; thus, been used throughout this work.

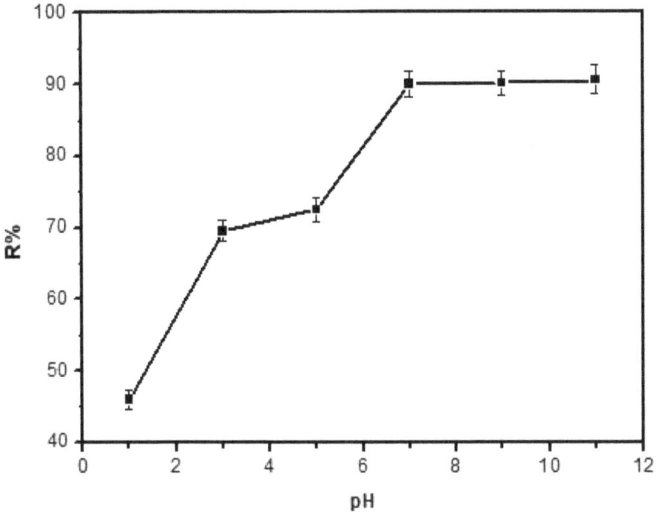

Figure 3. Effect of changing pH values on the efficiency of removal of Ni(II) ions.

3.5. *Effect of Adsorbent Particle Size*

Different particle sizes ranging from 100, 150, 250, 400, and 600 μm have been examined and the obtained results (Figure 4) showed that as the particle size decreases, the removal amount increases from 82% to 90%. This is due to the availability of more surface area obtainable for the removal of Ni(II) ions as the particle size decreases. However, no difference in adsorption was observed with particle sizes of 100 and 150 μm. Particle size of 100 μm was used throughout the study.

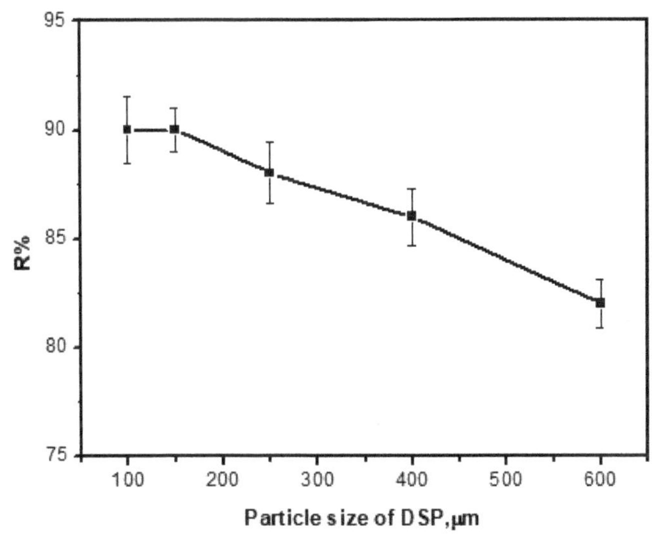

Figure 4. Effect of particle size of the DSP on the efficiency of removal of Ni(II) ions.

3.6. *Effect of Adsorbent Mass*

The dependence of the adsorption efficiency on DSP mass was studied to determine the optimal mass. Results displayed in Figure 5 show that the removal efficiency of Ni(II) ions increased as the DSP mass increases from 0.05 g to 0.30 g. Increasing the DSP mass to more than 0.30 g has no significant effect on adsorption effectiveness. This is due to the fact that the surface area of the adsorbent increases with its mass. The optimal DSP mass (0.30 g) was throughout this study.

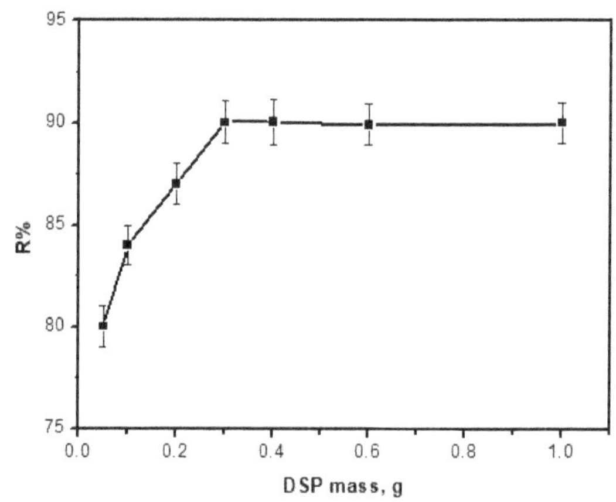

Figure 5. Effect of the DSP mass on the removal efficiency of Ni(II) ions.

3.7. *Effect of Contact Time*

In typical contaminant removal experiments, the contact time is considered a significant feature because it directly influences the adsorbent lifetime and the adsorption efficiency. Figure 6 shows the results obtained at different time intervals while all the other conditions (pH = 7.00, particle size = 100 μm, adsorbent mass = 0.30 g and temperature = 25.0 °C, revolutions per minute (rpm) = 150) were kept constant. It was found that the DSP reached the maximum adsorption of 90% for Ni(II) after 30 min.

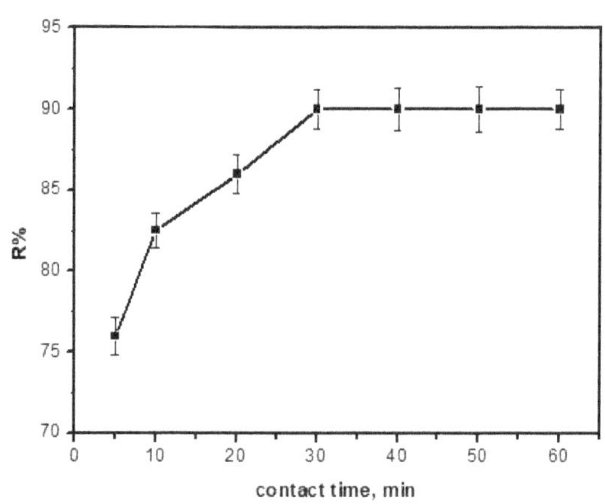

Figure 6. Effect of contact time on the efficiency of removal of Ni(II) ions.

3.8. Adsorption Kinetics

Kinetics of the adsorption process is the key feature for designing efficient adsorption experiments and this requires the use of proper kinetic model. Several kinetic parameters values are shown in Table 2. Adsorption kinetics control the rate, g the efficacy of DSP [16]. Several kinetic models were examined such as intraparticle diffusion, pseudo-second-order model, and pseudo-first-order.

Table 2. Kinetic parameters for the removal of Ni(II) ion by DSP.

Kinetic Models	Parameters	
Pseudo-first-order model	q_e (mg/g)	12.2
	k_1 (min^{-1})	4.0×10^{-4}
	R^2	0.6988
Pseudo-second-order model	q_e (mg/g)	3.1
	k_2 (g/mg min)	0.3
	R^2	0.9937
Intraparticle diffusion model	k_{id} (mg/g. $min^{(1/2)}$)	0.056
	I	2.6
	R^2	0.7978

3.8.1. Pseudo-First-Order Model

This model is denoted by Equation (2) [17].

$$\ln\left(q_e - q_t\right) = \ln q_e - k_1 t \quad (2)$$

The pseudo-second order kinetic model is denoted by Equation (3).

$$\frac{t}{q_t} = \frac{1}{k_2 q_e^2} + \frac{t}{q_e} \quad (3)$$

where q_e and q_t are the equilibrium and adsorption capacities at time (t) and equilibrium, and k_1, k_2 are rate constants for pseudo-first-order and pseudo-second-order, respectively. Moreover, 0.6977 and 0.9937 are values of correlation coefficients (R^2) for pseudo-first-order and pseudo-second-order models (Table 2), respectively. Figure 7 proves that the system obeys the pseudo-second-order kinetics model. This is in good agreement with previous studies [12,13,15].

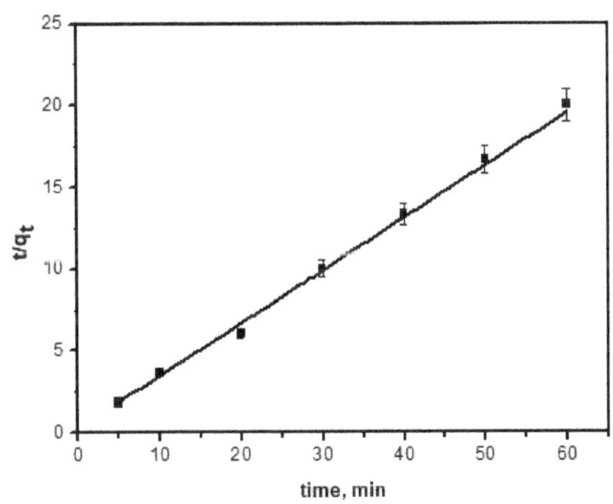

Figure 7. Pseudo-second-order kinetics.

3.8.2. Intra-Particle Diffusion Kinetic Model

Intra-particle diffusion kinetic model is displayed by Equation (4)

$$q_t = k_{id}\, t^{1/2} + I \tag{4}$$

k_{id} is the intra-particle diffusion rate constant (mg/g. min$^{(1/2)}$) and I is a constant that associated to the boundary layer thickness (mg/g). The value of (k_{id}) was determined from the slope of Equation (4) and presented in Table 2. The relationship between q_t and $t_{1/2}$ was non-linear, demonstrating that several processes are governing the adsorption process. The initial curved portion of the plot is due to the impact of boundary layer diffusion. The curved portion denotes that the intra-particle diffusion is controlled by the rate of constant k_{id}.

3.9. Adsorption Isotherm

Adsorption models are frequently exploited to explain the adsorbate/adsorbent interactions to determine the adsorption capacity of the adsorbent. To evaluate the adsorption isotherms for the DSP, Freundlich, Langmuir, Temkin, and Dubinin–Radushkevich (D–R) adsorption models were examined.

3.9.1. Langmuir Model

Equation (5) presents the Langmuir linear equation

$$\frac{C_e}{q_e} = \frac{C_e}{q_m} + \frac{1}{bq_m} \tag{5}$$

where q_e is the equilibrium quantity of Ni(II) ions adsorbed on the DSP surface at equilibrium (mg/g), C_e is the equilibrium concentration of Ni(II) ions in solution (mg/L), q_m is the maximum adsorption of Ni(II) ions (mg/g), and b (L/mg) is the Langmuir constant. According to Equation (4), values of C_e/q_e were plotted against C_e and the results are displayed in Figure 8a. Values of q_m and b were obtained from the slope and intercept, respectively. The b value refers to the adsorption binding energy [14], whereby a higher b value means more binding affinity between adsorbent and adsorbate. The parameters (q_m, b and R^2) are displayed in Table 3.

Table 3. Parameters obtained from various isotherm models.

Adsorption Model	Isotherm Constants	Value
Langmuir	q_{max} (mg/g)	40.8
	K_L (L/g)	0.029
	R^2	0.887
Freundlich	N	0.67
	K_f ((mg/g)/(mg/L) n)	4.6
	R^2	0.9981
Temkin	A (L/g)	2.1
	b_t (kJ/mol)	370.3
	R^2	0.899
D-R	β	2.0×10^{-7}
	q_m (mg/g)	7.18
	R^2	0.846

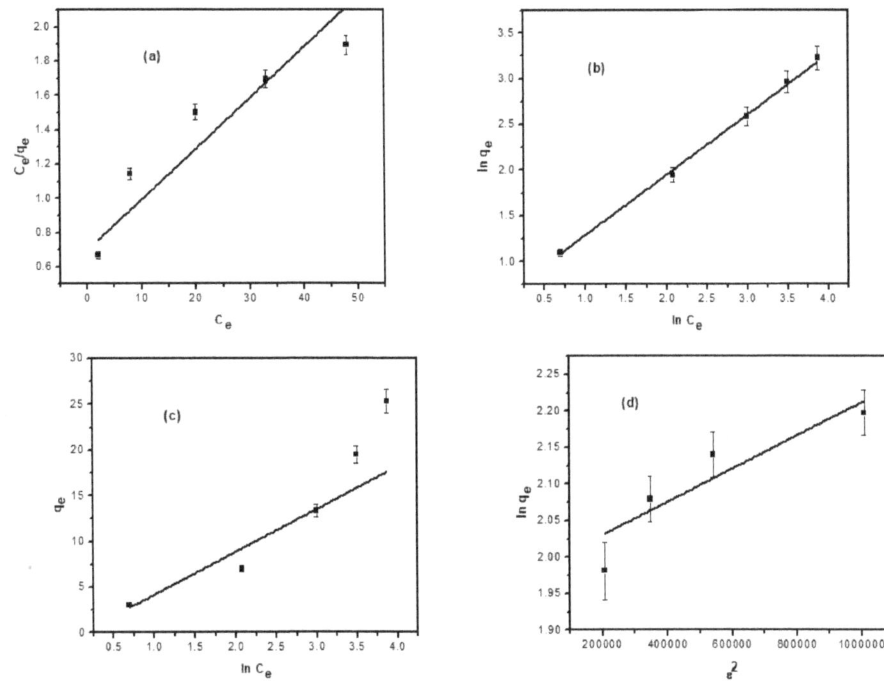

Figure 8. Isotherm models of adsorption of Ni(II) ions by the DSP (**a**), Langmuir, (**b**) Freundlich, (**c**) Temkin and (**d**) D-R.

3.9.2. Freundlich Model

This is an experimental association relating the adsorption of solutes from a liquid onto an adsorbent surface and adopts that various adsorption layers with a number of adsorption energies are involved. This model describes the affinity between the quantities of Ni(II) ions adsorbed per the dosage of the DSP, q_e, and the concentration of the Ni(II) ions at equilibrium, C_e. Linear Freundlich model is represented by Equation (6) [12].

$$\ln q_e = \ln K_f + \frac{1}{n}\ln C_e \tag{6}$$

where n and K_f represent Freundlich constants describe the process intensity capacity. Values of K_f and n are obtained from the intercept and slope of Figure 8b, respectively. Value of n is an indicator to the adsorption nature according to the following way: if $n < 1$, adsorption is classified as a physical process, if $n = 1$, adsorption is linear and if $n > 1$, adsorption is considered as a chemical process. The range of n values and K_f value are given in Table 3. Results indicate that adsorption of Ni(II) ion on the surface of the DSP is a physical process [13].

3.9.3. Temkin Isotherm

Temkin model proposes that the adsorption heat of all particles in the layer decrease sharply, rather than logarithmic with coverage [18]. The adsorption potential of DSP to Ni(II) ions can be verified by applying Temkin isotherm model. The linear formula of Temkin is shown in Equation (7):

$$q_e = \frac{RT}{b_t}\ln A + \frac{RT}{b_t}\ln C_e \tag{7}$$

where R is the universal gas constant, T is the absolute temperature, b_t is Temkin constant associated with adsorption heat, and A is a constant related to adsorption capacity. A and b_t values are found from the slope and intercept of Figure 8c and given along with R^2 in Table 3.

3.9.4. Dubinin–Radushkevich (D–R) Isotherm Model

This model estimates the energy of adsorption. It is commonly used to understand the mechanism of adsorption [19]. This isotherm is not proposed only for constant adsorption potential or homogeneous adsorbent but also for both heterogonous surfaces. The linear equation of this model is presented in Equation (8):

$$\ln q_e = \ln q_m - \beta\,\varepsilon^2 \tag{8}$$

ε is given by Equation (9):

$$\varepsilon = RT\ln\left(1 + \frac{1}{C_e}\right) \tag{9}$$

β is a constant associated with the adsorption free energy, q_m is the theoretical saturation capacity based on D–R isotherm (mg/g). Values of β, q_m and R^2 are obtained from Figure 8d and shown in Table 3. The free sorption energy E_s, is the change in free energy when one mole of adsorbate is stimulated to the solid surface and is calculated by Equation (10).

$$E_s = \frac{1}{\sqrt{2\beta}} \tag{10}$$

The adsorption type can be deduced from the E_s value. The adsorption process is considered chemical when E_s value is in the range 8.0 to16.0 kJ/mol and physical when E_s is less than 8.0 kJ/mol. In this work, the E_s value was determined as 2.24 kJ/mol concluding that the adsorption taking action is of physical nature.

3.10. Effect of Temperature

Figure 9 displays the influence of temperature on the adsorption of Ni(II) ions onto the DSP surface. The adsorption efficiency decreases as the temperature increases. This is justified due to the damage of some adsorption sites at elevated temperatures.

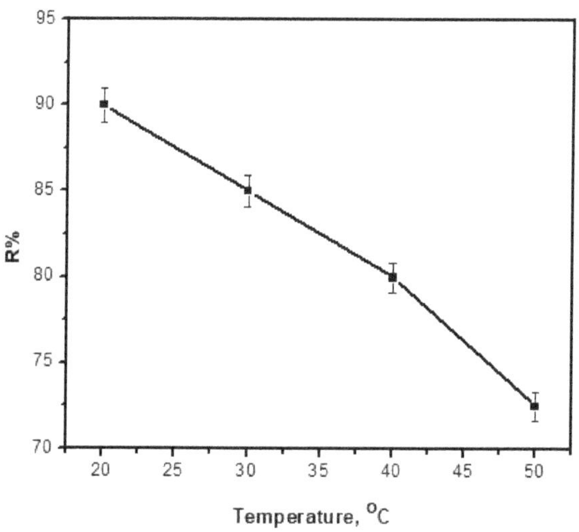

Figure 9. Effect of temperature on the removal efficiency of Ni(II) ions by the DSP.

3.11. Adsorption Thermodynamics

Thermodynamical parameters are vital in any adsorption investigations as the temperature is strongly connected to the kinetic energy of adsorbate. In this study, adsorption tests were performed at different temperatures, viz. 298, 308, 318 and 328 K for the sorption of initial Ni(II) ion concentration (50 mg/L) on DSP at their particular optimum pH values, DSP mass and contact time. Entropy (ΔS°), enthalpy (ΔH°), and Free energy (ΔG°) change are governed by Equations (11) and (12).

$$\Delta G^o = RT \ln K_D \tag{11}$$

$$\ln K_D = \frac{-\Delta H^o}{RT} + \frac{\Delta S^o}{R} \tag{12}$$

where K_D value were calculated by Equation (13)

$$K_D = \frac{q_e}{C_e} \tag{13}$$

The thermodynamical parameters are listed in Table 4. Negative signs of ΔG° indicate that the process is spontaneous. It is clear that the negative values of ΔG° decrease as the temperature increases. This is attributed to the fact that additional positions on the surface of the DSP are destroyed at elevated temperatures. Values of ΔG° for Ni(II) ions adsorption onto the DSP were found in the range of −3.5 to −5.7 kJ/mol. It is well established that physical adsorption free energy change (ΔG°) is ranging between −20 and 0 kJ/mol and chemical adsorption between −400 to −80 kJ/mol [13]. Thus, adsorption process is mainly a physical sorption process. This finding is in good agreement with the parameters obtained from Freundlich, Dubinin–Radushkevich (D-R) and Temkin models. ΔS° and ΔH° were calculated from the intercept and slope of Figure 10 and presented in Table 3. The negative value of ΔH° (−27.0 kJ/mol) proves that the adsorption is an exothermic process. The positive value of ΔS° (71.0 J/mol) designates the affinity of the DSP towards the Ni(II) ions.

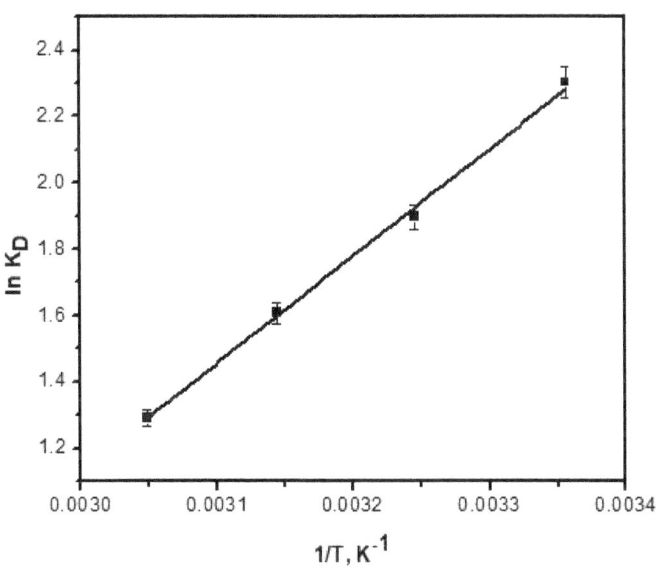

Figure 10. Relationship between ln K_D and 1/T.

Values of sticking probability (S*) and activation energy (Ea) were calculated by applying the modified Arrhenius equation that shown in Equations (14) and (15) [19]:

$$S^* = (1 - \theta)e^{\frac{-E_a}{RT}} \quad (14)$$

$$\theta = \left(1 - \frac{C_e}{C_i}\right) \quad (15)$$

where θ is the surface coverage, C_i represents the original concentration of Ni(II) ions, E_a is the activation energy of the system and S* is the sticking probability.

The sticking probability, S*, is a function of the adsorbent/adsorbate adsorption process but should fulfil the circumstance $0 < S^* < 1$ and depends on the temperature of the process. E_a and S* values were obtained from the slope and intercept of Figure 10, and are recorded in Table 4. The Ea value is very low (−2.01 kJ/mol) demonstrating the facile adsorption process and the negative charge of E_a value indicates an exothermic process. This is in good agreement with the negative value of ΔH° [15]. Since the value of S* <<< 1, therefore, the probability of Ni(II) ion to stick on the DSP is very high and thus the adsorption process is favorable [18].

Table 4. Values of thermodynamic parameters.

T, K	K_D	ΔG° (kJ/mol)	ΔS° (J/mol K)	ΔH° (kJ/mol)	E_a (kJ/mol)	S* (J K/mol)
298	10.0	−5.7				
308	6.7	−4.9	71.7	−27.0	−2.01	0.002
318	5.0	−4.2				
328	3.6	−3.5				

3.12. Comparisons with Other Adsorbents

Table 5 presents the results of this study compared with other biosorbents. Under the same experimental conditions in terms of optimum pH, temperature, DSP proved high adsorption efficiency in contrast to the chosen biosorbents. It is interesting to note that the DSP was used without any further chemical or thermal modifications. This reduces its cost and hazard to the minimum.

Table 5. Comparison of the DSP with other biosorbents used for the removal of Ni(II) ions.

Biosorbent	q_m (mg/g)	Isotherm Model	Kinetic Model	Optimal pH	Reference
Date seeds powder	41.0	Freundlich	2nd order	7.00	This study
Aloe barbadensis Miller leaves	10.0	Freundlich	2nd order	3.00	[20]
peel of Artocarpus nobilis fruit	12.1	Langmuir	**	4.00	[21]
Modified Aloe barbadensis leaves	29.0	Langmuir	2nd order	7.00	[22]
brown algae Sargassum sp.	1.3	Langmuir	2nd order	6.50	[23]
Calotropis procera roots	0.6	Langmuir	**	3.00	[24]
Peat	61.3	Langmuir	2nd order	9.00	[14]
Barely straw	35.8	Langmuir	**	4.85	[25]
Orange peel	162.6	Langmuir	2nd order	5.50	[26]
Tea factory waste	18.4	Langmuir	**	4.00	[27]

** not reported.

4. Conclusions

In this study, the powder obtained from date seeds has been tested as a competent (removal effectiveness > 90%), inexpensive, eco-friendly and natural material for adsorption of Ni(II) ions from artificial wastewater. The date seeds powder retained significant amounts of Ni(II) readily. The maximum adsorption capacity for DSP was found to be 41 mg/g. The optimal conditions for efficient removal of 50 ppm of Ni(II) ion were found to be 0.30 g of DSP after 30 min of shaking time at pH 7.00. This adsorbent has positive characteristics making it appropriate to purify wastewater. Such characteristics include (i) cost-effective material: date seeds powder is an expensive material. The only cost of this material would be for the collection, transportation and grinding. It does not need any further chemical or thermal modification (ii) availability: date seeds are abundantly existing in Saudi Arabia (iii) efficiency: date seeds powder can remove more than 90% of Ni(II) ions, which present in artificial wastewater.

Author Contributions: Conceptualization I.H.A. and A.E.; methodology, A.E.; validation, E.I.B.; formal analysis, E.I.B.; investigation A.E.; resources, A.B.E.; writing—original draft preparation, I.H.A. and A.E.; writing—review and editing, E.I.B.; supervision, B.K.; project administration, I.H.A.; funding acquisition, A.B.E. All authors have read and agreed to the published version of the manuscript.

Funding: This research was funded by DEANSHIP OF SCIENTIFIC RESEARCH, KING KHALID UNIVERSITY, grant number R.G.P.1/139/40.

Conflicts of Interest: The authors declare no conflict of interest.

References

1. Kulkarni, R.M.; Shetty, K.V.; Srinikethan, G. Cadmium (II) and nickel (II) biosorption by Bacillus laterosporus (MTCC 1628). *J. Taiwan Inst. Chem. Eng.* **2014**, *45*, 1628–1635. [CrossRef]
2. Putra, W.P.; Kamari, A.; Yusoff, S.N.M.; Ishak, C.F.; Mohamed, A.; Hashim, N.; Isa, I.M. Biosorption of Cu (II), Pb (II) and Zn (II) ions from aqueous solutions using selected waste materials: Adsorption and characterisation studies. *J. Encapsul. Adsorpt. Sci.* **2014**, *4*, 720–726.
3. Karatas, M. Removal of Pb(II) from water by natural zeolitic tuff: Kinetics and thermodynamics. *J. Hazard. Mater.* **2012**, *199*, 383–393. [CrossRef] [PubMed]
4. Ahmad, T.; Danish, M.; Rafatullah, M.; Ghazali, A.; Sulaiman, O.; Hashim, R.; Ibrahim, M.N.M. The use of date palm as a potential adsorbent for wastewater treatment: A review. *Environ. Sci. Pollut. Res.* **2012**, *19*, 1464–1468. [CrossRef]
5. Lee, C.G.; Lee, S.; Park, J.A.; Park, C.; Lee, S.J.; Kim, S.B.; An, B.; Yun, S.T.; Lee, S.H.; Choi, J.W. Removal of copper, nickel and chromium mixtures from metal plating wastewater by adsorption with modified carbon foam. *Chemosphere* **2017**, *166*, 203–211. [CrossRef]
6. Prithviraj, D.; Deboleena, K.; Neelu, N.; Noor, N.; Aminur, R.; Balasaheb, K.; Abul, M. Biosorption of nickel by *Lysinibacillus* sp. BA2 native to bauxite mine. *Ecotoxicol. Environ. Saf.* **2014**, *107*, 260–268. [CrossRef]
7. Salem, N.M.; Awwad, A.M. Biosorption of Ni(II) from electroplating wastewater by modified (*Eriobotrya japonica*) loquat bark. *J. Saudi Chem. Soc.* **2014**, *18*, 379–386. [CrossRef]
8. Onundi, Y.B.; Mamun, A.A.; Al Khatib, M.F.; Ahmed, M. Adsorption of copper, nickel and lead ions from synthetic semiconductor industrial wastewater by palm shell activated carbon. *Int. J. Environ. Sci. Technol.* **2010**, *7*, 751–758. [CrossRef]
9. Alomá, I.; Martín-Lara, M.A.; Rodríguez, I.L.; Blázquez, G.; Calero, M. Removal of nickel (II) ions from aqueous solutions by biosorption on sugarcane bagasse. *J. Taiwan Inst. Chem. Eng.* **2012**, *43*, 275–281. [CrossRef]
10. El Marouani, M.; Azoulay, K.; Bencheikh, I.; Elfakir, L.; Rghioui, L.; El Hajji, A.; Sebbahi, S.; El Hajjaji, S.; Kifani-Sahban, F. Application of raw and roasted date seeds for dyes removal from aqueous solution. *J. Mater. Environ. Sci.* **2018**, *9*, 2387–2396.
11. Chang, Y.S.; Au, P.I.; Mubarak, N.M.; Khalid, M.; Jagadish, P.; Walvekar, R.; Abdullah, E.C. Adsorption of Cu(II) and Ni(II) ions from wastewater onto bentonite and bentonite/GO composite. *Environ. Sci. Pollut. Res.* **2020**, *27*, 33270–33296. [CrossRef] [PubMed]
12. Mehrmad, N.; Moravegi, M.; Parvareh, A. Adsorption of Pb(II), Cu(II) and Ni(II) ions from aqueous solutions by functionalised henna powder (*Lawsonia Inermis*); isotherm, kinetic and thermodynamic studies. *Int. J. Environ. Anal. Chem.* **2020**. [CrossRef]
13. Khan, M.I.; Almesfer, M.K.; Danish, M.; Ali, I.H.; Shoukry, H.; Patel, R.; Gardy, J.; Nizami, A.S.; Rehan, M. Potential of Saudi natural clay as an effective adsorbent in heavy metals removal from wastewater. *Desalin. Water Treat.* **2019**, *158*, 140–151. [CrossRef]

14. Bartczak, P.; Norman, M.; Klapiszewski, L.; Karwan'ska, N.; Kawalec, M.; Baczyn´ska, M.; Wysokowski, M.; Zdarta, J.; Ciesielczyk, F.; Jesionowski, T. Removal of nickel(II) and lead(II) ions from aqueous solution using peat as a low-cost adsorbent: A kinetic and equilibrium study. *Arab. J. Chem.* **2018**, *11*, 1209–1222. [CrossRef]
15. Ali, I.H.; Al Mesfer, M.K.; Khan, M.I.; Danish, M.; Alghamdi, M.M. Exploring Adsorption Process of Lead (II) and Chromium (VI) Ions from Aqueous Solutions on Acid Activated Carbon Prepared from *Juniperus procera* Leaves. *Processes* **2019**, *7*, 217. [CrossRef]
16. Chang, C.F.; Chang, C.Y.; Chen, K.H.; Tsai, W.T.; Shie, J.L.; Chen, Y.H. Adsorption of naphthalene on zeolite from aqueous solution. *J. Colloid. Interface Sci.* **2004**, *277*, 29–34. [CrossRef]
17. Ali, I.H.; Sulfab, Y. Kinetics and mechanism of oxidation of cisdiaquabis(glycinato)chromium(III) by periodate ion in aqueous solutions. *Transit. Met. Chem.* **2013**, *38*, 79–84. [CrossRef]
18. Tempkin, M.I.; Pyzhev, V. Kinetic of Ammonia Synthesis on Promoted Iron Catalyst. *Acta Phys. Chim. USSR* **1940**, *12*, 327–356.
19. Günay, A.; Arslankaya, E.; Tosun, I. Lead removal from aqueous solution by natural and pretreated clinoptilolite: Adsorption equilibrium and kinetics. *J. Hazard. Mater.* **2007**, *146*, 362–371. [CrossRef]
20. Gupta, S.; Kumar, A. Removal of nickel (II) from aqueous solution by biosorption on *A. barbadensis Miller* waste leaves powder. *Appl. Water Sci.* **2019**, *9*, 96–107. [CrossRef]
21. Priyantha, N.; Kotabewatta, P.A. Biosorption of heavy metal ions on peel of *Artocarpus nobilis* fruit: 1—Ni(II) sorption under static and dynamic conditions. *Appl. Water Sci.* **2019**, *9*, 37–47. [CrossRef]
22. Gupta, S.; Sharma, S.K.; Kumar, A. Biosorption of Ni(II) ions from aqueous solution using modified *Aloe barbadensis* Miller leaf powder. *Water Sci. Eng.* **2019**, *12*, 27–36. [CrossRef]
23. Barquilha, C.E.R.; Cossich, E.S.; Tavares, C.R.; Silva, E.A. Biosorption of nickel(II) and copper(II) ions by *Sargassum* sp. in nature and alginate extraction products. *Bioresour. Technol. Rep.* **2019**, *5*, 43–50. [CrossRef]
24. Pandey, P.K.; Choubey, S.; Verma, Y.; Pandey, M.; Kamal, S.S.; Chandrashekhar, K. Biosorptive removal of Ni(Ii) from wastewater and industrial effluent. *Int. J. Environ. Res. Public Health* **2007**, *4*, 332–339. [CrossRef] [PubMed]
25. Thevannan, A.; Mungroo, R.; Niu, C. Biosorption of nickel with barley straw. *Bioresour. Technol.* **2010**, *101*, 1776–1780. [CrossRef]
26. Feng, N.; Guo, X.; Liang, S.; Zhu, Y.; Liu, J. Biosorption of heavy metals from aqueous solutions by chemically modified orange peel. *J. Hazard. Mater* **2011**, *185*, 49–54. [CrossRef]
27. Malkoc, E.; Nuhoglu, Y. Investigations of nickel(II) removal from aqueous solutions using tea factory waste. *J. Hazard. Mater* **2005**, *127*, 120–128. [CrossRef]

processes

MDPI

Article

Biosorption of Co^{2+} Ions from Aqueous Solution by K_2HPO_4-Pretreated Duckweed *Lemna gibba*

Jessica Lizeth Reyes-Ledezma, Eliseo Cristiani-Urbina and Liliana Morales-Barrera *

Departamento de Ingeniería Bioquímica, Escuela Nacional de Ciencias Biológicas, Instituto Politécnico Nacional, Av. Wilfrido Massieu s/n, Unidad Profesional Adolfo López Mateos, Ciudad de Mexico 07738, Mexico; jessicalrl22@hotmail.com (J.L.R.-L.); ecristianiu@yahoo.com.mx (E.C.-U.)

* Correspondence: lmoralesb@ipn.mx; Tel.: +52-55-5729-6000 (ext. 57827)

Received: 19 October 2020; Accepted: 23 November 2020; Published: 25 November 2020

Abstract: The wastewater of the many industries that use divalent cobalt (Co^{2+})-containing compounds has elevated levels of this metal. Thus, novel technology is needed to efficiently remove Co^{2+} ions from aqueous solutions. Biosorption is a low-cost technique capable of removing heavy metals from contaminated water. This study aims to evaluate the performance of KH_2PO_4-pretreated *Lemna gibba* (*PLEM*) as a biosorbent of Co^{2+} in aqueous solutions tested under different conditions of pH, particle size, and initial Co^{2+} concentration. Kinetic, equilibrium, and thermodynamic studies were conducted. The capacity of biosorption increased with a greater initial Co^{2+} concentration and was optimal at pH 7.0 and with small-sized biosorbent particles (0.3–0.8 mm). The pseudo-second-order sorption model best describes the experimental data on Co^{2+} biosorption kinetics. The Sips and Redlich-Peterson isotherm models best predict the biosorption capacity at equilibrium. According to the thermodynamic study, biosorption of Co^{2+} was endothermic and spontaneous. The effect of pH on the biosorption/desorption of Co^{2+} suggests that electrostatic attraction is the main biosorption mechanism. SEM-EDX verified the presence of Co^{2+} on the surface of the pretreated-saturated biosorbent and the absence of the metal after desorption.

Keywords: divalent cobalt; *Lemna gibba*; biosorption; desorption; SEM-EDX

1. Introduction

Excessive population growth, urbanization, and industrial development have increased the pollution of the planet and altered ecosystems. Of all environmental pollution, the contamination of water is the most worrisome because of affecting the primordial element on which life is based. The main source of water pollution is the discharge of industrial wastewater with diverse toxic substances, among which heavy metals are of particular concern [1].

Cobalt is a heavy metal found in the Earth's crust, being a natural component of volcanic emissions, as well as surface and subterranean water. It is released into the environment through anthropogenic activities: Burning fossil fuels, applying fertilizers, mining, electroplating, manufacturing batteries, and producing commodities with industrial processes involving cobalt-containing compounds, among others.

Although cobalt is an essential nutrient in human metabolism and the principal component of vitamin B12 [2], it is harmful to our health beyond trace levels, competing with other elements that constitute integral parts of a proper metabolic function [3]. In excess, it can give rise to skin irritation and problems in bone development, as well as respiratory, cardiac, thyroid, liver, and gastric disorders [4,5]. Due to being hazardous to humans and ecosystems [6], cobalt-contaminated wastewater should be treated prior to being released into the environment.

Since low concentrations of cobalt are difficult to remove from water by conventional physicochemical treatments, it is necessary to apply innovative technology characterized by safety, efficiency, and versatility. One alternative is biosorption, a process independent of cell metabolism [7]. This technique, which has been little studied as a remedy for Co^{2+} pollution, can be carried out by living, dead, or inactive biological material [8,9].

Biosorption is a process of capturing heavy metals by physical adsorption (physisorption), ionic interchange, chemisorption (e.g., complexation, coordination, and chelation), and microprecipitation [10]. Diverse biological materials are capable of biosorption, including agroindustrial waste, microbial biomass, and biopolymers. These economical materials are available in great quantities, and the respective processes are environmentally friendly [11,12]. Unlike physicochemical methods, biosorption techniques can efficiently remove low concentrations of metals from aqueous solutions. If biosorption is followed by desorption, the metals can be recovered and the biosorbents regenerated for later use [13].

The current contribution focuses on the biosorbent potential of *Lemna gibba*, a macrophyte of universal distribution commonly known as duckweed. This plant, which quickly proliferates to double its biomass in about two days, lends itself to the bioremediation of aquatic systems, due to its small size (2–4 mm) and ability to bioaccumulate toxic compounds (e.g., heavy metals) [14]. Because eutrophication has provoked an excessive spread of *Lemna gibba*, it is now a plague in many places. Its excessive growth in the form of a thick mat on the aquatic body leads to navigation problems, harbors harmful fauna, and prevents sunlight from reaching photosynthetic species in the water below, thus interrupting the correct oxygenation of its environment [15,16]. Apart from being abundantly available, the plant material holds promise as a sustainable biosorbent for treating wastewater contaminated with cobalt and other heavy metals.

According to a previous report, pretreatment of *L. gibba* with K_2HPO_4 substantially improves the availability of sorption sites on the surface of plant cells, and therefore, their capacity for Co^{2+} biosorption, which is achieved by removing salts and producing a higher negative charge (−35 vs. −26 mV). The zero point of charge (ζ_0) was 2.37 for unpretreated and 1.62 for K_2HPO_4-pretreated *Lemna gibba*, thus creating a greater attraction in the latter for positively charged Co^{2+}. The ATR-FTIR analysis of K_2HPO_4-pretreated *Lemna gibba* revealed an important role of its hydroxyl and carboxyl groups in the removal of Co^{2+} [17]. The aim of the present study was to analyze the performance of K_2HPO_4-pretreated *L. gibba* as a biosorbent under distinct conditions of pH, particle size, and the initial concentration of Co^{2+}. Various theoretical models were tested to find the best one for describing the experimental data on biosorption. To determine the best eluent solution for desorption, saturated *L. gibba* was processed with strong and weak acids, as well as some alkaline compounds. Considering that recyclability is a prerequisite for the practical application of biosorption technology, three biosorption/desorption cycles were herein evaluated.

2. Materials and Methods

2.1. Reagents

The reagents employed in the experiments were all of analytical grade (JT Baker®, Monterrey, Mexico). During the biosorption experiments, the pH of the solutions was maintained constant by adding HCl and NaOH in the solution at a concentration of 0.1 M and 0.01 M, respectively. The different concentrations of Co^{2+} were prepared by making dilutions of a stock solution of $CoCl_2{\cdot}6H_2O$ (>98% purity) containing 1 g L^{-1} of Co^{2+}.

2.2. Preparation of the Biosorbent

Lemna gibba was collected from the Xochimilco canals in Mexico City (19°15′31.8″ N 99°05′05.3″ W). It was cleaned with running tap water and then deionized water before being dried in a Luzeren® oven (Proveedor de Laboratorios, Mexico) at 60 °C for 48 h. Afterward, the material was ground in a hammer mill (Glen Creston, Ltd., London, UK) and sieved (U.S. ASTM) to obtain fractions of the biosorbent, each with a particular particle size between 0.3 and 2.0 mm (0.3–0.5, 0.5–0.8, 0.8–1.4, and 1.4–2.0 mm). The fractions were all pretreated with K_2HPO_4. Briefly, 5 g (dry weight) of *Lemna gibba* per liter were exposed to K_2HPO_4 (0.3 M) at 18 °C for 30 min. During the pretreatment, the material was agitated at 140 rpm in an orbital shaker (All Sheng™, Hangzhou Allsheng Instruments Co, Ltd., Hangzhou, China). Upon completion of the exposure time, the biosorbent was washed with deionized water. When the resulting wash water had a pH near the deionized water being used, the material was dried in an oven at 60 °C for 48 h [17]. Each fraction of dried, pretreated *Lemna gibba* (*PLEM*) was stored in a separate, well-labeled, hermetically-sealed bottle at room temperature (rt).

2.3. The Influence of Different Physicochemical Parameters on the Biosorption of Co^{2+} by PLEM

Experiments to evaluate the effect of several physicochemical variables on the biosorption of Co^{2+} by *PLEM* were carried out in 500 mL Erlenmeyer flasks. They contained 120 mL of a solution with a known concentration of Co^{2+} at a predetermined pH value. Subsequently, an addition was made of 0.12 g of PLEM at a certain particle size, thus achieving a biosorbent concentration of 1 g (dry weight) L^{-1}. The suspensions were left at 18 °C (rt) for 2 h under constant agitation at 140 rpm in an orbital shaker (All Sheng™, Hangzhou Allsheng Instruments Co, Ltd., Hangzhou, China). The pH of the solutions was adjusted to the desired value and maintained constant throughout the assay by adding 0.1 M HCl and 0.01 M NaOH.

Firstly, the pH varied (2, 3, 4, 5, 6, and 7), while maintaining the initial concentration of Co^{2+} (C_{ini}) at 100 mg L^{-1} and the particle size of *PLEM* at 0.3–0.5 mm. Later, distinct particle sizes (0.3–0.5, 0.5–0.8, 0.8–1.4, 1.4–2.0, and 0.3–0.8 mm) were utilized, while maintaining C_{ini} at 100 mg L^{-1} and the pH at 7.0. Finally, different initial values of C_{ini} (10, 20, 40, 60, 80, 100, 200, and 300 mg L^{-1}) were used, while maintaining the pH at 7.0 and the particle size at 0.3–0.8 mm.

During the experiment, samples were taken at various exposure times and filtered to afford a solution free of biosorbent. The filtrate of each flask was diluted properly for the posterior quantification of the cobalt concentration. From the values obtained, the biosorption capacity of Co^{2+} by *PLEM* was calculated at a series of exposure times using Equation (1):

$$q = \frac{V}{M}(C_{ini} - C) \tag{1}$$

where q (mg g^{-1}) is the capacity of biosorption of Co^{2+}, V (L) is the total volume of the solution, M (g) is the biosorbent mass, and C_{ini} and C (mg L^{-1}) correspond to the initial concentration of Co^{2+} in the solution and its concentration at time t (h), respectively. When the system reaches equilibrium, $t = t_{eq}$, $C = C_{eq}$ and $q = q_{eq}$. Based on the values of biosorption capacity found, the most suitable pH of the solution and the best particle size for the removal of Co^{2+} were selected for the rest of the biosorption experiments. For each of the parameters examined, controls free of biosorbent were established and analyzed for possible changes in the concentration of cobalt.

2.4. Kinetic Modeling of the Biosorption of Co^{2+} by PLEM

For the kinetic modeling of the biosorption of Co^{2+} by *PLEM*, the equations of pseudo-first-order, pseudo-second-order, and fractional power were employed (Table 1).

Table 1. Biosorption models were tested.

Kinetic Models	Equation	Parameters
Pseudo-first-order [18]	$q = q_{eq1}\left[1 - e^{(-k_1 t)}\right]$	k_1—pseudo-first-order sorption velocity constant (min^{-1}) q_{eq1}—equilibrium biosorption capacity predicted by the model ($mg\ g^{-1}$)
Pseudo-second-order [18]	$q = \frac{t}{\left(\frac{1}{k_2 \cdot q_{eq2}^2}\right)+\left(\frac{t}{q_{eq}}\right)}$	k_2—pseudo-second-order sorption velocity constant ($g\ mg^{-1}\ min^{-1}$) q_{eq2}—equilibrium biosorption capacity predicted by the model ($mg\ g^{-1}$)
Fractional power [18]	$q = k_{FP} t^{v}$	k_{FP}—constant of the model ($mg\ g^{-1}$) v—velocity constant (h^{-1})
Isothermal models	**Equation**	**Parameters**
Langmuir [19,20]	$q_{eq} = \frac{q_{mL}\ b_L C_{eq}}{1+b_L C_{eq}}$ $R_L = \frac{1}{1+b_L C_{ini}}$	q_{mL}—maximum biosorption capacity determined by Langmuir ($mg\ g^{-1}$) b_L—Langmuir constant, linked to affinity for the active sites ($L\ mg^{-1}$) C_{ini}—initial concentration ($mg\ L^{-1}$) R_L—separation factor
Freundlich [19]	$q_{eq} = k_F\ C_{eq}^{1/nF}$	k_F—Freundlich constant, related to the biosorption capacity ($mg\ g^{-1}\ (mg\ L^{-1})^{-1/nF}$) n_F—Freundlich constant, linked to the intensity of sorption
Sips [19]	$q_{eq} = \frac{q_{mSP}\ k_{SP}\ C_{eq}^{n_{SP}}}{1+k_{SP}\ C_{eq}^{n_{SP}}}$	q_{mSP}—maximum biosorption capacity, determined by Sips ($mg\ g^{-1}$) k_{SP}—constant of the model ($mg\ L^{-1})^{-nSP}$ n_{SP}—exponent of the model
Redlich-Peterson [19]	$q_{eq} = \frac{k_{RP}\ C_{eq}}{1+a_{RP}\ C_{eq}^{b_{RP}}}$	k_{RP}—constant of the model ($L\ g^{-1}$) a_{RP}—constant of the model ($mg\ L^{-1})^{-bRP}$ b_{RP}—exponent of the model

2.5. Biosorption Isotherm Studies at Different Temperatures

In 125 mL flasks were poured 30 mL of solutions of Co^{2+} at distinct concentrations (20, 40, 60, 80, 100, 200, and 300 mg L^{-1}), adjusting the pH to 7.0. Then 0.03 g of *PLEM* (particle size = 0.3–0.8 mm) was placed in each flask to ensure a concentration of 1 g L^{-1} of *PLEM*. The suspensions were left for 2 h at 18, 30, 40, 50, or 60 °C to reach biosorption equilibrium. Subsequently, the samples from each flask were filtered, and the residual concentration of Co^{2+} was quantified in each filtrate. With the experimental results of the biosorption capacity found at equilibrium (q_{eq}) and the residual concentration of cobalt at equilibrium (C_{eq}) for each initial concentration of metal assayed (C_{ini}), the isotherm for adsorption was calculated. It was then possible to select the best mathematical model for describing the experimental behavior. With this objective in mind, models of two (Langmuir and Freundlich) and three parameters (Sips and Redlich-Peterson) were used (Table 1).

2.6. Determination of the Thermodynamic Parameters

The thermodynamic parameters examined were the changes in Gibbs free energy (ΔG^0, J mol^{-1}), in standard entropy (ΔS^0, J mol^{-1} K^{-1}), and in standard enthalpy (ΔH^0, J mol^{-1}). With the data on the isotherms for biosorption at equilibrium, the coefficient of distribution (K_d, L g^{-1}) was obtained for each temperature and concentration assayed using Equation (2) [21]:

$$K_d = \frac{q_{eq}}{C_{eq}} \quad (2)$$

In the graph of Ln K_d vs. C_{eq} for each temperature, the point at which the ordinate crosses the origin corresponds to Ln K_0 (K_0 being the sorption constant at equilibrium, L g^{-1}). These values were substituted in Equation (3) to find the change in Gibbs free energy [22]:

$$\Delta G^0 = -RT\ Ln\ K_0 \tag{3}$$

where R is the constant of the ideal gases (8.315 J mol^{-1} K^{-1}), and T is the absolute temperature (K) during biosorption. The change in standard entropy (ΔS^0) was found by Equation (4):

$$\Delta S^0 = \frac{\partial \Delta G^0}{\partial T} \tag{4}$$

The slope of the graph of ΔG^0 vs. T indicates the mean value of ΔS^0. The change in the standard enthalpy was furnished by Equation (5):

$$\Delta G^0 = \Delta H^0 + T\Delta S^0 \tag{5}$$

2.7. Desorption of Co^{2+} from the Biosorbent

To evaluate desorption, the biosorbent was first saturated by exposing *PLEM* (1 g L^{-1}, with a particle size of 0.3–0.8 mm) to a solution of Co^{2+} (300 mg L^{-1}, pH 7.0, rt) under constant agitation at 140 rpm for 2 h. Upon completion of this time, the biosorbent was washed with deionized water several times to eliminate the excess cobalt and then dried in an oven at 60 °C for 48 h. Finally, it was stored in hermetically-sealed bottles until further use.

For the desorption of Co^{2+} from *PLEM*, diverse solutions were tested as eluents: Water at rt (H_2O rt, the control), water at 60 °C (H_2O 60 °C), various acidic solutions (HCl, H_2SO_4, HNO_3, $C_2H_2O_4$, KH_2PO_4, and NH_4Cl) and three alkaline compounds (NaOH, $NaHCO_3$, and K_2HPO_4). The concentration of all compounds was 0.1 M. Desorption was carried out by placing 120 mL of one of the distinct eluent solutions in each Erlenmeyer flask and adding the saturated biosorbent at a concentration of 1 g L^{-1}. The material was maintained under constant agitation at 140 rpm and 18 °C for 2 h, collecting and filtering samples from each of the flasks at different times. The concentration of desorbed metal on each filtrate was quantified. The percentage of desorption at time t was calculated with Equation (6) [23]:

$$E_D(\%) = \frac{V(C_D - C_{ini})}{M\ q_{eq}} \times 100 \tag{6}$$

where C_{ini} and C_D (mg L^{-1}) are the initial concentration of metal in the solution (t = 0 h) and the concentration of Co^{2+} eluted from the solution at time t, respectively, and q_{eq} (mg g^{-1}) is the amount of Co^{2+} retained per gram of biosorbent (determined experimentally). The results of the percentage of the desorption were compared to select the adequate solution for eluting Co^{2+} from *PLEM*.

2.8. Biosorption-Desorption Cycles

PLEM was saturated with Co^{2+} for 2 h, as described in the previous section. Upon completion of this time, samples of the solution were taken to assess the biosorption capacity of *PLEM* in the first stage (Equation (1)). Subsequently, the saturated biosorbent was washed, dried, and subjected to the desorption of Co^{2+} (as already explained) by putting 1 g L^{-1} of the material in a solution with the selected eluent and leaving it under constant agitation at 140 rpm and rt for 2 h. Samples were then taken to quantify the concentration of Co^{2+} in the solution and calculate the percentage of desorption for the first cycle (Equation (6)). *PLEM* was washed with deionized water and dried at 60 °C for 48 h to be submitted to posterior cycles. Three cycles of biosorption/desorption were carried out under the same conditions, allowing for the comparison of the capacity of biosorption and percentage of desorption from one cycle to another.

2.9. Scanning Electron Microscope Coupled to Energy-Dispersive X-ray Spectroscopy (SEM-EDX)

The possible changes in the structure and composition of the surface of *PLEM*, due to the process of biosorption and the posterior desorption of Co^{2+} were explored on a scanning electron microscope (SEM). The three types of samples of *PLEM* (unexposed to Co^{2+}, saturated, and desorbed in the first cycle) were dried for 24 h at 60 °C. Subsequently, they were covered with carbon to be later observed with a JEOL high-resolution scanning electron microscopy (HR-SEM) (model JSM7800F, Jeol Ltd., Tokyo, Japan) with an acceleration voltage of 5 kV.

2.10. Analytical Methods

Co^{2+} was quantified by the dimethylglyoxime (DMG) method, with which a compound is formed with an intensity of color proportional to the concentration of Co^{2+} present in the solution [24]. The measurement of absorbance was conducted in a Spectronic Genesys UV/Vis 10 spectrophotometer (Thermo Electron Scientific Instruments Corp, Madison, WI, USA) at 400 nm. The concentration of Co^{2+} was established by constructing metal-type curves with at least 10 distinct known concentrations.

2.11. Statistical Analysis

Each experiment was performed independently at least twice, and the determinations of residual cobalt were made at least three times, with the aim of attaining the appropriate statistical power. Data are expressed as the mean ± standard deviation (SD) of the values obtained experimentally. Regarding the values from the kinetics of biosorption and the experimental biosorption capacity at equilibrium (q_{eq}), differences between groups were examined with two-way ANOVA and Tukey's test (with a confidence interval of $\alpha = 0.05$) on the GraphPad Prism® Ver 8.4 program 2020 (GraphPad Software Inc, San Diego, CA, USA). The kinetic and equilibrium parameters were scrutinized by nonlinear regression on the same software, selecting the best model in accordance with a variety of error functions: The correlation coefficient (R^2), the absolute sum of squares (*ASE*), the standard deviation of the residuals (*Sy.x*) and Akaike's information criterion (*AICc*). The data from the three cycles of biosorption/desorption were compared with one-way ANOVA and Dunnett's test (confidence interval, $\alpha = 0.05$) on the GraphPad Prism® Ver 8.4 program 2020 (GraphPad Software Inc., San Diego, CA, USA).

3. Results and Discussion

No change in the concentration of Co^{2+} was found for the *PLEM*-free solutions, used as controls for the evaluation of the influence of the physicochemical conditions herein tested. Thus, the removal of Co^{2+} from the aqueous solution can be fully attributed to the effect of biosorption produced by *PLEM*.

3.1. The Effect of pH

The level of pH is one of the physicochemical factors that most influence the biosorption of heavy metals [25]. The pH values of 2–7 were presently employed because the precipitation of cobalt was observed experimentally as of pH 8, likely due to the formation of cobalt hydroxide [26,27]. At each pH value, the biosorption capacity was evaluated with respect to time (Figure 1a). With the pH at 2 or 3, the cobalt removal capacity was near 0.

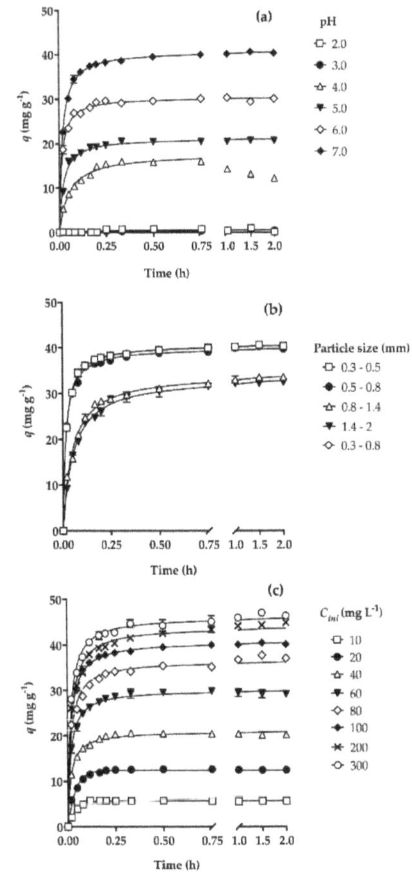

Figure 1. Capacity of biosorption of Co^{2+} by pretreated *Lemna gibba (PLEM)*: (**a**) At various pH values of the solution (C_{ini} = 100 mg L^{-1}, particle size = 0.3–0.5 mm), (**b**) with distinct particle sizes of *PLEM* (C_{ini} = 100 mg L^{-1}, pH = 7.0), and (**c**) at different initial concentrations of the metal (pH = 7.0, particle size = 0.3–0.8 mm). The continuous lines were predicted by the pseudo-second-order kinetic model.

The sorption capacity was enhanced with each increment in pH from 4 to 7, which can be easily explained by considering the pH of the plant material (1.67), which results in zero point of charge (ζ_0) [17]. When the pH of a solution is less than that found at ζ_0, the net charge of the surface of the biosorbent is positive. Hence, an electrostatic repulsion exists between the positive charge of both the metal ions and the surface of the biosorbent [28]. In contrast, when a solution has a pH value above that at ζ_0, the net charge of the surface of the biosorbent is negative, and there is an attraction with the positively charged metal ion [29]. A pH value of 5–7 herein afforded the fastest biosorption of Co^{2+} during the first 10 min (0.17 h) of the experiment. After this time, however, the velocity of removal of the metal decreased until reaching equilibrium, at which point the velocity of net transfer was 0. The initial rapid biosorption was due to the greater number of sites on *PLEM* available for the uptake of the sorbate and the higher concentration of Co^{2+} in the aqueous solution. As time passed, the available sites and the concentration of free cobalt ions were both diminished, leading to a gradual decline in the velocity of the removal of Co^{2+} until reaching the equilibrium dynamic. It was observed that as the pH increased, the biosorbent removed more Co^{2+}, and therefore, required more time to reach equilibrium

(t_{eq}). The same phenomenon has been reported for the effect of pH on the biosorption of other divalent metal ions [29].

A summary of the of Co^{2+} removal capacity at experimental equilibrium (q_{eq}), the time to reach equilibrium (t_{eq}), and the values of the parameters and error functions for each model and at each pH value assayed are provided in Table 2. None of the kinetic models employed fit the experimental results at pH 2 or 3, probably owing to the minimal biosorption of Co^{2+} under these conditions. At pH 4, a reduction in the removal capacity was only found after 0.75 h (Figure 1a), a time period not included in the process of biosorption. Hence, the corresponding data was not considered when determining the values of the parameters for the kinetic models. With a pH of 4–7, the pseudo-second-order model had the highest correlation coefficient (R^2) and the lowest values for *ASE, Sy.x*, and *AICc* compared to the other two models (pseudo-first-order and fractional power). The Elovich model was also evaluated, but is not listed in the tables because the R^2 was too small, and the parameters obtained had exaggerated *SD* values. Given that a pH of 7 produced the greatest biosorption capacity at equilibrium, this value was used for further testing.

Table 2. Kinetic parameters of the biosorption of Co^{2+} by *PLEM* at various pH values of the solution (C_{ini} = 100 mg L^{-1}, particle size = 0.3–0.5 mm).

Parameter	pH 4.0	pH 5.0	pH 6.0	pH 7.0
q_{eq} (mg g^{-1})	12.35 ± 0.09	17.85 ± 0.08	29.78 ± 0.18	40.13 ± 0.18
t_{eq} (h)	0.25	0.25	0.25	0.5
Pseudo-first-order				
q_{eq1} (mg g^{-1})	12.19 ± 0.15	17.51 ± 0.13	29.20 ± 0.22	38.80 ± 0.26
k_1 (h^{-1})	18.81 ± 0.98	25.03 ± 1.11	41.79 ± 2.32	34.96 ± 1.59
R^2	0.9740	0.9757	0.9699	0.9777
ASE	17.570	33.360	104.200	139.400
Sy.x	0.6468	0.7860	1.389	1.607
AICc	−33.79	−22.54	41.25	57.55
Pseudo-second-order				
q_{eq2} (mg g^{-1})	13.69 ± 0.22	18.61 ± 0.11	30.57 ± 0.16	40.86 ± 0.09
k_2 (g mg^{-1} h^{-1})	1.88 ± 0.17	2.28 ± 0.10	2.51 ± 0.13	1.50 ± 0.03
R^2	0.975	0.990	0.989	0.998
ASE	16.81	13.75	35.23	12.31
Sy.x	0.6327	0.5046	0.8077	0.4775
AICc	−35.73	−72.18	−19.50	−78.37
Fractional power				
k_{FP} (mg g^{-1})	15.03 ± 0.56	18.37 ± 0.33	30.61 ± 0.39	40.85 ± 0.52
v (h^{-1})	0.211 ± 0.02	0.107 ± 0.01	0.073 ± 0.007	0.086 ± 0.008
R^2	0.758	0.676	0.667	0.7293
ASE	68.31	133.90	190.00	342.90
Sy.x	1.341	1.637	1.949	2.619
AICc	55.70	80.49	73.88	104.6

3.2. The Effect of Particle Size

Particle size is a physical property that affects the surface area of contact between a sorbent and the liquid phase, thus playing a key role in biosorption [30,31]. When the particle size is reduced, the area of contact is amplified, and the sites of sorption are more accessible, generating a better capacity, efficiency, and velocity of biosorption and a decrease in the time to reach equilibrium (Figure 1b). The present results are in agreement with previous reports of an enhanced biosorption capacity as the particle size diminishes, considering particles from 0.3 to 2.0 mm (Table 3).

Table 3. Kinetic parameters of the biosorption of Co^{2+} by *PLEM*, using different particle sizes (C_{ini} = 100 mg L^{-1}, pH = 7.0).

Parameter	Particle Size (mm)				
	0.3–0.5	**0.5–0.8**	**0.8–1.4**	**1.4–2**	**0.3–0.8**
q_{eq} (mg g^{-1})	40.13 ± 0.18	39.77 ± 0.25	33.38 ± 0.17	32.25 ± 0.07	40.05 ± 0.16
t_{eq} (h)	0.5	0.75	1.0	1.0	0.5
			Pseudo-first-order		
q_{eq1} (mg g^{-1})	38.80 ± 0.26	38.01 ± 0.29	31.74 ± 0.32	30.91 ± 0.31	38.76 ± 0.25
k_1 (h^{-1})	34.96 ± 1.59	36.02 ± 1.98	13.66 ± 0.62	12.41 ± 0.52	35.08 ± 1.56
R^2	0.978	0.968	0.969	0.973	0.979
ASE	139.4	191.9	154.2	130.8	134.0
Sy.x	1.607	1.885	1.690	1.556	1.575
AICc	57.55	75.43	63.20	53.97	55.32
			Pseudo-second-order		
q_{eq2} (mg g^{-1})	40.86 ± 0.09	40.07 ± 0.10	34.45 ± 0.23	33.64 ± 0.19	40.80 ±0.09
k_2 (g mg^{-1} h^{-1})	1.50 ± 0.03	1.55 ± 0.04	0.61 ± 0.02	0.56 ± 0.02	1.52 ± 0.09
R^2	0.998	0.998	0.991	0.994	0.998
ASE	12.31	14.66	46.04	27.98	11.73
Sy.x	0.478	0.521	0.923	0.720	0.466
AICc	−78.37	−68.58	−4.504	−32.39	−81.09
			Fractional power		
k_{FP} (mg g^{-1})	40.85 ± 0.52	40.18 ± 0.45	33.06 ± 0.58	32.02 ± 0.59	40.78 ± 0.52
v (h^{-1})	0.09 ± 0.01	0.09 ± 0.01	0.17 ± 0.01	0.18 ± 0.01	0.08 ± 0.01
R^2	0.729	0.781	0.819	0.814	0.723
ASE	342.9	256.4	422.2	453.9	348.1
Sy.x	2.619	2.265	2.906	3.013	2.638
AICc	104.6	89.47	115.4	119.2	105.4

The biosorption of Co^{2+} was not significantly different ($p > 0.05$) between the size intervals of 0.3–0.5 mm and 0.5–0.8 mm. Therefore, a kinetic study was carried out to remove Co^{2+} by *PLEM* at a particle size of 0.3–0.8 mm. The statistical analysis with two-way ANOVA and Tukey's test indicated the lack of significant difference ($p > 0.05$) between the equilibrium biosorption capacity q_{eq} values of the samples with the following three particle sizes: 0.3–0.5, 0.5–0.8 mm, and 0.3–0.8 mm. The Co^{2+} biosorption rate was slightly faster (as expected) at the smaller particle size range (0.3–0.5 mm), reaching equilibrium at 0.5 h. The particle size range of 0.5–0.8 mm achieved equilibrium in a longer period of time (0.75 h), probably due to the greater surface area available with a smaller particle size, leading to faster binding of Co^{2+} ions to the surface of the biosorbent. With a particle size range of 0.3–0.8 mm, the time required to reach equilibrium (t_{eq}) of Co^{2+} biosorption by PLEM was 0.5 h, similar to the time found for the smallest particles tested (0.3–0.5 mm).

One advantage of employing a particle size of 0.3–0.8 mm is that it is possible to utilize fixed-bed columns packed with the material. Volesky [32] suggested using a particle size of 0.4–0.7 mm, since smaller sizes could obstruct the bed and provoke a drop in pressure. Additionally, particles of 0.3–0.8 mm (but not smaller) allow for the application of more biosorbent material. If the particle size range is under 0.3 mm, pretreatment is more difficult. Hence, a particle size of 0.3–0.8 mm was chosen for the rest of the experiments. The experimental results of the Co^{2+} removal capacity at equilibrium (q_{eq}) were compared to the parameters of the kinetic models assayed (Table 3). As can be appreciated, the equation of the pseudo-second-order model shows a higher correlation coefficient (R^2) and lower error functions (*ASE*, *Sy.x*, and *AICc*) than the other two models.

3.3. The Effect of the Initial Co^{2+} Concentration

The initial concentration of metallic ions is an important variable because it significantly affects the biosorption capacity and the time to reach equilibrium [33]. A boost in the initial concentration of the metal from 10 to 300 mg L^{-1} generated an 8.46-fold rise (from 5.46 to 46.17 mg g^{-1}) in the biosorption capacity at equilibrium (Figure 1c). Increasing the initial concentration of the sorbate, while maintaining the concentration of the biosorbent constant likely amplified the driving force behind sorption (the transfer of the cobalt ions from the aqueous solution to the surface of the biosorbent), a consequence of the higher gradient of concentration. Moreover, there is a greater probability of Co^{2+} binding to the active sites available in the sorbent, which would bring about a better biosorption capacity [34]. The experimental data on biosorption capacity at equilibrium (q_{eq}), the time required to reach equilibrium (t_{eq}), and the values of the parameters of the kinetic models and their corresponding error functions are listed in Table 4. Of the theoretical models applied to the data, the pseudo-second-order model gave the values closest to those found experimentally (as occurred with the other environmental variables) for the distinct initial concentrations of Co^{2+}.

The sorption velocity (k_2) is a kinetic parameter known to be related to the time to reach equilibrium, and therefore, depends on the initial concentration of the metal. The analysis of the kinetic parameters with two-way ANOVA and multiple comparisons by Tukey's test revealed a significant difference in relation to t_{eq} and k_2 between two initial concentrations of Co^{2+} (C_{ini}): 10 and 300 mg L^{-1}. The corresponding values for t_{eq} were 0.05 and 0.75 h, while those for k_2 were 6.847 and 1.402 g mg^{-1} h^{-1}, respectively (Table 4). Thus, an increase in the initial concentration of cobalt led to a decrease in k_2 and a longer time necessary to reach equilibrium, which is in agreement with previous reports on the biosorption of metallic ions [33,35].

Table 4. Kinetic parameters of the biosorption of Co^{2+} by *PLEM* at various initial concentrations of the metal (particle size = 0.3–0.8 mm, pH = 7.0).

Parameter	C_{ini} (mg L^{-1})							
	10	**20**	**40**	**60**	**80**	**100**	**200**	**300**
q_{eq} (mg g^{-1})	5.46 ± 0.16	12.18 ± 0.20	20.20 ± 0.13	29.22 ± 0.19	36.44 ± 0.48	40.05 ± 0.16	44.22 ± 0.31	46.17 ± 0.41
t_{eq} (h)	0.05	0.08	0.16	0.2	0.5	0.5	0.75	0.75
Pseudo-first-order								
q_{eq1} (mg g^{-1})	5.71 ± 0.03	12.44 ± 0.07	19.95 ± 0.14	28.64 ± 0.23	34.53 ± 0.41	38.76 ± 0.25	41.56 ± 0.46	43.81 ± 0.42
k_1 (h^{-1})	24.01 ± 0.85	25.00 ± 0.81	32.93 ± 1.58	34.58 ± 1.91	35.81 ± 2.94	35.08 ± 1.56	33.50 ± 2.53	37.68 ± 2.58
R^2	0.9853	0.9871	0.9741	0.9666	0.9288	0.9785	0.9371	0.9514
ASE	2.324	8.894	43.08	113.9	356.2	134.0	456.9	384.9
Sy.x	0.2075	0.4058	0.8932	1.452	2.568	1.575	2.909	2.670
AICc	−171.7	−96.58	−8.221	46.20	110.1	55.32	124.0	114.4
Pseudo-second-order								
q_{eq2} (mg g^{-1})	6.05 ± 0.09	13.15 ± 0.09	20.97 ± 0.10	30.13 ± 0.15	36.52 ± 0.27	40.80 ± 0.09	44.03 ± 0.26	46.19 ± 0.23
k_2 (g mg^{-1} h^{-1})	6.85 ± 0.76	3.37 ± 0.20	2.85 ± 0.13	2.06 ± 0.09	1.65 ± 0.11	1.52 ± 0.03	1.27 ± 0.07	1.40 ± 0.07
R^2	0.9456	0.9830	0.9915	0.9915	0.9804	0.9981	0.9882	0.9912
ASE	8.617	11.74	14.17	29.02	98.25	11.73	85.78	69.72
Sy.x	0.3995	0.4663	0.5122	0.7331	1.349	0.4660	1.260	1.136
AICc	−98.35	−81.02	−70.50	−30.34	37.94	−81.09	30.34	18.74
Fractional power								
k_{FP} (mg g^{-1})	5.89 ± 0.16	12.92 ± 0.26	20.84 ± 0.32	30.05 ± 0.42	36.85 ± 0.36	40.78 ± 0.53	44.33 ± 0.41	46.49 ± 0.46
v (h^{-1})	0.099 ± 0.02	0.099 ± 0.01	0.083 ± 0.009	0.083 ± 0.008	0.095 ± 0.005	0.085 ± 0.007	0.098 ± 0.005	0.088 ± 0.006
R^2	0.4581	0.5935	0.6402	0.6845	0.8463	0.7226	0.8675	0.8184
ASE	31.46	84.08	128.5	218.8	157.8	348.1	206.0	269.9
Sy.x	0.7932	1.297	1.603	2.092	1.777	2.638	2.030	2.323
AICc	−19.63	31.48	53.53	81.21	64.23	105.4	78.08	92.13

3.4. Biosorption Isotherm Studies at Various Temperatures

To understand the sorbate-sorbent interaction, it is crucial to assess the isotherm of biosorption and model it at several temperatures. This approach also allows for the prediction of the maximum biosorption capacity of the sorbent (q_m) and consequently a comparison of distinct sorbents (a prerequisite for the design of an adsorption system) [36,37]. Biosorption at equilibrium was established by examining the variation of the biosorption capacity at equilibrium (q_{eq}) with respect to the concentration of the sorbent at equilibrium (C_{eq}). The relation between the experimental isotherms and those predicted by the theoretical models for the biosorption of Co^{2+} by *PLEM* at different temperatures is shown in Figure 2.

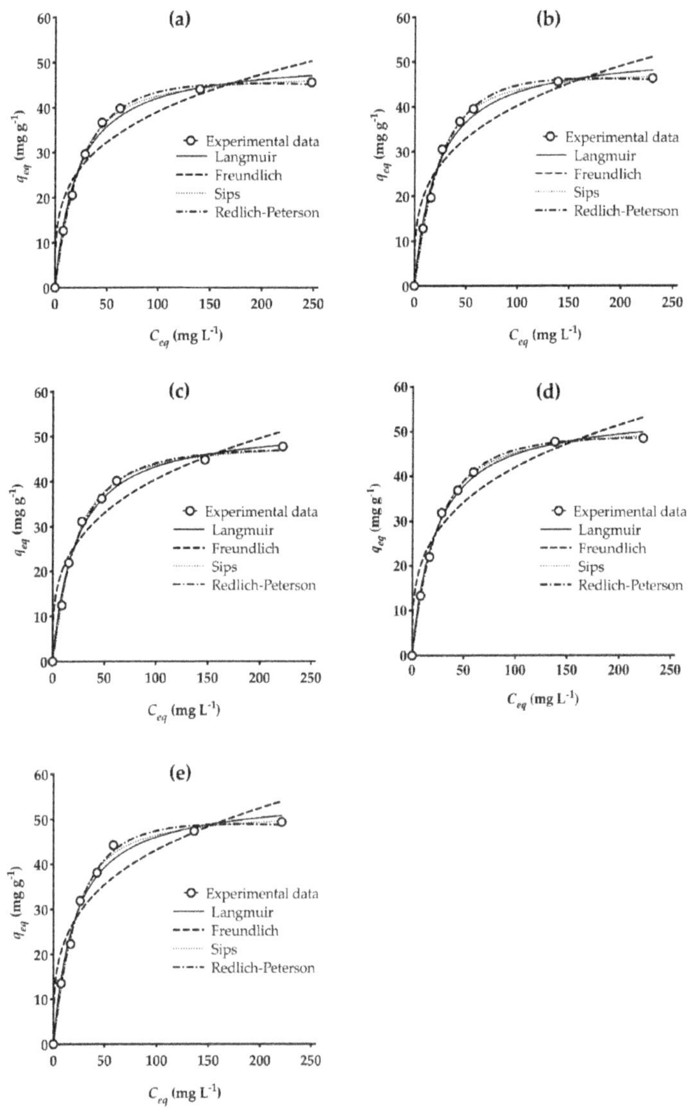

Figure 2. Isotherms for the adsorption of Co^{2+} by *PLEM* at the following temperatures: (**a**) 18 °C, (**b**) 30 °C, (**c**) 40 °C, (**d**) 50 °C, and (**e**) 60 °C (pH = 7.0, particle size = 0.3–0.8 mm).

The maximum experimental sorption capacity ($q_{m\,exp}$) was determined at each temperature, as were the values of the other parameters and the error functions (R^2, *ASE*, *Sy.x*, and *AICc*) for the models of isotherms (Table 5).

Table 5. Isotherms for the biosorption of Co^{2+} by *PLEM*.

Parameter	Temperature (°C)				
	18	30	40	50	60
$q_{m\,exp}$ (mg g^{-1})	46.17 ± 0.41	46.33 ± 0.19	47.73 ± 0.12	48.49 ± 0.21	49.35 ± 0.22
			Langmuir		
b_L (L mg^{-1})	0.047 ± 0.005	0.046 ± 0.005	0.046 ± 0.004	0.045 ± 0.003	0.048 ± 0.006
q_{mL} (mg g^{-1})	51.12 ± 1.49	52.66 ± 1.82	52.78 ± 1.48	54.97 ± 1.27	55.57 ± 1.91
R_L	0.508 – 0.068	0.509 – 0.073	0.508 – 0.075	0.513 – 0.076	0.498 – 0.071
R^2	0.993	0.990	0.993	0.996	0.990
ASE	13.65	19.18	12.9	8.856	21.46
Sy.x	1.508	1.788	1.466	1.215	1.891
AICc	16.28	18.99	15.82	12.81	19.89
			Freundlich		
k_F (mg g^{-1} (L g^{-1})$^{1/nF}$)	11.05 ± 2.65	10.82 ± 2.66	10.6 ± 2.39	10.94 ± 2.43	11.71 ± 2.80
n_F	3.64 ± 0.70	3.504 ± 0.67	3.433 ± 0.59	3.422 ± 0.58	3.53 ± 0.67
R^2	0.924	0.922	0.938	0.939	0.926
ASE	140.7	150.4	120.7	126	161
Sy.x	4.842	5.006	4.484	4.582	5.179
AICc	34.94	35.47	33.71	34.05	36.01
			Sips		
k_{SP} (L g^{-1})	0.022 ± 0.005	0.018 ± 0.005	0.022 ± 0.005	0.026 ± 0.004	0.023 ± 0.008
q_{mSP} (mg g^{-1})	47.55 ± 1.026	48.25 ± 1.065	48.79 ± 1.122	51.57 ± 0.975	51.55 ± 1.677
n_{SP}	1.295 ± 0.086	1.367 ± 0.095	1.294 ± 0.092	1.224 ± 0.066	1.304 ± 0.131
R^2	0.998	0.998	0.998	0.999	0.996
ASE	3.713	4.38	4.055	2.504	9.674
Sy.x	0.8617	0.936	0.900	0.7077	1.391
AICc	15.19	16.51	15.9	12.04	22.85
			Redlich-Peterson		
k_{RP} (L g^{-1})	1.822 ± 0.096	1.789 ± 0.126	2.006 ± 0.222	1.964 ± 0.099	1.996 ± 0.178
a_{RP} (L mg^{-1})^{bRP}	0.017 ± 0.003	0.014 ± 0.004	0.023 ± 0.009	0.019 ± 0.004	0.015 ± 0.006
b_{RP}	1.142 ± 0.029	1.171 ± 0.044	1.095 ± 0.057	1.116 ± 0.028	1.162 ± 0.054
R^2	0.999	0.998	0.996	0.999	0.997
ASE	2.111	4.197	7.774	1.824	7.069
Sy.x	0.6498	0.9162	1.247	0.604	1.189
AICc	10.68	16.17	21.1	9.507	20.34

Regarding the isotherm models of two parameters, the Langmuir model afforded the best correlation coefficient ($R^2 > 0.99$) and the smallest error functions. The value of the separation factor (R_L) reflects the nature of biosorption, which is considered unfavorable with $R_L \geq 1$, favorable with $0 < R_L < 1$, an irreversible if $R_L = 0$ [38]. The values of R_L calculated presently indicate that biosorption is favorable ($0.07 < R_L < 0.5$) at all temperatures assayed.

On the other hand, each of the models of three parameters (Sips and Redlich-Peterson) provided a higher correlation coefficient ($R^2 > 0.996$) and lower error functions than the models of two parameters. Overall, the Redlich-Peterson model gave the lowest error functions. The values of maximum biosorption capacity predicted by the isotherm of Sips (q_{mSP} = 47.55 to 51.55 mg g^{-1}) at the five temperatures herein employed were very close to the experimental data ($q_{m\,exp}$ = 46.17 to 49.35 mg g^{-1}). Compared to the capacity for the biosorption of Co^{2+} previously reported for diverse biosorbents, the value found in the current study reveals an excellent capacity for *PLEM* (Table 6). Thus, it is an attractive biosorbent for the detoxification of water contaminated with Co^{2+}.

Table 6. Capacity for the biosorption of Co^{2+} by different materials.

Material	Biosorption Capacity (mg g^{-1})	pH	Reference
Cocos nucifera leaf	3.69	5	[39]
Spent coffee	5.37	6	[40]
Rhytidiadelphus squarrosus	7.38	6	[41]
Natural hemp fibers	13.58	4.5–5	[26]
Activated carbon from hazelnut shells	13.88	6	[42]
Alginate from *Callithamnion corymbosum* sp.	18.79	4.4	[43]
Sargassum wightii	20.63	4.5	[44]
Carbonized lemon peel	22.00	6	[45]
Watermelon rind	23.30	5	[46]
Prunus dulcis bio-char	27.86	7	[36]
Teak leaves	29.48	5	[47]
$MgCl_2$-pretreated *Ficus carica* leaves	33.9	6	[48]
Acacia nilotica	35.45	5	[37]
NaOH-treated lemon peels	35.71	6	[49]
Almond green hull	45.50	ND	[50]
PLEM	46.47	7	The present study
Corn silk	82.04	6	[25]
NaOH-pretreated *Mangifera indica* leaves	114	5	[51]

ND, no data.

3.5. Thermodynamic Parameters

Graphs were constructed to find the thermodynamic parameters, ΔG^0 (Figure 3a), ΔH^0, and ΔS^0 (Figure 3b), and the corresponding values were determined (Table 7).

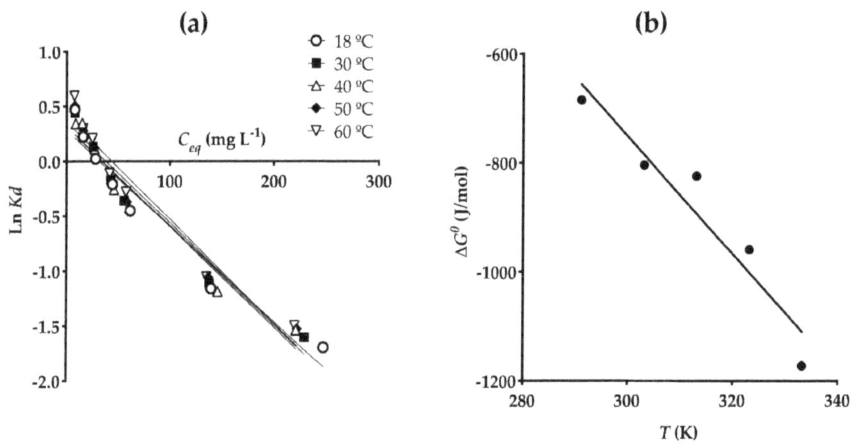

Figure 3. Graphs based on the values of (**a**) Ln K_d vs. C_{eq} and (**b**) ΔG^0 vs. T, which were used to calculate the thermodynamic parameters of biosorption of Co^{2+} by *PLEM*.

Table 7. Thermodynamic parameters of biosorption of Co^{2+} by *PLEM*.

Temperature °C	ΔG^0 (J mol^{-1})	ΔH^0 (J mol^{-1})	ΔS^0 (J mol^{-1} K^{-1})
18	−684.8	2461.3	10.81 ± 1.89
30	−804.3	2471.6	
40	−825.0	2558.8	
50	−959.4	2532.6	
60	−1172.1	2427.9	

The Gibbs free energy (ΔG^0) values are negative for the biosorption of Co^{2+} by *PLEM* (Table 7), suggesting a spontaneous process. The biosorption has been reported to improve as the temperature rises [22]. The positive values of ΔH^0 show an endothermic biosorption, which is consistent with the enhanced biosorption capacity ($q_{m\,exp}$) presently found at higher temperatures (Table 5). The change in the mean calculated standard enthalpy was $\Delta H^0{}_{prom}$ = 2.49 KJ mol^{-1}. A value below 40 kJ mol^{-1} is indicative of a process of physisorption [21]. The positive value of standard entropy (ΔS^0) reveals a high affinity of Co^{2+} for *PLEM* [22], and thus, the probability that the metal promotes structural changes in the biosorbent. Hence, the process of biosorption likely increases the degree of disorder of the whole system [25,52]. According to the values of the thermodynamic parameters, adsorption of Co^{2+} by *PLEM* is spontaneous and favorable, allowing this material to be utilized for the removal of metal from polluted water.

3.6. Desorption

The elution of Co^{2+} after its sorption by *PLEM* was tested with various acids and bases (Figure 4). Overall, the strong acids (HCl, HNO_3, and H_2SO_4) were the best eluent solutions, giving superior desorption percentages (>94%) compared to the weak acids (<65%) or alkaline compounds (<20%). Water, whether at rt or 60 °C, was not capable of eluting more than 10% of Co^{2+}.

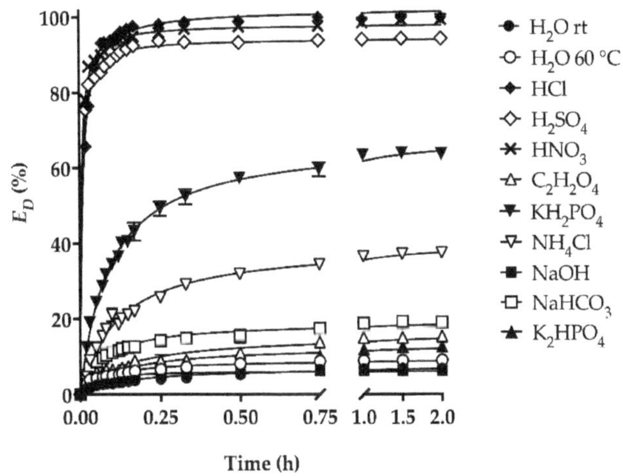

Figure 4. Kinetics of desorption of Co^{2+} from *PLEM* with distinct eluent solutions.

Thus, the biosorbent was positively charged at the pH of acid solutions, resulting in an electrostatic repulsion with the sorbate [53]. Accordingly, physisorption seems to play a key role in the process of biosorption of Co^{2+} by *PLEM*. On the other hand, a high concentration of H^+ ions in the acid solutions could cause competition with Co^{2+} for these sorption sites, favoring ionic interchange, and consequently, the desorption process [54]. Since 0.1 M HCl was the eluent with the greatest percentage of desorption (100%), the biosorbent was eluted with this solution in posterior assays.

The effect of pH on the biosorption/desorption of Co^{2+} suggests that the main biosorption mechanism is electrostatic attraction, a physical process between negatively charged groups of the biosorbent and the positive charge of Co^{2+}. The thermodynamic value of $\Delta H^0{}_{prom}$ (2.49 KJ mol^{-1}) indicates a physisorption process, which reinforces the idea of electrostatic attraction being the principal mechanism of biosorption.

3.7. Biosorption-Desorption Cycles

Considering the indispensable requirement of recyclability for the practical application of a biosorbent, an evaluation of the cycles of biosorption/desorption is necessary to assure that the material can be regenerated in a cost-effective manner [23]. Additionally, insights are provided as to the best way to dispose of the biosorbent once it is no longer useful. Few such studies have been reported for the biosorption/desorption of Co^{2+} [9,43].

The biosorption capacity of *PLEM* in the first cycle (46.17 ± 0.41 mg g^{-1}) was diminished 8.53% in the second cycle and a cumulative 17.89% by the end of the third cycle (Figure 5a), representing significant differences. Hence, the eluent herein employed (0.1 M HCl) could have damaged the composition and structure of the biosorbent, affecting the sorption sites and reducing the capacity of Co^{2+} removal from one cycle to the next [55]. However, *PLEM* maintained an elevated capacity of Co^{2+} removal throughout the three cycles. During all three cycles, moreover, Co^{2+} was completely desorbed (E_D = 100%) from the biosorbent (Figure 5b), evidencing its recyclability. After the end of its useful life, *PLEM* can be integrated into compost with null impact on the environment because of not containing any Co^{2+}.

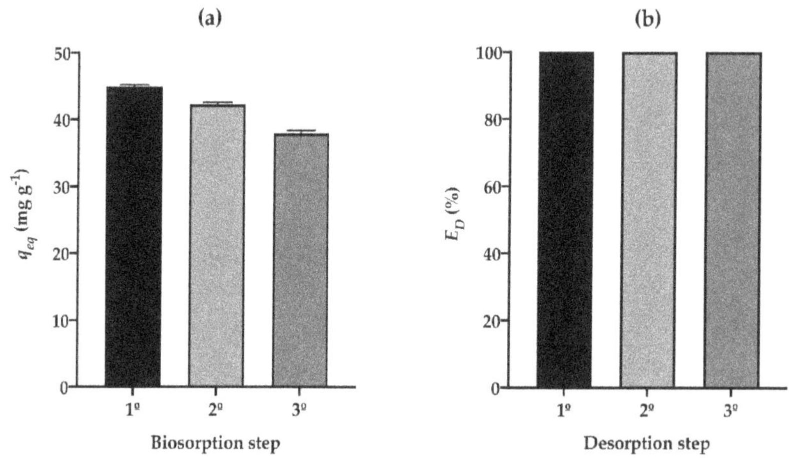

Figure 5. (**a**) Capacity of biosorption and (**b**) the percentage of desorption of Co^{2+} by *PLEM* during three cycles of biosorption/desorption.

3.8. Scanning Electron Microscopy Coupled with Energy-Dispersive X-ray Spectroscopy (SEM-EDX)

The SEM-EDX analysis of *PLEM* before exposure to Co^{2+} (Figure 6a) reveals a course and porous surface with agglomerations of the biosorbent. Hence, the surface is characterized by an ample exposure of the active sites for the capture of Co^{2+}. The EDX spectra of *PLEM* evidences a surface free of Co^{2+}.

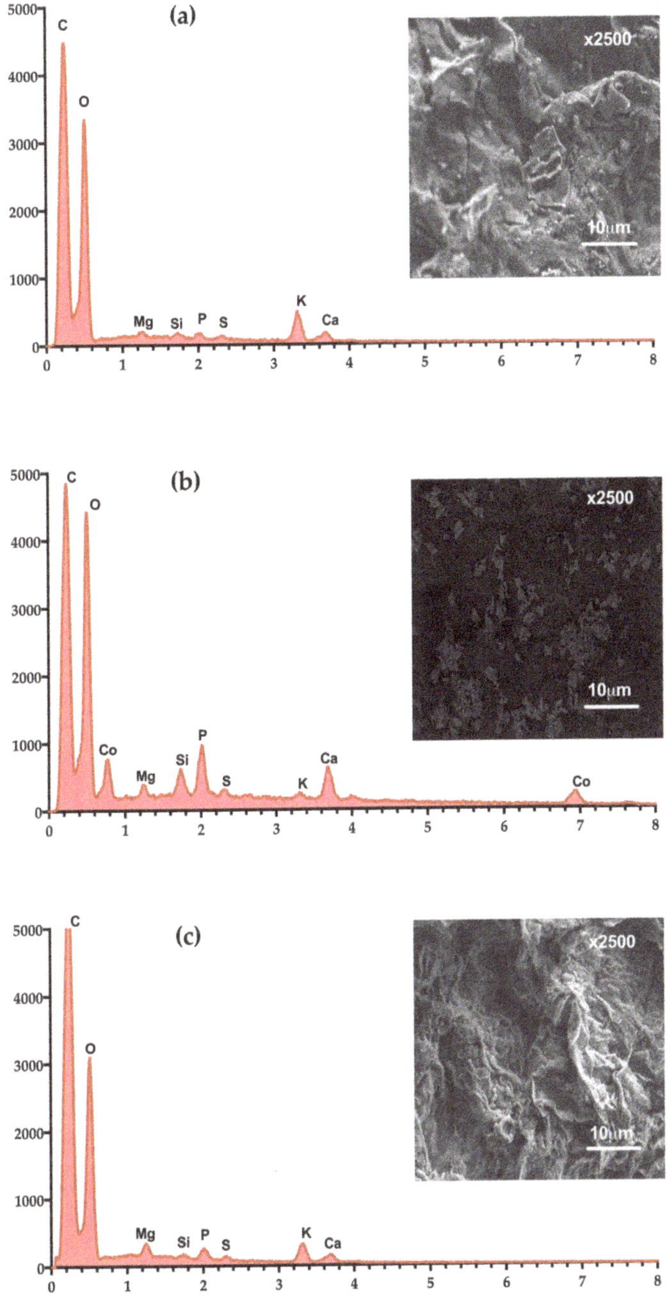

Figure 6. SEM and EDX micrograph of *PLEM* during the first biosorption/desorption cycle: (**a**) Before exposure to Co^{2+}, (**b**) saturated with Co^{2+}, and (**c**) subsequent to desorption of the biosorbent with 0.1 M HCl.

The micrograph of *PLEM* saturated with Co^{2+} (Figure 6b) shows a more homogenous surface (on which rectangular particles are dispersed) than *PLEM* prior to contact with Co^{2+}. The following desorption with 0.1 M HCl (Figure 6c), the appearance of the surface of *PLEM* is similar to that observed before exposure to Co^{2+}. In the EDX spectrum, two peaks corresponding to Co^{2+} indicate its presence after the biosorption step (Figure 6b). The absence of such peaks after desorption (Figure 6c) evidenced the adequacy of the eluent solution for the total recovery of the metallic ion. Consequently, HCl was able to regenerate the biosorbent for posterior cycles of biosorption/desorption.

4. Conclusions

The results demonstrate that *PLEM* is an attractive, economical, sustainable, and environmentally friendly material for the removal of Co^{2+} from aqueous solutions. The capacity of biosorption of Co^{2+} by *PLEM* was enhanced by smaller particle size, a greater pH of the solution, and a higher initial concentration of the metal. The main mechanism of removal of Co^{2+} from the aqueous solution is physisorption based on electrostatic attraction. While the kinetics of the experimental biosorption data were adequately described by the pseudo-second-order model, the isotherms of biosorption at equilibrium at different temperatures were properly predicted by the Sips and Redlich-Peterson models. According to the thermodynamic study, the biosorption of Co^{2+} by *PLEM* is an endothermic and spontaneous process. The best eluent solution for the recovery of both the metal and the biosorbent material was 0.1 M HCl. *PLEM* can be used for at least three cycles of biosorption/desorption, with a high capacity of biosorption and complete desorption in each cycle, revealing the recyclability of the material, and therefore, the possibility of its economical use. The SEM-EDX analysis confirmed the biosorption of Co^{2+} by *PLEM* and the posterior desorption of the plant material by means of its exposure to 0.1 M HCl.

Author Contributions: Conceptualization, L.M.-B. and E.C.-U.; methodology, J.L.R.-L., L.M.-B. and E.C.-U.; software, J.L.R.-L.; validation, J.L.R.-L. and L.M.-B.; formal analysis, L.M.-B. and E.C.-U.; investigation, J.L.R.-L.; resources, L.M.-B. and E.C.-U.; writing—original draft preparation, review and editing, L.M.-B. and E.C.-U.; visualization, L.M.-B.; supervision, L.M.-B. and E.C.-U.; project administration, L.M.-B. and E.C.-U.; funding acquisition, L.M.-B. and E.C.-U. All authors have read and agreed to the published version of the manuscript.

Funding: This research was funded by the Instituto Politécnico Nacional, Secretaría de Investigación y Posgrado, project number: 20201814.

Acknowledgments: The authors are grateful for the technical support provided by the Centro de Nanociencias y Micro y Nanotecnologías, IPN. The CONACyT awarded a graduate scholarship to the coauthor J.L.R.-L., L.M.-B. and E.C.-U. hold grants from EDI-IPN, COFAA-IPN, and SNI-CONACYT. The authors thank Bruce Allan Larsen for proofreading the manuscript.

Conflicts of Interest: The authors declare that they have no conflict of interest.

References

1. Vilvanathan, S.; Shanthakumar, S. Modeling of fixed-bed column studies for removal of cobalt ions from aqueous solution using *Chrysanthemum indicum*. *Res. Chem. Intermed.* **2017**, *43*, 229–243. [CrossRef]
2. World Health Organization (WHO). Cobalt and Inorganic Cobalt Compounds. In *Concise International Chemical Assessment Document 69*; Kim, J.H., Gibb, H.J., Howe, P.D., Eds.; WHO: Geneva, Switzerland, 2006; pp. 1–84. Available online: https://apps.who.int/iris/handle/10665/43426 (accessed on 3 August 2020).
3. He, E.-K.; Baas, J.; Van Gestel, C.A. Interaction between nickel and cobalt toxicity in *Enchytraeus crypticusis* due to competitive uptake. *Environ. Toxicol. Chem.* **2015**, *34*, 328–337. [CrossRef] [PubMed]
4. Oguz, E.; Ersoy, M. Biosorption of cobalt(II) with sunflower biomass from aqueous solutions in a fixed bed column and neural networks modelling. *Ecotoxicol. Environ. Saf.* **2014**, *99*, 54–60. [CrossRef] [PubMed]
5. Anirudhan, T.; Deepa, J.; Christa, J. Nanocellulose/nanobentonite composite anchored with multi-carboxyl functional groups as an adsorbent for the effective removal of Cobalt(II) from nuclear industry wastewater samples. *J. Colloid Interface Sci.* **2016**, *467*, 307–320. [CrossRef]

6. Fawzy, M.A.; Hifney, A.F.; Adam, M.S.; Al-Badaani, A.A. Biosorption of cobalt and its effect on growth and metabolites of *Synechocystis pevalekii* and *Scenedesmus bernardii*: Isothermal analysis. *Environ. Technol. Innov.* **2020**, *19*, 100953. [CrossRef]
7. Pipíška, M.; Trajteľová, Z.; Horník, M. Compartmentalization of Co and Mn in live cells of *Escherichia coli*: Investigation using 60Co and 54Mn as radioindicators. *J. Radioanal. Nucl. Chem.* **2017**, *314*, 1197–1205. [CrossRef]
8. Pipíška, M.; Richveisová, B.M.; Frišták, V.; Horník, M.; Remenárová, L.; Stiller, R.; Soja, G. Sorption separation of cobalt and cadmium by straw-derived biochar: A radiometric study. *J. Radioanal. Nucl. Chem.* **2017**, *311*, 85–97. [CrossRef]
9. Pipíška, M.; Zarodňanská, S.; Horník, M.; Ďuriška, L.; Holub, M.; Safarik, I. Magnetically Functionalized Moss Biomass as Biosorbent for Efficient Co^{2+} Ions and Thioflavin T Removal. *Materials* **2020**, *13*, 3619. [CrossRef] [PubMed]
10. Robalds, A.; Naja, G.M.; Klavins, M. Highlighting inconsistencies regarding metal biosorption. *J. Hazard. Mater.* **2016**, *304*, 553–556. [CrossRef]
11. Saman, N.; Tan, J.-W.; Mohtar, S.S.; Kong, H.; Lye, J.W.P.; Mat, H.; Hassan, H.; Mat, H. Selective biosorption of aurum(III) from aqueous solution using oil palm trunk (OPT) biosorbents: Equilibrium, kinetic and mechanism analyses. *Biochem. Eng. J.* **2018**, *136*, 78–87. [CrossRef]
12. Moradi, P.; Hayati, S.; Ghahrizadeh, T. Modeling and optimization of lead and cobalt biosorption from water with Rafsanjan pistachio shell, using experiment based models of ANN and GP, and the grey wolf optimizer. *Chemom. Intell. Lab. Syst.* **2020**, *202*, 104041. [CrossRef]
13. Muller da Cunha, J.; Klein, L.; Moro Bassaco, M.; Hiromitsu Tanabe, E.; Bertuol, D.A.; Dotto, G.L. Cobalt recovery from leached solutions of lithium-ion batteries using waste materials as adsorbents. *Can. J. Chem. Eng.* **2015**, *93*, 2198–2204. [CrossRef]
14. Greenberg, B.M.; Huang, X.-D.; Dixon, D.G. Applications of the aquatic higher plant Lemna gibba for ecotoxicological assessment. *J. Aquat. Ecosyst. Health* **1992**, *1*, 147–155. [CrossRef]
15. Arroyave, M.D. La lenteja de agua (*Lemna minor* L.): Una planta acuática promisoria. *Rev. EIA* **2004**, *1*, 33–38. Available online: http://www.scielo.org.co/scielo.php?script=sci_arttext&pid=S1794-12372004000100004&lng=en&tlng=es (accessed on 9 November 2020).
16. Canales-Gutiérrez, Á. Evaluación de la biomasa y manejo de *Lemna gibba* (Lenteja de agua) en la bahía interior del lago titicaca, puno. *Ecol. Apl.* **2010**, *9*, 91–99. [CrossRef]
17. Reyes-Ledezma, J.L. Biosorción de Co(II) a Partir de Soluciones Acuosas por *Lemna* sp. Master's Thesis, Instituto Politécnico Nacional, Mexico City, Mexico, 2017.
18. Reyes-Ledezma, J.L.; Uribe-Ramírez, D.; Cristiani-Urbina, E.; Morales-Barrera, L. Biosorptive removal of acid orange 74 dye by HCl-pretreated *Lemna* sp. *PLoS ONE* **2020**, *15*, e0228595. [CrossRef]
19. Morales-Barrera, L.; Flores-Ortiz, C.M.; Cristiani-Urbina, E. Single and Binary Equilibrium Studies for Ni^{2+} and Zn^{2+} Biosorption onto *Lemna Gibba* from Aqueous Solutions. *Processes* **2020**, *8*, 1089. [CrossRef]
20. Gupta, S.; Sharma, S.; Kumar, A. Biosorption of Ni(II) ions from aqueous solution using modified *Aloe barbadensis* Miller leaf powder. *Water Sci. Eng.* **2019**, *12*, 27–36. [CrossRef]
21. Guo, Z.; Liu, X.; Huang, H. Kinetics and Thermodynamics of Reserpine Adsorption onto Strong Acidic Cationic Exchange Fiber. *PLoS ONE* **2015**, *10*, e0138619. [CrossRef]
22. Zhao, Y.; Zhao, X.; Deng, J.; He, C. Utilization of chitosan–clinoptilolite composite for the removal of radiocobalt from aqueous solution: Kinetics and thermodynamics. *J. Radioanal. Nucl. Chem.* **2016**, *308*, 701–709. [CrossRef]
23. Ramírez-Rodríguez, A.E.; Reyes-Ledezma, J.L.; Chávez-Camarillo, G.M.; Cristiani-Urbina, E.; Morales-Barrera, L. Cyclic biosorption and desorption of acid red 27 onto *Eichhornia crassipes* leaves. *Rev. Mex. Ing. Química* **2018**, *17*, 1121–1134. [CrossRef]
24. Gramlich, A.; Moradi, A.B.; Robinson, B.H.; Kaestner, A.; Schulin, R. Dimethylglyoxime (DMG) staining for semi-quantitative mapping of Ni in plant tissue. *Environ. Exp. Bot.* **2011**, *71*, 232–240. [CrossRef]
25. Yu, H.; Pang, J.; Ai, T.; Liu, L. Biosorption of Cu^{2+}, Co^{2+} and Ni^{2+} from aqueous solution by modified corn silk: Equilibrium, kinetics, and thermodynamic studies. *J. Taiwan Inst. Chem. Eng.* **2016**, *62*, 21–30. [CrossRef]
26. Tofan, L.; Teodosiu, C.; Paduraru, C.; Wenkert, R. Cobalt (II) removal from aqueous solutions by natural hemp fibers: Batch and fixed-bed column studies. *Appl. Surf. Sci.* **2013**, *285 Pt A*, 33–39. [CrossRef]

27. Lee, M.-Y.; Hong, K.-J.; Kajiuchi, T.; Yang, J.-W. Determination of the efficiency and removal mechanism of cobalt by crab shell particles. *J. Chem. Technol. Biotechnol.* **2004**, *79*, 1388–1394. [CrossRef]
28. Xu, H.; Liu, Y.; Tay, J.-H. Effect of pH on nickel biosorption by aerobic granular sludge. *Bioresour. Technol.* **2006**, *97*, 359–363. [CrossRef]
29. Suazo-Madrid, A.; Morales-Barrera, L.; Aranda-García, E.; Cristiani-Urbina, E. Nickel(II) biosorption by *Rhodotorula glutinis*. *J. Ind. Microbiol. Biotechnol.* **2011**, *38*, 51–64. [CrossRef]
30. Abbas, M.; Kaddour, S.; Trari, M. Kinetic and equilibrium studies of cobalt adsorption on apricot stone activated carbon. *J. Ind. Eng. Chem.* **2014**, *20*, 745–751. [CrossRef]
31. Vijayaraghavan, K.; Yun, Y.-S. Bacterial biosorbents and biosorption. *Biotechnol. Adv.* **2008**, *26*, 266–291. [CrossRef]
32. Volesky, B. Detoxification of metal-bearing effluents: Biosorption for the next century. *Hydrometallurgy* **2001**, *59*, 203–216. [CrossRef]
33. Bulgariu, D.; Bulgariu, L. Potential use of alkaline treated algae waste biomass as sustainable biosorbent for clean recovery of cadmium(II) from aqueous media: Batch and column studies. *J. Clean. Prod.* **2016**, *112*, 4525–4533. [CrossRef]
34. Mishra, A.; Tripathi, B.D.; Rai, A.K. Packed-bed column biosorption of chromium(VI) and nickel(II) onto Fenton modified Hydrilla verticillata dried biomass. *Ecotoxicol. Environ. Saf.* **2016**, *132*, 420–428. [CrossRef] [PubMed]
35. Vilvanathan, S.; Shanthakumar, S. Biosorption of Co(II) ions from aqueous solution using *Chrysanthemum indicum*: Kinetics, equilibrium and thermodynamics. *Process. Saf. Environ. Prot.* **2015**, *96*, 98–110. [CrossRef]
36. Kılıç, M.; Kırbıyık, Ç.; Çepelioğullar, Ö.; Putun, A.E. Adsorption of heavy metal ions from aqueous solutions by bio-char, a by-product of pyrolysis. *Appl. Surf. Sci.* **2013**, *283*, 856–862. [CrossRef]
37. Thilagavathy, P.; Santhi, T. Kinetics, Isotherms and Equilibrium Study of Co(II) Adsorption from Single and Binary Aqueous Solutions by Acacia nilotica Leaf Carbon. *Chin. J. Chem. Eng.* **2014**, *22*, 1193–1198. [CrossRef]
38. Rangabhashiyam, S.; Anu, N.; Nandagopal, M.S.G.; Selvaraju, N. Relevance of isotherm models in biosorption of pollutants by agricultural byproducts. *J. Environ. Chem. Eng.* **2014**, *2*, 398–414. [CrossRef]
39. Hymavathi, D.; Prabhakar, G. Optimization, equilibrium, and kinetic studies of adsorptive removal of cobalt(II) from aqueous solutions using *Cocos nucifera* L. *Chem. Eng. Commun.* **2017**, *204*, 1094–1104. [CrossRef]
40. Imessaoudene, D.; Hanini, S.; Bouzidi, A.; Ararem, A. Kinetic and thermodynamic study of cobalt adsorption by spent coffee. *Desalin. Water Treat.* **2015**, *57*, 6116–6123. [CrossRef]
41. Marešová, J.; Pipíška, M.; Rozložník, M.; Horník, M.; Remenárová, L.; Augustín, J. Cobalt and strontium sorption by moss biosorbent: Modeling of single and binary metal systems. *Desalination* **2011**, *266*, 134–141. [CrossRef]
42. Demirbaş, E. Adsorption of Cobalt(II) Ions from Aqueous Solution onto Activated Carbon Prepared from Hazelnut Shells. *Adsorpt. Sci. Technol.* **2003**, *21*, 951–963. [CrossRef]
43. Lucaci, A.R.; Bulgariu, D.; Ahmad, I.; Bulgariu, L. Equilibrium and Kinetics Studies of Metal Ions Biosorption on Alginate Extracted from Marine Red Algae Biomass (*Callithamnion corymbosum* sp.). *Polymers* **2020**, *12*, 1888. [CrossRef] [PubMed]
44. Vijayaraghavan, K.; Jegan, J.; Palanivelu, K.; Velan, M. Biosorption of cobalt(II) and nickel(II) by seaweeds: Batch and column studies. *Sep. Purif. Technol.* **2005**, *44*, 53–59. [CrossRef]
45. Bhatnagar, A.; Minocha, A.; Sillanpää, M. Adsorptive removal of cobalt from aqueous solution by utilizing lemon peel as biosorbent. *Biochem. Eng. J.* **2010**, *48*, 181–186. [CrossRef]
46. Lakshmipathy, R.; Sarada, N. Application of watermelon rind as sorbent for removal of nickel and cobalt from aqueous solution. *Int. J. Miner. Process.* **2013**, *122*, 63–65. [CrossRef]
47. Vilvanathan, S.; Shanthakumar, S. Removal of Ni(II) and Co(II) ions from aqueous solution using teak (*Tectona grandis*) leaves powder: Adsorption kinetics, equilibrium and thermodynamics study. *Desalin. Water Treat.* **2016**, *57*, 3995–4007. [CrossRef]
48. Dabbagh, R.; Moghaddam, Z.A.; Ghafourian, H. Removal of cobalt(II) ion from water by adsorption using intact and modified *Ficus carica* leaves as low-cost natural sorbent. *Desalin. Water Treat.* **2016**, *57*, 19890–19902. [CrossRef]

49. Singh, S.A.; Shukla, S.R. Adsorptive removal of cobalt ions on raw and alkali-treated lemon peels. *Int. J. Environ. Sci. Technol.* **2016**, *13*, 165–178. [CrossRef]
50. Ahmadpour, A.; Tahmasbi, M.; Bastami, T.R.; Besharati, J.A. Rapid removal of cobalt ion from aqueous solutions by almond green hull. *J. Hazard. Mater.* **2009**, *166*, 925–930. [CrossRef]
51. Nadeem, R.; Zafar, M.N.; Afzal, A.; Hanif, M.A.; Saeed, R. Potential of NaOH pretreated *Mangifera indica* waste biomass for the mitigation of Ni(II) and Co(II) from aqueous solutions. *J. Taiwan Inst. Chem. Eng.* **2014**, *45*, 967–972. [CrossRef]
52. Bhattacharyya, K.G.; Gupta, S.S. Removal of Cu(II) by natural and acid-activated clays: An insight of adsorption isotherm, kinetic and thermodynamics. *Desalination* **2011**, *272*, 66–75. [CrossRef]
53. Ye, W.-M.; He, Y.; Chen, Y.-G.; Chen, B.; Cui, Y.J. Adsorption, Desorption and Competitive Adsorption of Heavy Metal Ions from Aqueous Solution onto GMZ01 Bentonite. In *Engineering Geology for Society and Territory*; Lollino, G., Giordan, D., Thuro, K., Carranza-Torres, C., Wu, F., Marinos, P., Delgado, C., Eds.; Springer International Publishing: Cham, Switzerland, 2015; Volume 6, pp. 533–536.
54. Zhang, X.; Wang, X. Adsorption and Desorption of Nickel(II) Ions from Aqueous Solution by a Lignocellulose/Montmorillonite Nanocomposite. *PLoS ONE* **2015**, *10*, e0117077. [CrossRef] [PubMed]
55. Ronda, A.; Calero, M.; Blázquez, G.; Pérez, A.; Martín-Lara, M.Á. Optimization of the use of a biosorbent to remove heavy metals: Regeneration and reuse of exhausted biosorbent. *J. Taiwan Inst. Chem. Eng.* **2015**, *51*, 109–118. [CrossRef]

processes

MDPI

Article

Cu(II) and As(V) Adsorption Kinetic Characteristic of the Multifunctional Amino Groups in Chitosan

Byungryul An

Department of Civil Engineering, Sangmyung University, Cheonan 31066, Korea; bran@smu.ac.kr; Tel.: +82-41-550-5497

Received: 29 July 2020; Accepted: 19 September 2020; Published: 21 September 2020

Abstract: Amino groups in the chitosan polymer play as a functional group for the removal of cations and anions depending on the degree of protonation, which is determined by the solution pH. A hydrogel beadlike porous adsorbent was used to investigate the functions and adsorption mechanism of the amino groups by removal of Cu(II) as a cation and As(V) as an anion for a single and mixed solution. The uptakes of Cu(II) and As(V) were 5.2 and 5.6 μmol/g for the single solution and 5.9 and 3.6 μmol/g for the mixed solution, respectively. The increased total capacity in the presence of both the cation and anion indicated that the amino group (NH_2 or NH_3^+) species was directly associated for adsorption. The application of a pseudo second-order (PSO) kinetic model was more suitable and resulted in an accurate correlation coefficient (R^2) compared with the pseudo first-order (PFO) kinetic model for all experimental conditions. Due to poor linearization of the PFO reaction model, we attempted to divide it into two sections to improve the accuracy. Regardless of the model equation, the order of the rate constant was in the order of As(V)-single > Cu(II)-single > As(V)-mixed > Cu(II)-mixed. Also, the corresponding single solution and As(V) showed a higher adsorption rate. According to intraparticle and film diffusion applications displaying two linear lines and none passing through zero, the rate controlling step in the chitosan hydrogel bead was determined by both intraparticle and film diffusion.

Keywords: adsorption; amino group; kinetic; multifunction; cation; anion

---

1. Introduction

The application of natural biopolymers, chitosan, has been reported for water and wastewater treatment as adsorbents owing to their natural abundance, nontoxicity, hydrophilicity, and biodegradability [1,2]. Chitosan has been utilized in various ways according to their required use, such as (nano)particles [3], beads [1], or membranes [4] in solution, and have demonstrated high effective removal efficiencies and uptakes for cations and anions. Similar to alginate, chitosan bead was formed by covalent crosslinking [5] and H-bonds interaction [6]. Chitosan exhibits the ability to remove cation heavy metals [7]. Moreover, chitosan is useful for removing anions, such as As(V), phosphate [8], dyes [9], and toluene [10]. This is because of the unique characteristic linked to the amino group. The removal by the amino group can be explained by Equation (1) for cations and Equation (2) for anions. The amino functional groups exist as NH_3^+ and NH_2 via protonation and deprotonation, respectively, and these bind with an anion via electrostatic forces and with a cation heavy metal via surface complexation [11].

$$\text{R-NH}_2 + \text{Cu(II)} \rightarrow \text{R-NH}_2\text{Cu(II) by coordination} \quad (1)$$

$$\text{R-NH}_3^+ + \text{HAsO}_4^{2-} \rightarrow \text{R-NH}_3^+\text{HAsO}_4^{2-} \text{ by electrostatic force} \quad (2)$$

The adsorption process in water and wastewater treatment has been recently regarded as one of the most effective technologies for pollutant removal from gas or liquid to solid media, owing to the flexible design and easy operation with high removal efficiency [12]. However, because of the presence of cation and anion pollutants in solution, each process was separately installed and operated as a dual or multiple column in the full process to remove cations and anions, respectively. For example, Cr(VI) and Cr(III) have been treated by anion exchange and cation exchange [13]. Otherwise, chemical modification involves preparing a multifunctional adsorbent by grafting [14], coating to inorganic or organic materials, or both [12,15]. Therefore, interest in the simultaneous removal of cations and anions has increased. The degree of protonation is defined by the pK_a value, which usually ranges from 6.0 to 7.0 [11]. The preparation of a granular type is possible with a porous adsorbent, which leads to the application of an adsorption process [16]. Moreover, the chemical stability (insoluble chitosan) under acidic conditions is increased by the crosslinking reaction, resulting in an expansion of its use in the field of water and wastewater treatment [17].

Generally, for porous adsorbents, like granular activated carbon (GAC), the adsorption process occurs through four consecutive stages [18]. The migration of the molecular adsorbate is divided into (1) transferring from the bulk solution to a layer of thin film (bulk diffusion), which is excluded when three stages are described; (2) solute transport by diffusion through a liquid boundary layer to the external surface of the adsorbent; (3) penetration from the external (sorbent) surface of the sorbent to the intraparticle pores; and (4) adsorption on both the external and internal available surfaces via interactions of the solute. Because the first and fourth stages are relatively fast, the second and third stages determine the rate of the entire adsorption process, which is called the rate limiting or controlling step [19–21].

Several mathematical models have been introduced and reported to describe the optimized fitting of adsorption kinetics and diffusion. Based on these models, the models can be classified with adsorption reaction and adsorption diffusion [22]. The most common adsorption reaction models are the pseudo first-order (PFO) and pseudo second-order (PSO) models that describe the interaction of pollutants present as ions and molecules on the adsorbent surface, although sometimes erroneous overestimation of the rate constant occurs due to the underestimation of instantaneous driving force for sorption [23]. The nonlinear PFO (NPFO) and PSO (NPSO) should be linearized to calculate the parameters, which can be compared to determine the adsorption mechanism.

Although the adsorption reaction model provides adsorption information, such as uptake, rate, and nature of the adsorption, more valuable adsorption mechanism would be provided from an adsorption diffusion model that describes what controls the overall rate. The adsorption diffusion model has been classified as having two parts: an internal diffusion and external mass transfer model. The internal diffusion model was introduced by Crank [24], Weber and Morris [25], and Bangham [23]. Among the three internal diffusion models, the model by Weber and Morris is the simplest and is widely used to explain the adsorption mechanism. The external mass transfer model was proposed by Spahn and Schlunder (1975) [26] and Boyd (1947) [27].

Since the amino group interacts with Cu(II) and As(V) by coordination and electrostatic force, respectively, each interaction is influenced by the presence of counter-ion. Therefore, it is required to determine the effect of counter-ion on the adsorption characteristic. In the present study, the overall goal was to investigate the multifunctional behavior of the amino group and determine the adsorption rate and sorption mechanism of Cu(II) and As(V) on hydrogel chitosan beads in batch and kinetic sorption tests. The specific objectives were to (1) determine the removal efficiency of Cu(II) and As(V) under single and mixed conditions, (2) observe the sorption kinetics using PFO and PSO, (3) compare the linear and nonlinear models from the adsorption kinetics, and (4) investigate intraparticle and external mass transfer diffusion and the effect of the presence of a counter-ion.

2. Material and Methods

2.1. Chemicals

Chitosan was purchased from Sigma-Aldrich (St. Louis, MI, USA) as flakes with a medium molecular weight of ~250,000 g/mol and a 75–85% degree of deacetylation from chitin. The glutaraldehyde solution (25 wt.%) was obtained from SHOWA (Tokyo, Japan). The solution of Cu(II) and As(V) was prepared using $CuCl_2{\cdot}2H_2O$ and arsenate ($NaHAsO_4H_2O$) purchased from Sigma-Aldrich (St. Louis, MI, USA). All other chemicals, including HCl and NaOH, were purchased from Sigma-Aldrich (St. Louis, MI, USA) and were ACS grade and used without further purification.

2.2. Preparation of Chitosan Beads

Chemically stable chitosan beads were prepared in three steps [12,28]. Briefly, we first prepared a 2.5% chitosan solution using 1% HCl from chitosan flakes and beads of chitosan were prepared by dropping the chitosan solution in 1M NaOH during mild stirring, and crosslinking was achieved by adding chitosan beads to a 0.5 M glutaraldehyde (GA) solution. Finally, the cross-linked hydrogel chitosan beads were washed several times and stored in Deionized water (DI) until use.

2.3. Batch Adsorption Tests

A series of batch tests were performed to determine the uptake of Cu(II) and As(V) and determine the adsorption rate for the single and mixed solutions. The adsorption experiment to find the uptake of Cu(II) and As(V) was conducted with 0.15 ± 0.01 g of chitosan bead in 50 mL of 0.031 mmol/L for Cu(II) or 0.021 mmol/L for As(V) at single solution, and 0.032 mmol/L for Cu(II) and 0.027 mmol/L for As(V) at mixed solution. Each sample was rotated at 90 rpm for 48 h.

To determine the adsorption rate, batch kinetic tests were carried out in a 0.5 L glass bottle with 0.75 ± 0.015 g of chitosan bead, including 0.055 and 0.049 mmol/L of Cu(II) and As(V) at single solution, and 0.038 and 0.055 mmol/L of Cu(II) and As(V) at mixed solution, respectively. For both experiments, the initial solution pH of 5.0–5.5 for was adjusted with weak HCl or NaOH solution at a desired time. An aliquot of the sample solution was removed at desired intervals and stored in a refrigerator (4 °C). Cu(II) and As(V) in single and mixed solutions are referred to as Cu(II)-single and As(V)-single and Cu(II)-mixed and As(V)-mixed, respectively. The concentration of Cu(II) and As(V) were analyzed with ICP–OES (Prodigy, Reemanlabs, Mason, OH, USA). The uptake of adsorbate was calculated by

$$q_t = \frac{(C_0 - C_t)V}{M} \tag{3}$$

2.4. Adsorption Reaction Model

The experimental data were analyzed by using the following four kinetic models.

2.4.1. Pseudo First-Order Model

The assumption of pseudo first-order (PFO) is that (1) the rate of occupation of sorption sites is proportional to the number of unoccupied sites and (2) sorption only occurs on localized sites and there is no interaction between the sorbed ions, which correspond to the monolayer of adsorbates on the adsorbent surface [29,30]. The equation of PFO, which was first proposed by Lagergreen (1907), can be represented as follows [31]:

$$\frac{dq_t}{dt} = K_1(q_e - q_t) \tag{4}$$

2.4.2. Pseudo Second-Order Model

The pseudo second-order (PSO) model was first described by Ho and McKay (1998) for the kinetic process of the adsorption of divalent metal ions onto peat [32]. While the PFO model estimated the initial stage of adsorption, the adsorption behavior was described during the entire range or final stage of adsorption process via the PSO model [18,33]. The following PSO Equation (5) can be presented:

$$\frac{dq_t}{dt} = K_2(q_e - q_t)^2 \tag{5}$$

2.5. Adsorption Diffusion Model

2.5.1. Intraparticle Diffusion

The intraparticle diffusion model, which considers pore diffusion, was developed and proposed by Weber and Morris (1963) as follows [25]:

$$q_t = K_i t^{0.5} + C \tag{6}$$

The internal diffusion model assumes that internal diffusion of the adsorbate is the slowest step, resulting in the rate-controlling step during the adsorption process, and the adsorption is instantaneous [33].

2.5.2. External Mass Transfer

To identify the external mass transfer, it is assumed that the diffusion of the adsorbate is controlled by the liquid film around the adsorbent. Spahn and Schlunder (1975) suggested the following equation [26]:

$$V\frac{dC}{dt} = -K_s A(C - C_s) \tag{7}$$

3. Results and Discussion

3.1. Batch Test

A series of batch tests was conducted for the single and mixed solutions of Cu(II) and As(V) at pH ~5.5, and the uptakes are shown in Figure 1. The individual uptakes were ~5.2–5.6 μmol/g for Cu(II) and As(V) for the single solution, indicating the difference of two uptakes is less than 10%. For the mixed solution, the uptakes for Cu(II) and As(V) were 3.6 (30% decrease) and 5.9 (5% increase) μmol/g, respectively, resulting in a total capacity increase to 9.4 μmol/g, which reached 87% of the sum of each ion. This increased total capacity of chitosan can be explained by the hybrid characteristic of the amino group in chitosan. The available sorption sites are separated for Cu(II) and As(V) based on Equations (1) and (2), respectively. As a result, Cu(II) and As(V) can independently interact with NH_2 and NH_3^+, respectively. However, Figure 2, calculated based on the Katchalsky equation [34], shows that over 80% of amino groups are protonated as NH_3^+ at a pK_a of 6.0–7.0, which enhances the adsorption for As(V) via electrostatic forces. Theoretically, when the pK_a (depending on the molecular weight and the degree of deacetylation) is estimated to be 5.5 for the inner chitosan polymer, the reaction is proportionally related to the presence of amino groups. Another assumption is additional association with Cu(II) except for the amino groups. Domard (1987) suggested that the oxygen on the OH group in the chitosan polymer also contributes to chelation with heavy cation metals [35].

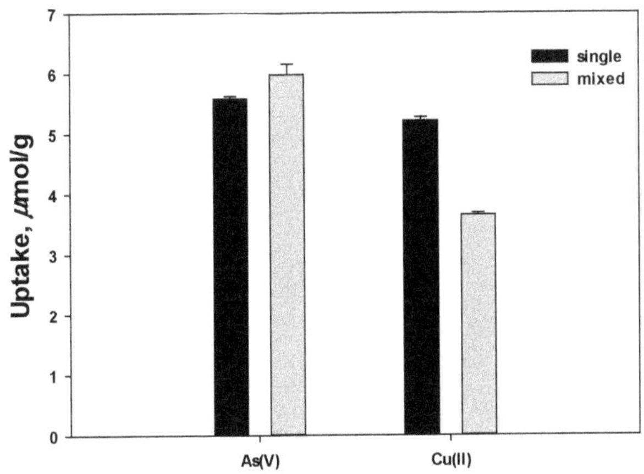

Figure 1. Removal uptake of Cu(II) and As(V) in a presence of single and/or mixed solution.

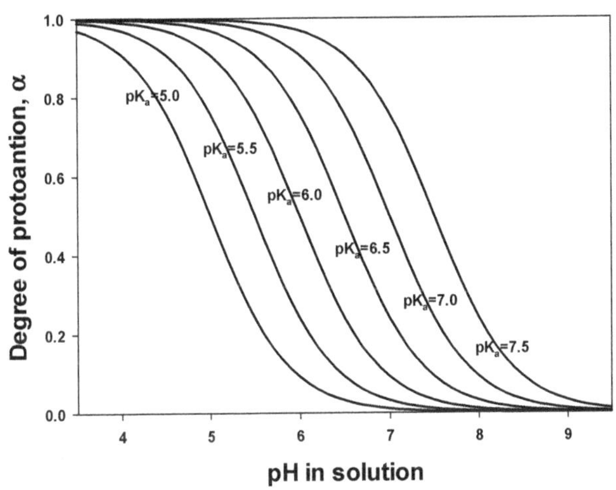

Figure 2. Degree of protonation (α) of chitosan as a function of pH and pK_a.

3.2. Removal Efficiency

The adsorption rates of Cu(II) and As(V) in the form of removal efficiency (%) vs. time (h) for single and mixed solutions are shown in Figure 3. The dash lines indicate the point of 90% uptake. For all cases, it is likely that the adsorption rate was initially faster and then became slower. This phenomenon may be governed by the chance of collision between ions and the adsorbent. At the beginning of the reaction, the higher concentration in the experiment led to surface adsorption with the higher driving force, which reduced the mass transfer resistance in the bulk and film layer [21]. Then, 90% of the sorption removal for Cu(II) was reached at 6 and 24 h, and for As(V), at 0.5 and 6 h for the single and mixed solutions, respectively. The sorption equilibrium for As(V) was more quickly achieved than for Cu(II), and the single solution was at least four times faster than the mixed solution. The relative long equilibrium of Cu(II) is similar to that in Kannamba et al. (2010) [36], who reported that 12 h is required for equilibrium using chitosan flakes. As(V) reached equilibrium at 5 h using 0.25–0.35 mm-sized chitosan [37].

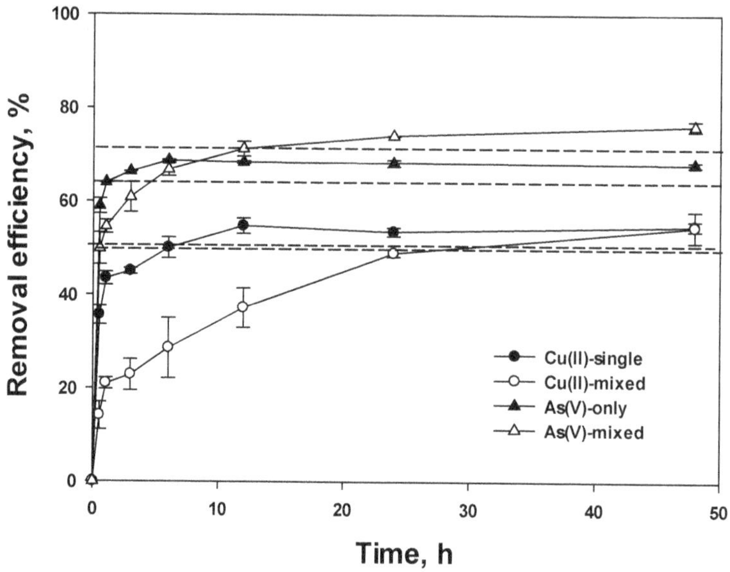

Figure 3. Removal efficiency (%) of Cu(II) and As(V) in the presence of single and/or mixed solutions (dash lines indicate 90% removal efficiency (%), and the dotted line for Cu(II)-single and Cu(II)-mixed were very close).

A longer equilibrium time was required for the mixed solution regardless of the species. This was because of the competition derived from the increased mass of the adsorbate for the limited absorbable sites despite their separation. According to prior studies [38], increasing the initial concentration retards the equilibrium despite the higher concentration gradient.

3.3. *Adsorption Reaction Kinetics*

3.3.1. Nonlinear Pseudo First-Order (PFO) and Second-Order (PSO) Models

From Equation (4) for PFO, the integrating form for the boundary conditions of $t = 0$ to $t = t$ and for $q_t = 0$ to $q_e = q_t$ is expressed as nonlinear (Equation (8)).

$$q_t = q_e\left(1 - e^{-K_1 t}\right) \tag{8}$$

After separating the variables from Equation (5) for PSO and by applying the boundary condition ($t = 0$ to $t = t$ and $q_t = 0$ to $q_e = q_t$), the nonlinear PSO follows Equation (9).

$$q_t = \frac{K_2 q_e t}{1 + K_2 q_e^2 t} \tag{9}$$

Nonlinear PFO (NPFO) and nonlinear PSO (NPSO) were modeled using Equations (8) and (9), and the calculated variables, including $q_e$cal., K, and R^2, are listed in Figure 4 and Table 1. Upon comparing the R^2, although both NPFO and NPSO showed good agreement (>0.94), except for Cu(II)-mixed, NPSO achieved higher accuracy for all experimental conditions. Additionally, the $q_e$cal. was closer to $q_e$exp. for NPSO than for NPFO. As shown in Figure 4, the simulated NPFO line did not likely represent a sudden change in q_e at 3–6 h leading to a lower R^2. Moreover, the pseudo n order equation, Equation (10), based on adsorption capacity was simulated to determine the accurate kinetic order (n) [39]. The calculated n values were 1.98, 2.01, 1.84, and 1.98 for Cu(II)-single, As(V)-single, Cu(II)-mixed, and As(V)-mixed, respectively.

$$q_t = q_e\{1 - [\frac{1}{\left(1 + K_n(n-1)q_e^{n-1}t\right)}]^{\frac{1}{n-1}}\} \quad (10)$$

Although methodological bias is introduced for PSO [40], the NPSO is the more appropriate kinetic model with a higher applicability for PSO, indicating that the adsorption process is mainly associated with chemisorption [18,41], whereas PFO proceeds by diffusion through a boundary [22].

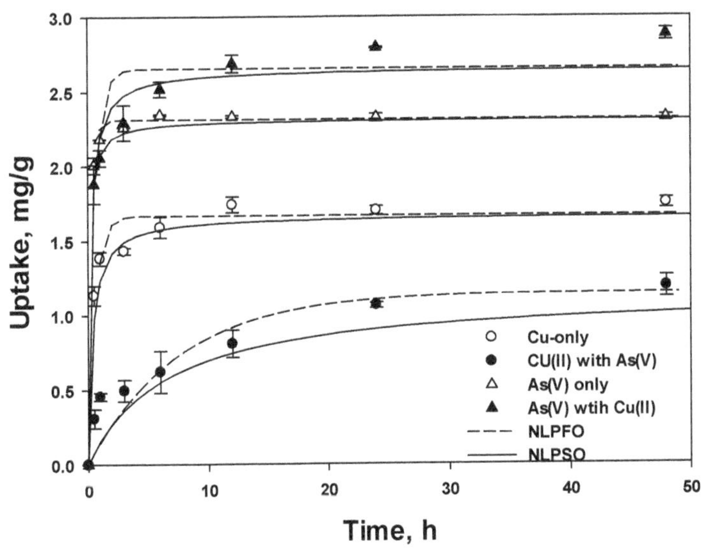

Figure 4. Nonlinear pseudo first-order (dots) and second-order (solid lines) kinetic models.

Table 1. Kinetic parameters using nonlinear PFO and PSO models for Cu(II) and As(V) in single and mixed solutions.

		Cu(II)-Single	Cu(II)-Mixed	As(V)-Single	As(V)-Mixed
	$q_e$exp. [(b)]	1.85	1.25	2.4	2.92
PFO	$q_e$cal. [(c)]	1.64	1.07	2.30	2.62
	K	2.16	0.187	3.96	2.10
	R^2	0.969	0.819	0.997	0.942
PSO	$q_e$cal.	1.71	1.17	2.34	2.74
	K	2.19	0.258	5.28	1.28
	R^2	0.986	0.891	0.999	0.979
PNO [(a)]	$q_e$cal.	1.45	0.813	2.03	2.32
	K	15.7	4.04	36.3	12.4
	n	1.98	1.84	2.01	1.98

[(a)] indicates pseudo n order. [(b)] indicates experimental q_e. [(c)] indicates calculated q_e.

The rate constant K was considered to determine the adsorption equilibrium [42]. Both absolute K_1 and K_2 are in the order of As(V)-single > Cu(II)-single > As(V)-mixed > Cu(II)-mixed. This observation shows that a higher K was calculated for the single solution and As(V), regardless of the kinetic model.

3.3.2. Linear PFO and PSO

Linearization of PFO and PSO was attempted to determine the best-fitting model by minimizing the error distribution between experimental and predicted values. Linear PFO (LPFO) and PSO (LPSO) were derived from Equations (8) and (9), respectively, and expressed by Equations (11) and (12), respectively.

$$\ln(q_e - q_t) = \ln q_e - K_1 t \tag{11}$$

$$\frac{t}{q_t} = \frac{1}{K_2 q_e^2} + \frac{1}{q_e} t \tag{12}$$

Equations (11) and (12) are frequently applied to fit experiment data instead of nonlinear PFO and PSO models. The calculated parameters are listed in Table 2 and shown in Figures 5 and 6 for LPFO and LPSO, respectively. Note that Equation (12) is one of the linearized PSO expressions among the five equations derived from Equation (9) and is the most valid and widely used expression, showing the highest R^2 and similarity between $q_e$exp. and $q_e$cal. [43,44]. Our study also obtained the most valid R^2 from Equation (12) (data not shown).

Table 2. Kinetic parameters for the linear PFO and PSO models for Cu(II) and As(V) in single and mixed solutions.

		PFO										PSO		
		Without Separation			$t \leq 1$			$t \geq 3$						
	q_e-exp.	q_e-cal.	K_1	R^2	q_e-cal.	K_1	R^2	q_e-cal.	K_1	R^2	q_e-cal.	K_2	R^2
Cu(II)single	1.85	0.529	0.0437	0.526	1.69	1.38	0.952	0.268	0.0231	0.514	1.75	5.37	0.999
Cu(II)mixed	1.25	1.03	0.102	0.990	1.17	0.489	0.979	1.11	0.104	0.992	1.23	1.86	0.981
As(V)single	2.4	0.259	0.0371	0.240	1.95	2.39	0.918	0.0816	0.00212	0.0129	2.32	12.5	0.999
As(V)mixed	2.92	0.959	0.0700	0.820	2.54	1.21	0.860	0.552	0.0530	0.954	2.89	24.2	0.999

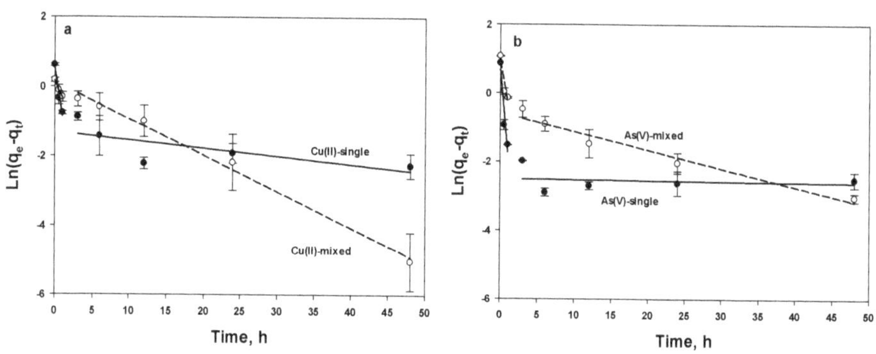

Figure 5. Linear PFO model for (**a**) Cu(II) and (**b**) As(V) with two sections.

Table 2 and Figures 5 and 6 clearly verify the effect of linearization. While the fitting accuracy by R^2 increased from 0.891 to 0.981 for Cu(II)-mixed and others were enhanced to 0.999 for LPSO, the accuracy decreased from ~0.9 to 0.526 and 0.240 for Cu(II)-single and As(V)-single, respectively, for the LPFO model. Similarly, poor values from linearization were reported in several studies [45,46]. Attempts have been made to overcome the poor relation for the LPFO model. Na and Park (2011) [47] and Simonin (2016) [40] separated the times in terms of the initial from the other times. Both results show that the calculated parameters at the initial time can be relatively compared to the nonlinear equation and PSO; however, the rest of the region should be ignored, indicating that a different adsorption mechanism is expected [47]. In this study, Figure 5 is shown with two regions and Table 2 includes the separate parameters clarified from previous studies. For the condition of $1 \leq h$, the R^2 and $q_e$cal. are comparable to NLFO, but at $3 \geq h$, both R^2 and $q_e$cal. were not comparable Thus, we concluded that linearization of the PFO model included unreasonable data that were influenced by the failure to represent the kinetics and limited by the initial adsorption time [40].

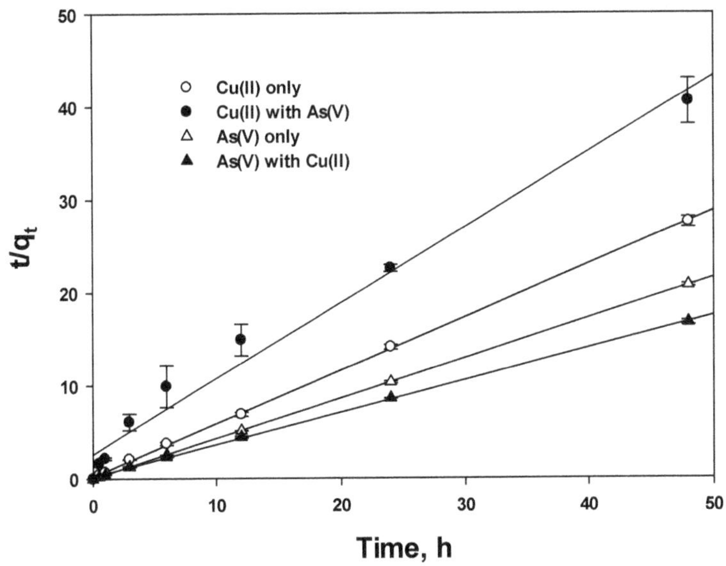

Figure 6. Linear PSO model for Cu(II) and As(V).

Upon comparison of nonlinear and linear PFO and PSO models, the highest R^2 was acquired for LPSO followed by NPSO and NPFO, and the relative lowest was for the LPFO model. This trend is the same as shown by Kumar (2006), who studied methylene blue on activated carbon [44]. Moreover, the $q_e$cal. was compared with $q_e$exp. for all conditions. Based on the results of comparison with R^2 and q_e-cal., to estimate the appropriate parameters, the nonlinear application was more suitable and acceptable in this study. Although the comparison of R^2 cannot describe a sufficient criterion [48], Simonin (2016) suggested that the PSO model better describes a diffusion-controlled process than the PFO model [40].

Before discussing the value of K, the effect of the initial concentration should be first considered because a higher initial concentration contributes to a higher concentration gradient, which can enhance the initial sorption rate and require more time to reach equilibrium [49]. However, it can cause a decrease in the entire sorption rate due to the higher competition between adsorbate and adsorbent active sites [50]. In this study, Cu(II) and As(V) in the single and mixed solutions were 0.055 and 0.049 mmol/L and 0.038 and 0.055 mmol/L, respectively, where ~31% of the Cu(II) concentration was decreased in the mixed solution and a ~10% difference in concentration between Cu(II) and As(V) in single solution was found. According to the K_1 and K_2 from NPFO, NPSO, and LPSO (LPFO was excluded), the K value was in the following order regardless of kinetic model: As(V)-single > Cu(II)-single > As(V)-mixed >> Cu(II)-mixed. The effect of the initial concentration difference for Cu(II) was negligible because the K value for Cu(II)-mixed showed 3–11 times lower than for other conditions, although the concentration of Cu(II) in the mixed solution showed a 31% decrease.

Based on the order of K, As(V) and single solution had higher rate constants than Cu(II) and the mixed solution. Generally, the value of K increased with increased initial concentration as chemical conditions [49] and shaking speed as physical conditions [50]. The phenomena indicate that for the mixed condition, As(V) and Cu(II) competed with each other at limited activated sites, and the rate declined. As mentioned in Section 3.1, the functional group in chitosan, the amino group, is present in the form of NH_2 or NH_3^+ depending on the solution pH, and it has been suggested that

Cu(II) can coordinate with NH_2 and As(V) can react with NH_3^+ via electrostatic forces [15]. Therefore, electrostatic forces initially occurred and led to the rapid adsorption rate of As(V).

3.4. *Diffusion Model*

Both diffusion and adsorption on the activated site were influenced by the surface area, the reactivity of the surface, and liquid film thickness for external diffusion, and by the reaction of surface and pore structure for internal diffusion. Generally, film diffusion is controlled in a specific system followed at poor mixing, low concentration, small particle sizes of the adsorbent, and a higher affinity of the adsorbate for the adsorbent; whereas, good mixing, larger particle size of the adsorbent, high concentration of adsorbate, and a low affinity of the adsorbate for the adsorbent are controlled at intraparticle diffusion [21,49,51].

3.4.1. Intraparticle Diffusion

A plot of adsorbate uptake versus the square root of time ($t^{0.5}$) is shown in Figure 7, based on Equation (6), and some calculated parameters are listed in Table 3. There are theoretical interpretations of the intraparticle diffusion equation. C is an arbitrary constant representing the boundary layer thickness, and a larger value of C represents a thicker boundary layer [18,21,52]. If the value of C is zero, which corresponds to no boundary layer, the linear line should pass through the origin. Consequently, film diffusion could be ignored due to no or less thickness, and thus, intraparticle diffusion would remain as the rate-controlling step through the entire adsorption kinetic process. However, this is only the theoretical explanation using Equation (6). Many studies have reported nonzero intercepts, indicating that the rate-limiting step involves both intraparticle and film diffusion in most adsorption processes. As shown in Figure 7 and Table 3, owing to the very low R^2 with the single regression line, it is unable to predict one linear line for the experiment data; instead, it could be divided with two different segments for all cases. Thus, Figure 7 was segregated with two linear regressions, illustrating that both film and intraparticle diffusion control the adsorption diffusion. Moreover, another study [53] suggested that intraparticle diffusion has three regressions in the macro, meso, and micro pores with a horizontal line as the equilibrium. Another study observed three linear regions, including initially rapid surface loading, then pore diffusion, and finally horizontal equilibrium [21]. However, it was not easy to differentiate three or four regions in this study. Consequently, the two linear regions were primarily observed and clearly explained by the adsorption process between the film and intraparticle diffusion [49,54]. Additionally, the initial rapid increase was represented by film diffusion [21,55]. As a result, two linear lines and the value of C (y-axis) both revealed diffusion was coincidentally involved in the adsorption of Cu(II) and As(V).

Some observations were made based on the data in Table 2. The highest and lowest values of K_i were for As(V)-single and Cu(II)-mixed in the first regression line, respectively, and the order was reversed to Cu(II)-mixed and As(V)-single in the second line. As mentioned regarding the first rapid adsorption related to film diffusion, the higher K_{i1} for As(V) compared with that of Cu(II) indicated a higher surface adsorption reaction with an active site leading to an electrostatic force between NH_3^+ and $HAsO_4^-$. According to the assumption where close to the zero of slope (K) in the second section represents intraparticle diffusion, which stands for the equilibrium process [21], the single solutions of Cu(II) and As(V) quickly accomplished equilibrium, whereas mixed solutions required more time to reach equilibrium, which was the same result as with the PSO model.

The order of C was As(V), As(V)-Cu(II), Cu(II), and Cu(II)-As(V) regardless of separation, which is related to boundary diffusion or surface adsorption. As a result, the higher C for As(V) demonstrated that As(V) removal was primarily associated with surface adsorption [22].

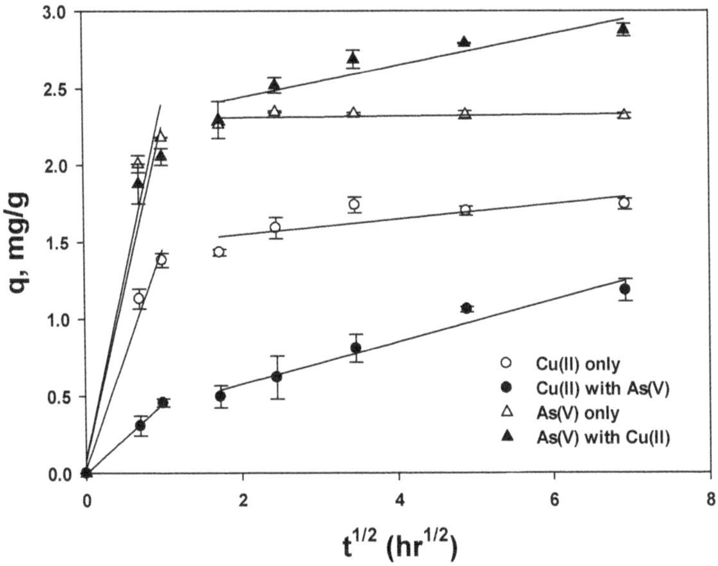

Figure 7. Weber and Morris intraparticle diffusion model with two sections.

Table 3. Diffusion parameters for the Weber and Morris intraparticle model for Cu(II) and As(V) in single and mixed solutions.

	First Section			Second Section			Without Separation		
	C	K	R^2	C	K	R^2	C	K	R^2
Cu(II)-single	0.0286	1.42	0.986	1.45	0.0497	0.605	2.23	0.103	0.850
Cu(II)-mixed	−0.00285	0.452	0.999	0.301	0.137	0.954	1.41	0.00916	0.482
As(V)-single	0.0861	2.30	0.953	2.30	0.00483	0.0994	2.21	0.00345	0.238
As(V)-mixed	0.0787	2.17	0.956	2.23	0.103	0.850	0.470	0.0175	0.852

3.4.2. External Mass Transfer

To identify the external mass transfer, Equation (7) was expressed with Equation (13) by plotting the values between C_t/C_o and ln (C_t/C_o), respectively.

$$-ln\frac{C_t}{C_0} = K_s\frac{A}{V}t \quad (13)$$

Like the intraparticle diffusion model, when the plot using Equation (13) is linear and passes through zero regardless of K_s, A, or V, it suggests that film diffusion governs the adsorption kinetic process [56]. To carefully predict the external diffusion, Figure 8a was segregated with two regions in Figure 8b, and then showed that the slope sharply increased at the beginning ($\leq$ 1 h) and gradually decreased after 1 h, which is a similar trend as that found in a prior study [47]. Note that the simplified Equation (13) is comparable with the LPFO model in Equation (9); additionally, Figure 8 is very similar to Figure 5 in the current study. The rapid slope ($K_s \cdot A/V$) of the regression line indicates that external mass transfer occurred [57], and the lowest slope for the Cu(II) mixed solution can be assumed to be relatively associated with a strong effect of intraparticle diffusion. Therefore, both intraparticle and external diffusion simultaneously influence the adsorption kinetics.

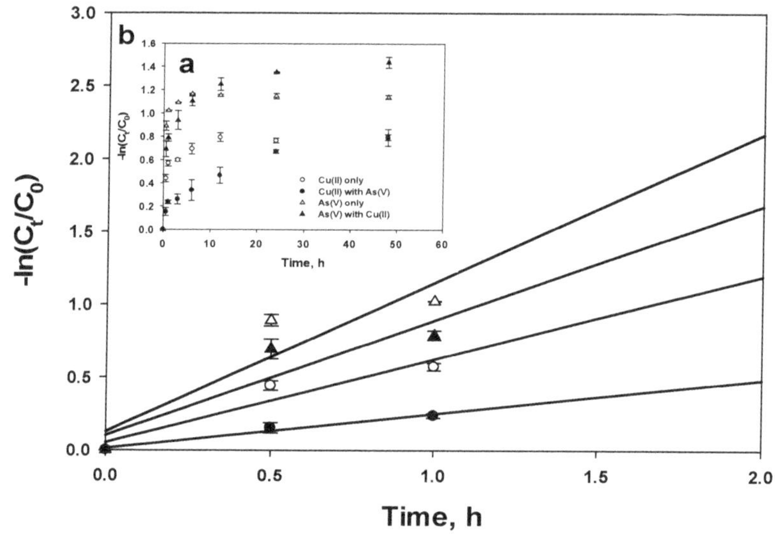

Figure 8. Spahn and Schlunder film diffusion model shown for (**a**) the entire experimental time scale and (**b**) less than 2 h.

4. Conclusions

In this study, we described the interaction between the amino group in the chitosan polymer and the cation of Cu(II) and anion of As(V) to determine the adsorption mechanism. Additionally, the effect of each ion on adsorption was also studied for single and mixed solutions. According to the increased capacity of the Cu(II) and As(V) mixed solution, the amino group in chitosan was in the form of NH_2 and NH_3^+, and each functional group directly participated in the removal of the anion and cation, respectively. The PSO model is more suitable than PFO, indicating that surface chemisorption occurred primarily for both Cu(II) and As(V) adsorption. Because the linearization of PFO expanded the error, lowering the correlation coefficient (R^2), two segregation regression lines dependent on time were effective at describing the adsorption. Because the two types of diffusion models showed no passing through of the zero regression line, both intraparticle and film diffusion were considered as rate-controlling steps for Cu(II) and As(V) adsorption. A higher rate constant (K) for As(V) than for Cu(II) was obtained from the electrostatic interaction between NH_3^+ and $HAsO_4^{2-}$.

Funding: This research was supported by the National Research Foundation of Korea (NRF-2019R1A2C1009129).

Conflicts of Interest: The author declares no conflict of interest.

Nomenclature

A	Surface area of adsorbent (m^2)
C_0	Initial concentration in the solution, mg/L
C_t	Liquid phase concentration at any time, mg/L
K_1	Sorption rate constant (min^{-1}) at PFO
K_2	Sorption rate constant (g/mg·min) at PSO
K_i	Intraparticle diffusion rate constant (mg/g· $min^{1/2}$)
K_n	n order diffusion rate constant
K_s	The liquid-film mass transfer coefficient (m^2/s)
M	Mass of adsorbent (g)
n	Kinetic order

q_e Adsorbate uptake at equilibrium (mg/g)
q_t Adsorbate uptake at any time (mg/g)
t Reaction time (min)
V Volume of batch experiment (L)

References

1. An, B.; Son, H.; Chung, J.; Choi, J.W.; Lee, S.H.; Hong, S. Calcium and hydrogen effects during sorption of copper onto an alginate-based ion exchanger: Batch and fixed-bed column studies. *Chem. Eng. J.* **2013**, *232*, 51–58. [CrossRef]
2. Sessarego, S.; Rodrigues, S.C.G.; Xiao, Y.; Lu, Q.; Hill, J.M. Phosphonium-enhanced chitosan for Cr(VI) adsorption in wastewater treatment. *Carbohydr. Polym.* **2019**, *211*, 249–256. [CrossRef] [PubMed]
3. Morisi, R.E.; Alsabagh, A.M.; Nasr, S.; Zaki, M.M. Multifunctional nanocomposites of chitosan, silver nanoparticles, copper nanoparticles and carbon nanotubes for water treatment: Antimicrobial characteristics. *Int. J. Biol. Macromol.* **2017**, *97*, 264–269. [CrossRef]
4. Wan, Y.; Creber, K.A.M.; Peppley, B.; Bui, V.T. Ionic conductivity of chitosan membranes. *Polymer* **2003**, *44*, 1057–1065. [CrossRef]
5. Philippova, O.E.; Korchagina, E.V. Chitosan and its hydrophobic derivatives: Preparation and aggregation in dilute aqueous solutions. *Polym. Scl. Ser. A* **2012**, *54*, 552–572. [CrossRef]
6. Wang, Z.; Nie, J.; Qin, W.; Hu, Q.; Tang, B.Z. Gelation process visualized by aggregation-induced emission fluorogens. *Nat. Commun.* **2016**, *7*, 12033. [CrossRef]
7. Wan Ngah, W.S.; Endud, C.S.; Mayanar, R. Removal of copper(II) ions from aqueous solution onto chitosan and cross-linked chitosan beads. *React. Funct. Polym.* **2002**, *50*, 181–190. [CrossRef]
8. Kwok, K.C.M.; Koong, L.F.; Al Ansari, T.; Mckay, G. Adsorption/desorption of arsenite and arsenate on chitosan and nanochitosan. *Environ. Sci. Pollut. Res.* **2018**, *25*, 14734–14742. [CrossRef]
9. Wang, X.; Tang, R.; Zhaong, Y.; Yu, Z.; Qi, C. Preparation of a novel chitosan based bioplymer dye and application in wood dyeing. *Polymers* **2016**, *8*, 338. [CrossRef]
10. Mohamed, M.; Ouki, S. Removal mechanism of toluene from aqueous solutions by chitin and chitosan. *Ind. Eng. Chem. Res.* **2011**, *50*, 9557. [CrossRef]
11. An, B.; Jung, K.Y.; Zhao, D.; Lee, S.H.; Choi, J.W. Preparation and characterization of polymeric ligand exchanger based on chitosan hydrogel for selective removal phosphate. *React. Funct. Polym.* **2014**, *85*, 45–53. [CrossRef]
12. An, B.; Choi, J.W. An experimental application of four types of chitosan bead for removal of cationic and anionic pollutants. *Water Air Soil pollut.* **2019**, *230*, 314. [CrossRef]
13. Zhang, J.; Guo, W.; Guo, Q.; Jin, L.; Liua, Z.; Hu, S. On-site separation of Cr(VI) and Cr(III) in natural waters by parallel cartridge ion-exchange columns. *RSC Adv.* **2017**, *7*, 50657. [CrossRef]
14. Dou, Q.; Liu, L.; Yang, B.; Lang, J.; Yan, X. Silica-grafted ionic liquids for revealing the respective charging behaviors of cations and anions in supercapacitors. *Nat. Commun.* **2017**, *8*, 2188. [CrossRef] [PubMed]
15. Guo, X.; Chen, F. Removal of arsenic by bead cellulose loaded with iron oxyhydroxide from groundwater. *Environ. Sci.Technol.* **2005**, *39*, 6808–6818. [CrossRef] [PubMed]
16. Wahba, M.I. Porous chitosan beads of superior mechanical properties for the covalent immobilization of enzymes. *Int. J. Biol. Macromol.* **2017**, *105*, 894–904. [CrossRef]
17. Li, N.; Bai, R. Copper adsorption on chitosan–cellulose hydrogel beads: Behaviors and mechanisms. *Sep. Purif. Technol.* **2005**, *42*, 237–247. [CrossRef]
18. Magdy, Y.H.; Altaher, H. Kinetic analysis of the adsorption of dyes from high strength wastewater on cement kiln dust. *J. Environ. Chem. Eng.* **2018**, *6*, 834–841. [CrossRef]
19. McKay, G. The Adsorption of basic dye onto silica from aqueous solution-solid diffusion model. *Chem. Eng. Sci.* **1984**, *39*, 129–138. [CrossRef]
20. Altaher, H.; Khalil, T.E.; Abubeah, R. The effect of dye chemical structure on adsorption on activated carbon: A comparative study. *Color. Technol.* **2014**, *30*, 205–214. [CrossRef]
21. Hameed, B.H.; El-Khaiary, M.I. Malachite green adsorption by rattan sawdust: Isotherm, kinetic and mechanism modeling. *J. Hazad. Mater.* **2008**, *159*, 574–579. [CrossRef]

22. Pholosi, A.; Naidoo, E.B.; Ofomaja, A.E. Intraparticle diffusion of Cr(VI) through biomass and magnetite coated biomass: A comparative kinetic and diffusion study. *S. Afr. J. Chem. Eng.* **2020**, *32*, 39–55. [CrossRef]
23. Huang, Y.; Farooq, M.U.; Lai, S.; Feng, X.; Sampranpiboon, P.; Wang, X.; Huang, W. Model fitting of sorption kinetics data: Misapplications overlooked and their rectifications. *AIChE* **2018**, *64*, 1793–1805. [CrossRef]
24. Crank, J. *The Mathematics of Diffusion*, 2nd ed.; Clarendon Press, Oxford Science Publications: London, UK, 1975.
25. Weber, W.J.; Morris, J.C. Kinetics of adsorption on carbon from solution. *J. Sanit. Eng. Div.* **1963**, *89*, 31–59.
26. Spahn, H.; Schlunder, U. The scale-up of activated carbon columns for water purification based on results from batch test. I. Theoretical and experimental determination of adsorption rates of single organic solutes in batch test. *Chem. Eng. Sci.* **1975**, *30*, 529–537. [CrossRef]
27. Boyd, G.E.; Adamson, A.W.; Myers, L.S. The exchange adsorption of ions from aqueous solutions by organic zeolites. II. Kinetics. *J. am. Chem. Soc.* **1947**, *69*, 2836–2848. [CrossRef]
28. Kim, T.; An, B. Adsorption characteristic of Cu(II) and phosphate using non-linear and linear isotherm equation for chitosan bead. *J. Korean Soc. Water Wastewater* **2020**, *34*, 201–210. [CrossRef]
29. Salam, M.A.; Al-Zhrani, G.; Kosa, S.A. Removal of heavy metal ions from aqueous solution by multi-walled carbon nanotubes modified with 8-hydroxyquinoline: Kinetic study. *J. Ind. Eng. Chem.* **2014**, *20*, 572–580. [CrossRef]
30. Largitte, L.; Pasquier, R. A review of the kinetics adsorption models and their application to the adsorption of lead by an activated carbon. *Chem. Eng. Res. Des.* **2016**, *109*, 495–504. [CrossRef]
31. Lagergreen, S. Zur theorie der sogenannten adsorption geloster stoffe. *Z. fur Chem. Ind. der Kolloide* **1907**, *2*, 15.
32. Ho, Y.S.; McKay, G. Sorption of dye from aqueous solution by peat. *Chem. Eng. J.* **1998**, *70*, 115–124. [CrossRef]
33. Wang, J.; Guo, X. Adsorption kinetic models: Physical meanings, applications, and solving methods. *J. Hazar. Mater.* **2020**, *390*, 122156. [CrossRef] [PubMed]
34. Katchalsky, A. Problems in the physical chemistry of polyelectrolytes. *J. Polym. Sci.* **1954**, *12*, 159–184. [CrossRef]
35. Domard, A. Determination, of N-acetyl content in chitosan samples by c.d. measurements. *Int. J. Biol. Macromol.* **1987**, *9*, 333–336. [CrossRef]
36. Kannamba, B.; Laxma Reddy, K.L.; AppaRao, B.V. Removal of Cu(II) from aqueous solutions using chemically modified chitosan. *J. Hazad. Mater.* **2010**, *175*, 939–948. [CrossRef]
37. Gérente, C.; Andrès, Y.; McKay, G.; Le Cloirec, P. Removal of arsenic(V) onto chitosan: From sorption mechanism explanation to dynamic water treatment process. *Chem. Eng. J.* **2010**, *158*, 593–598. [CrossRef]
38. Klapiszewski, L.; Siwińska-Stefańska, K.; Kołodyńska, D. Development of lignin based multifunctional hybrid materials for Cu(II) and Cd(II) removal from the aqueous system. *Chem. Eng. J.* **2017**, *330*, 518–530. [CrossRef]
39. Özer, A. Removal of Pb(II) ions from aqueous solutions by sulphuric acid- treated wheat bran. *J. Hazard Mater.* **2007**, *141*, 753–761. [CrossRef]
40. Simonin, J.P. On the comparison of pseudo-first order and pseudo-second order rate laws in the modeling of adsorption kinetics. *J. Chem. Eng.* **2016**, *300*, 254–263. [CrossRef]
41. Konggidinata, M.I.; Chao, B.; Lian, Q.; Subramaniam, R.; Zappi, M.; Gang, D.D. Equilibrium, kinetic and thermodynamic studies for adsorption of BTEX onto ordered mesoporous carbon (OMC). *J. Hazad. Mater.* **2017**, *336*, 249–259. [CrossRef]
42. Plazinski, W.; Rudzinski, W.; Plazinska, A. Theoretical models of sorption kinetics including a surface reaction mechanism: A Review. *Adv. Colloid Interface Sci.* **2009**, *152*, 2–13. [CrossRef] [PubMed]
43. Blanchard, G.; Manuaye, M.; Martin, G. Removal of heavy metals from waters by means of natural zeolites. *Water Res.* **1984**, *18*, 1501–1507. [CrossRef]
44. Kumar, K.V. Linear and non-linear regression analysis for the sorption kinetics of methylene blue onto activated carbon. *J. Hazard. Mater.* **2006**, *B137*, 1538–1544. [CrossRef] [PubMed]
45. Moussout, H.; Ahlafi, J.; Aazza, M.; Maghat, H. Critical of linear and nonlinear equations of pseudo-first order and pseudo-second order kinetic models. *Karbala Int. J. Mod. Sci.* **2018**, *4*, 244–254. [CrossRef]

46. Lin, J.; Wang, L. Comparison between linear and non-linear forms of pseudo-first-order and pseudo-second-order adsorption kinetic models for the removal of methylene blue by activated carbon. *Front. Environ. Sci. En.* **2009**, *3*, 320–324. [CrossRef]
47. Na, C.K.; Park, H.J. Applicability of theoretical adsorption models for studies on adsorption properties of adsorbents (II). *J. Korean Soc. Environ. Eng.* **2011**, *33*, 804–811. [CrossRef]
48. Guthrie, W. Engineering Statistics Handbook. Section 4.4 Model Validation; 2013. Available online: https://www.itl.nist.gov/div898/handbook/ (accessed on 12 October 2019).
49. Vadivelan, V.; Kumar, K.V. Equilibrium, kinetics, mechanism, and process design for the sorption of methylene blue onto rice husk. *J. Colloid Interf. Sci.* **2005**, *286*, 90–100. [CrossRef]
50. Albadarin, A.B.; Mangwandi, C.; Al-Muhtaseb, A.A.H.; Walker, G.M.; Allen, S.J.; Ahmad, M.N.M. Kinetic and thermodynamics of chromium ions adsorption onto low-cost dolomite adsorbent. *Chem. Eng. J.* **2012**, *179*, 193–202. [CrossRef]
51. Mohan, D.; Singh, K.P. Single- and multi-component adsorption of cadmium and zinc using activated carbon derived from bagasse—An agricultural waste. *Water Res.* **2002**, *36*, 2304–2318. [CrossRef]
52. Wang, W.; Maimaiti, A.; Shi, H.; Wu, R.; Wang, R.; Li, Z.; Qi, Q.; Yu, G.; Deng, S. Adsorption behavior and mechanism of emerging perfluoro-2- propoxypropanoic acid (GenX) on activated carbons and resins. *Chem. Eng. J.* **2019**, *364*, 132–138. [CrossRef]
53. Ho, Y.S.; McKay, G. The kinetics of sorption of basic dyes from aqueous solution by sphagnum moss peat. *Can. J. Chem. Eng.* **1998**, *76*, 822–827. [CrossRef]
54. Doke, K.M.; Khan, E.M. Equilibrium, kinetic and diffusion mechanism of Cr(VI) adsorption onto activated carbon derived from wood apple shell. *Arab. J. Chem.* **2017**, *10*, s252–s260. [CrossRef]
55. Hamdaoui, O. Batch study of liquid-phase adsorption of methylene blue using cedar sawdust and crushed brick. *J. Hazard. Mater.* **2006**, *B135*, 264–273. [CrossRef] [PubMed]
56. Gupta, S.S.; Bhattacharyya, K.G. Kinetics of adsorption of metal ions on inorganic materials: A review. *Adv. Colloid Interface Sci.* **2011**, *162*, 39–58. [CrossRef] [PubMed]
57. Fernadez, N.A.; Chacin, E.; Gutierrez, E.; Alastre, N. Adsorption of lauryl benzyl sulphonate on algae. *Bioresour. Technol.* **1995**, *54*, 111. [CrossRef]

processes

MDPI

Article

In Vitro Bioadsorption of Cd^{2+} Ions: Adsorption Isotherms, Mechanism, and an Insight to Mycoremediation

Raman Kumar [1,*], Priyanka Sharma [2], Ahmad Umar [3,4,*], Rajeev Kumar [2], Namita Singh [5], P. K. Joshi [6,7], Fahad A. Alharthi [8], Abdulaziz Ali Alghamdi [8] and Nabil Al-Zaqri [8]

1 Department of Biotechnology, Maharishi Markandeshwar (Deemed to be University), Mullana (Ambala), Haryana 133207, India
2 Department of Environment Studies, Panjab University, Chandigarh 160014, India; priyanka_prabhakar@ymail.com (P.S.); answalrajeev@gmail.com (R.K.)
3 Department of Chemistry, Faculty of Science and Arts, Najran University, Najran 11001, Saudi Arabia
4 Promising Centre for Sensors and Electronic Devices (PCSED), Najran University, Najran 11001, Saudi Arabia
5 Department of Bio and Nano Technology Guru Jambheshwar University of Science and Technology, Hisar 125001, India; namitasingh71@gmail.com
6 Division of Soil and Crop Management, Central Soil Salinity Research Institute, Karnal 132001 India; pkjoshi54@gmail.com
7 Department of Agri-Business Management, Maharishi Markandeshwar (Deemed to be University), Mullana (Ambala), Haryana 133207, India
8 Department of Chemistry, College of Science, King Saud University, P.O. Box 2445, Riyadh 11451, Saudi Arabia; fharthi@ksu.edu.sa (F.A.A.); aalghamdia@ksu.edu.sa (A.A.A.); nalzaqri@ksu.edu.sa (N.A.-Z.)
* Correspondence: ramankumar@mmumullana.org (R.K.); ahmadumar786@gmail.com (A.U.)

Received: 1 August 2020; Accepted: 29 August 2020; Published: 2 September 2020

Abstract: The objective of this paper is to establish the significance of the mycoremediation of contaminants such as Cd^{2+} to achieve sustainable and eco-friendly remediation methods. Industries such as electroplating, paint, leather tanning, etc. release an enormous amount of Cd^{2+} in wastewater, which can drastically affect our flora and fauna. Herein, we report on the in vitro bioadsorption of Cd^{2+} ions using fungal isolates obtained from different contaminated industrial sites. The detailed studies revealed that two fungal species, i.e., *Trichoderma fasciculatum* and *Trichoderma longibrachiatum*, were found to be most effective against the removal of Cd^{2+} when screened for Cd^{2+} tolerance on potato dextrose agar (PDA) in different concentrations. Detailed adsorption studies were conducted by exploring various experimental factors such as incubation time, temperature, pH, inoculum size, and Cd^{2+} salt concentrations. Based on optimum experimental conditions, *T. fasciculatum* exhibited approximately 67.10% removal, while *T. longibrachiatum* shows 76.25% removal of Cd^{2+} ions at pH 5.0, 120 h incubation time, at 30°C. The inoculum sizes for *T. fasciculatum* and *T. longibrachiatum* were 2.5% and 2.0%, respectively. Finally, the morphological changes due to Cd^{2+} accumulation were examined using scanning electron microscopy (SEM). Further, Fourier transform infrared spectroscopy (FTIR) spectroscopy reveals the presence of various functional groups (-CH, –C=O, NH and –OH), which seem to be responsible for the efficient binding of Cd^{2+} ions over the fungal surfaces.

Keywords: *T. longibrachiatum*; *T. fasciculatum*; bioadsorption; cadmium; heavy metals; isotherms; bioadsorption mechanism; mycoremediation

1. Introduction

The emerging worldwide environmental problems of the past decades are mostly related to the superfluous increase in metallic contaminants in the environment [1–6]. Inorganic contaminants such as heavy metals released into the environment through various industrial, agricultural, and domestic practices [5,6]. Heavy metals may also enter into the environment by natural sources, but the extent of such exposures is non-significant. Heavy metals are often present together with organic pollutants in industrial wastewater [7]. The industrial effluents are rich in dissolved salts of Cr, Cu, Ni, Zn, and Cd, etc., and can cause a serious problem to the surrounding environment. Heavy metals pollution causes serious problems due to their non-biodegradable nature [8,9]. Although these metals are vital for the proper functioning of biological systems but only in trace concentrations, higher concentrations may lead to disturbed functioning of bio-geo-chemical cycles [10–12]. In addition, serious health issues may arise if such metal ions can enter into the food chains and their corresponding products. Thus, even 1.0–10 mg/L can cause health problems in human beings such as jaundice, facial edema, blue lungs, kidney damage, hearing disorder, skin cancer, asthma, protein metabolism, and bronchial cancer, etc. [13,14]. Therefore, a number of traditional physiochemical treatment techniques have been reported to remove the heavy metals from wastewater, especially from the industrial wastewater, to name a few, electro-coagulation, solvent extraction, ion-exchange, electro-reduction, reverse osmosis, adsorption, membrane separation, chemical precipitation, and so on [15–29]. Even though used widely, these above-mentioned techniques exhibited several disadvantages which include the utilization of high-cost equipment and monitoring systems, the use of various expensive chemicals, the discharge of toxic sludge, long processing time, the production of toxic waste products that need further processing, and so on [23–31]. Thus, it is the necessity of the time to develop a new, simple, and inexpensive method for heavy metal removal from wastewater. Recently, by utilizing living/non-living microorganisms and their derivatives to remove the heavy metals from industrial wastewater is considered a promising technique [32–34]. Accordingly, the bioremediation can be considered as one of the most important and advantageous process that can efficiently remove the heavy metals from industrial wastewater. The entrapment of the heavy metal ions and their corresponding sorption on the binding cites of the cellular structures leads to the bioadsorption of heavy metal ions from the industrial wastewater [35,36]. As a result of the faster adsorption process, the microbial cells are considered as an important and advantageous scaffold for the biosorption of heavy metals and wastewater treatment.

In this study, different fungal isolates extracted from various contaminated sites (Haryana, India) were cultured and screened for tolerance to Cadmium (Cd^{2+}). Various process factors, i.e., inoculum size, pH, initial metal ion concentration, temperature, and incubation time on the Cd^{2+} removal by highly efficient Cd^{2+} tolerant fungal isolates have been investigated and presented in this article. In addition, mechanisms of Cd^{2+} removal by efficient Cd^{2+} tolerant fungal isolates have also been studied.

2. Materials and Methods

2.1. Chemicals Used

Potato dextrose broth (PDB) and nutrient media, potato dextrose agar (PDA), were procured from Hi-Media, Mumbai India. $Cd(NO_3)_2$, was provided by Fine-Chem Limited, Mumbai, India. The solutions were prepared in triply distilled water (conductivity =0.5 μS cm^{-1}), and the reagents were used without any further purification.

2.2. Sample Collection and Isolation of Fungi

Samples from various sites such as industrial effluents, sewerage discharge, and sludge were collected from Karnal, Panipat, and Sonepat districts of Haryana (India). The samples were stored at 4 °C prior to further processing. Fungal isolates were isolated from samples by the PDA serial dilution method. The serial dilutions were made up to 10^6 and 1 mL of each dilution i.e., 10^4 and 10^6

were poured over PDA plates. An incubation of cultured petri plates was performed at 28 °C for 96 h. The prominently grown fungal colonies were regenerated and purified using the streak plate method.

2.3. Screening and Identification of Efficient Cd^{2+}-Tolerant Fungal Isolate

The isolated fungal isolates were tested for their potential to remove Cd^{2+} at 25, 50, and 75 mg/L concentrations. The concentrations were prepared from a sterilized 100 mg/L solution of $Cd(NO_3)_2$ containing PDA. A loop full of purified isolates was positioned at the Petri plates' center comprising of PDA (pH 5.0) and the above-mentioned Cd^{2+} concentrations. The growth of fungal isolates on the pure PDA medium was labeled 'Control'. All plates were sealed with parafilm and subjected to incubation of 5 days. Based on the visual observation, the growth of fungal isolates was labeled as absent growth (−), least growth (+), less than normal growth (++), slightly less than normal growth (+++), and normal growth (++++). The highly efficient Cd^{2+}-tolerant fungal isolates were sent for identification to the Indian Agricultural Research Institute, New Delhi, India.

2.4. Optimization of Batch Cultures

The highly efficient Cd^{2+}-tolerant fungal culture was tested at different pH, i.e., 3.5, 4.0, 4.5, 5.0, 5.5, and 6.0 for uptake and percentage removal of Cd^{2+} from PDA containing 20 mg/L of Cd^{2+}. First, 1M HCl and 1M NaOH were used to adjust the pH of PDB. Then, 1mL spore suspension of efficient fungal culture containing 10^4 spores (1%) was inoculated into 250 mL Erlenmeyer conical flasks containing 100 mL of PDB enriched with 20 mg/L of Cd^{2+} followed by shaking these flasks for 120 h at 30 °C. For the control, a non-inoculated flask having 20 mg/L of Cd^{2+} and PDB was used. After 120h, fungal growth was collected through filtration using Whatman Filter Paper No. 1, which was later dried for 48 h at 80 °C in an oven and then weighed. Atomic absorption spectrophotometer (AAS, GBC 932, Dandenong, Australia) was used to evaluate the Cd^{2+} concentration left in the filtrate. To evaluate the incubation time effect, the incubation time was varied between 24 and 144h under optimum pH. Under optimum pH and incubation time, the effect of inoculum size was investigated by performing batch experiments at different inoculum sizes i.e., 0.5%, 1.0%, 1.5%, 2.0%, 2.5%, and 3.0%. Temperatures of 20, 25, 30, and 35 °C were studied to examine the temperature effect at optimum pH, incubation time, and inoculum size. Under optimum pH, the effect of initial Cd^{2+} concentration, inoculum size, and incubation time was investigated by changing the concentrations of Cd^{2+} ions from 5 to 50 mg/L. The process to perform all the experiments was the same as mentioned above. All experiments were performed in triplicate, and data were statistically examined by *t*-test with the help of SPSS 11.5.

The Cd^{2+} uptake by fungal biomass was calculated as follows:

$$Q = \frac{v(C_0 - C_f)}{M} \tag{1}$$

where Q is the Cd^{2+} uptake (mg/g); C_0 and C_f are the initial and leftover concentrations of Cd^{2+} (mg/L), respectively; v is the total volume of the solution (L); and M the dried biomass of fungus (g).

The removal of Cd^{2+} (%) was calculated using following equation:

$$\%\ \text{Removal of Cd}^{2+} = \frac{C(initial) - C(final)}{C(initial)} \times 100 \tag{2}$$

2.5. Equilibrium Isotherms

2.5.1. Langmuir Isotherm

Langmuir isotherm is a mathematical deduction of kinetic equilibrium between the condensation and loss of adsorbed molecules.

$$Ce/qe = 1/q_{max}\, b + Ce/q_{max} \quad (3)$$

where Ce is equilibrium concentration of Cd^{2+} (mg/L), qe is amount of Cd^{2+} adsorbed per gram of fungal biomass (mg/g), q_{max} is Langmuir constant, and b is the energy of adsorption. The values of b and q_{max} were evaluated by the intercept and slope of the graph, respectively.

2.5.2. Freundlich Isotherm

This isotherm is the result of empirical consideration and expressed as

$$qe = K_f\, Ce^{1/n} \quad (4)$$

where qe is the Cd^{2+} adsorbed (mg/g), Ce is the equilibrium concentration of Cd^{2+} (mg/L), K_f is the adsorption coefficient (mg/g) and directly related to the standard free energy change, and n is the empirical constant.

The linear logarithmic form of the isotherm is:

$$\text{Log } qe = (1/n) \log Ce + \log K_f \quad (5)$$

From the straight-line plot between log qe and log Ce, the values of of K_f and 1/n were found.

2.6. *Scanning Electron Microscopy (SEM) and Fourier Transform Infrared Spectroscopy (FTIR) Analysis*

The morphological changes in the fungi without and after incubating with Cd^{2+} were visualized using Scanning Electron Microscopy (SEM; JSM-6100, JEOL, Tokyo, Japan). The changes in surface chemical profiling were analyzed and interpreted by Fourier transform infrared spectroscopy (FTIR) (Perkin Elmer Spectrum BXII, Waltham, MA, USA, in 4000–400 cm^{-1} frequency range with a resolution of 1 cm^{-1}).

3. Results and Discussion

3.1. *Isolation, Screening, and Identification of Cd^{2+}-Resistant Fungal Biomass*

Twenty-five sludge, industrial effluents and sewage samples were collected from Panipat, Sonepat and Karnal districts of Haryana, India to isolate fungal isolates. All the fungi were maintained on PDA slants (pH 4.5) at 30 °C. Screening of these fungal isolates on PDA containing 25, 50, and 75 mg/L of Cd^{2+} showed that 13 fungal isolates were tolerant to 25 mg/L Cd^{2+}, 7 fungal isolates were tolerant to 50 mg/L, and 3 fungal isolates were tolerant to 75 mg/L (Table 1). Such phenomena indicate that some fungal isolates exhibited inhibition at higher metal ions concentrations. Out of three fungal isolates tolerant to 75 mg/L Cd^{2+}, the highly Cd^{2+}-tolerant fungal isolate FS7 and FS16 were identified as *Trichoderma fasciculatum* (ITCC 7547) and *Trichoderma longibrachiatum* (ITCC 7062) by the Indian Culture Collection, Indian Agricultural Research Institute. IARI, New Delhi, India.

Table 1. Growth profile of all the fungal isolates on potato dextrose agar (PDA) (pH 5.0, temperature 30 °C) containing 25, 50, and 75 mg/L of Cd^{2+} after five days.

Sr. No.	Fungal Isolate Code	Control	Cd^{2+} Ion Concentration (mg/L)		
			25	50	75
1	FS1	++++	++	++	+
2	FS2	++++	++	++	+
3	FS3	++++	++	++	+
4	FS4	++++	++++	+++	++
5	FS5	++++	++	++	+
6	FS6	++++	++	++	+
7	FS7	++++	++++	++++	+++
8	FS8	++++	–	–	–
9	FS9	++++	++++	+++	++
10	FS10	++++	++++	+++	++
11	FS11	++++	–	–	–
12	FS12	++++	++++	+++	+++
13	FS13	++++	–	–	–
14	FS14	++++	++	+	–
15	FS15	++++	–	–	–
16	FS16	++++	++++	++++	+++
17	FS17	++++	+	–	–
18	FS18	++++	+	–	–
19	FS19	++++	-	–	-
20	FS20	++++	+++	++	+
21	FS21	++++	+++	++	+
22	FS22	++++	++++	+++	++
23	FS23	++++	+++	++	+
24	FS24	++++	–	–	–
25	FS25	++++	++	+	–

Note: Absent growth (–), least growth (+), less than normal growth (++), slightly less than normal growth (+++), and normal growth (++++).

3.2. *Optimization of Process Parameters*

Various surface and physiochemical properties—for instance, pH, incubation time, inoculum size, temperature, metal ion concentrations, quantity of biosorbents, etc. are affecting the process for the bioaccumulation of heavy metal ions in fungal biomass.

3.2.1. pH Effect

In the process to remove the heavy metal ions from fungi, the pH plays an important and significant role. The observed results demonstrated that there was an enhancement in the Cd^{2+} ion uptake and percentage removal up to pH 5.0 from the liquid medium by *T. fasciculatum* and *T. longibrachiatum*. After pH 5.0, the uptake and percentage removal was significantly decreased with the increasing pH (Figure 1a). The outcomes showed that the ideal pH for uptake and percentage removal of Cd^{2+} + from the liquid medium by *T. fasciculatum* and *T. longibrachiatum* was 5.0.

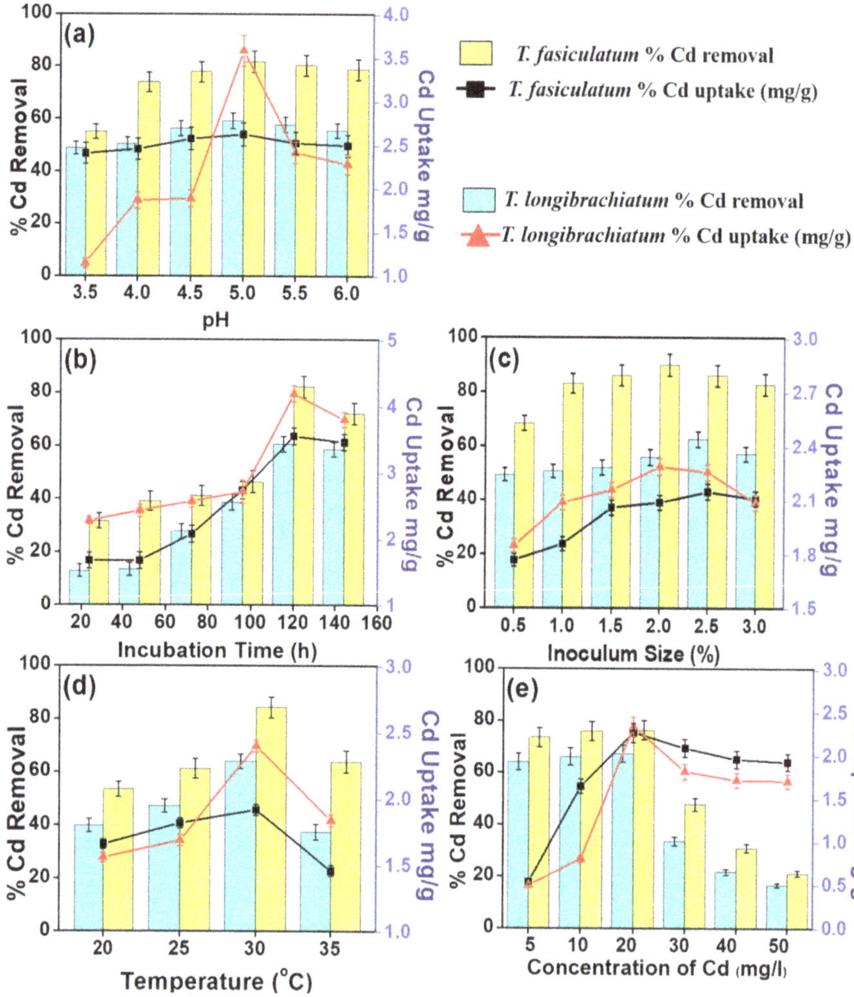

Figure 1. Effect of operational parameters on the Cd^{2+} removal efficiency by *T. fasciculatum* and *T. longibrachiatum* (**a**) pH (**b**) Incubation Time (**c**) Inoculum Size (**d**) Temperature (**e**) Concentration of Cd.

The observed uptake and percentage removal of Cd^{2+} from the liquid medium by *T. fasciculatum* at pH 5.0 were 2.63 mg/g and 59.20%, respectively. In addition, the uptake and percentage removal of Cd^{2+} by *T. longibrachiatum* were 3.59 mg/g and 81.60%, respectively. Interestingly, it has been reported that the optimum pH is important to get better results regarding the removal of heavy metal ions using fungal isolates [37,38]. In many cases, the optimal value of pH was 5.0, which shows a maximum removal of heavy metal ions (Pb(II), Cd(II), As(III), and Hg(II)) by fungus isolates such as *Penicillium purpurogenum* [37,38].

3.2.2. Incubation Time Effect

Figure 1b exhibits the incubation time effect on the Cd uptake and percentage removal from the liquid medium by *T. fasciculatum* and *T. longibrachiatum*. It was observed that at optimum pH 5.0, the uptake and percentage removal of Cd^{2+} ions from the liquid medium by *T. fasciculatum* and *T. longibrachiatum* were increased with increasing the incubation time. Interestingly, it was seen the uptake and percentage removal were enhanced from 24 h to 120 h; however, they reduce after 120 h. The maximum uptake and percentage removal of Cd^{2+} by *T. fasciculatum* at 120 h incubation time were 3.56 mg/g and 60.55%, respectively. Nevertheless, at 120 h incubation time, the uptake and percentage removal of Cd^{2+} ion by *T. longibrachiatum* were 4.2 mg/g and 82.2%, respectively.

It is believed by most of the researchers that ion exchange or physical adsorption at the surface of the cell might relate with the initial rapid phase [39]. However, in the presence of salt, the metal ion transport into the cytoplasm across the cell membrane via an active metabolism-dependent transport is responsible for the slower phase [40]. Moreover, in the absence of salt, other biosorption mechanisms that consist of complexation, microprecipitation, etc. due to the diffusion of metal ions into the cell debris are responsible for the second slower phase [41].

3.2.3. Inoculum Size Effect

Under the optimized conditions (pH 5.0, incubation time 120 h), with increasing the inoculum size from 0.5% to 3.0%, the Cd^{2+} ion uptake and percentage removal was enhanced up to a particular size and then reduced. Figure 1c depicts the relation between the Cd^{2+} ion uptake and percentage removal and inoculum size. The experimental observations confirmed that the uptake and percentage removal of the Cd^{2+} ion was increased with increasing the inoculum size from 0.5% to 2.5% for *T. fasciculatum* and 2.0% for *T. longibrachiatum,* while it declined afterwards. The maximum uptake and percentage removal of Cd^{2+} from the liquid medium by *T. fasciculatum* was approximately 2.15 mg/g and 62.60%, respectively. Moreover, under the same optimized experimental conditions, the Cd^{2+} ion uptake and percentage removal in the liquid medium by *T. longibrachiatum* were 2.29 mg/g and 89.85%, respectively. Interestingly, after the inoculum size 2.5% for *T. fasciculatum* and 2.0% for *T. longibrachiatum*, the uptake and percentage removal of the Cd^{2+} ions were decreased significantly. It has been reported that the percentage removal of heavy metal ions increases with enhancing the adsorbent dose due to the fact that it increases the adsorption sites, which leads to high adsorption and high removal efficiency [42–44].

3.2.4. Temperature Effect

Interestingly, the adsorption medium temperature is also substantially important in the biosorption of the metal ions by microbial cells for energy-dependent mechanisms. As reported in the literature, mostly, the adsorption is considered as an exothermal process [45]. To understand the effect of temperature on the uptake and percentage removal of Cd^{2+} ions, temperature-dependent experiments were performed by keeping all other optimized experimental conditions the same, i.e., pH 5, incubation time 120h, inoculum size 2% for *T. longibrachiatum.* Figure 1d exhibits the relation between the Cd^{2+} ion uptake and percentage removal and adsorption temperature. The observed results revealed that under optimized conditions, the maximum Cd^{2+}ion uptake and percentage removal from the liquid medium by both fungi were obtained at 30 °C. It was interesting to note that for *T. fasciculatum*, the maximum Cd^{2+} uptake and percentage removal was 1.92 mg/g and 63.85%, respectively while for the *T. longibrachiatum,* the uptake and percentage removal was 2.40 mg/g and 84.25%, respectively.

3.2.5. Initial Metal Concentration Effect

Initial metal ion concentrations also have a particular effect on the adsorption characteristics of heavy metal ions. Figure 1e shows the effect of initial metal ion concentration on the uptake and percentage removal of Cd^{2+} ions using fungal isolates. To check the initial ion concentrations, several concentrations (5–50 mg/L) were tested, and all the experiments were performed at optimized experimental conditions, i.e., pH 5.0, incubation time (120 h), inoculum size (2.5% for *T. fasciculatum*, and 2.0% for *T. longibrachiatum*), and temperature 30 °C. The observed results confirmed that under optimized reaction conditions, for both fungal isolates, the Cd^{2+} uptake and percentage removal increased with the increasing the Cd^{2+} concentrations from 5 to 20 mg/L, and afterwards, a decline was seen. At an optimum concentration of 20 mg/L, the uptake and percentage removal of Cd^{2+}

from liquid medium was 2.27 mg/g and 67.10% for *T. fasciculatum* and 2.34 mg/g and 76.25% for *T. longibrachiatum*, respectively. This enhancement may be because of an increase in the electrostatic associations (comparative with covalent interactions), including sites of continuously lower affinity for metal ions [46].

3.3. Adsorption Isotherms

The availability of a finite number of binding sites can be expected in the Langmuir model, which equally distributed to the sorbent surface. It presents the same attraction for single-layer sorption with no interaction between sorbed species. The Freundlich equation provides the adsorption data for a limited concentration range and proposed an adsorption sites heterogeneity on the biomass [47]. By changing the pH from 3.5 to 6.0 for Cd^{2+} ion removal from the liquid medium using *T. fasciculatum* and *T. longibrachiatum*, the obtained results were examined, and sorption isotherms were calculated. Figure 2 exhibits the calculated Langmuir and Freundlich isotherms for Cd^{2+} removal by *T. fasciculatum* and *T. longibrachiatum*. As a result of the higher value of correlation coefficients $R^2 > 0.90$, the observed results revealed that the Cd^{2+} removal using both the fungal isolates, i.e., *T. fasciculatum* and *T. longibrachiatum*, better fit with the Langmuir isotherm models compared with the Freundlich isotherm. Various parameters of Langmuir and Freundlich models for the Cd^{2+} ion sorption on the live cells of *T. fasciculatum* and *T. longibrachiatum* are presented in Table 2. As observed, the maximum Cd^{2+} ion adsorption capacities (q_{max}) for *T. fasciculatum* and *T. longibrachiatum* fungal isolates were 1.90 mg/g and 0.80 mg/g, while the calculated correlation coefficients (R^2) were 0.99 and 0.99 for both the fungal isolates, respectively.

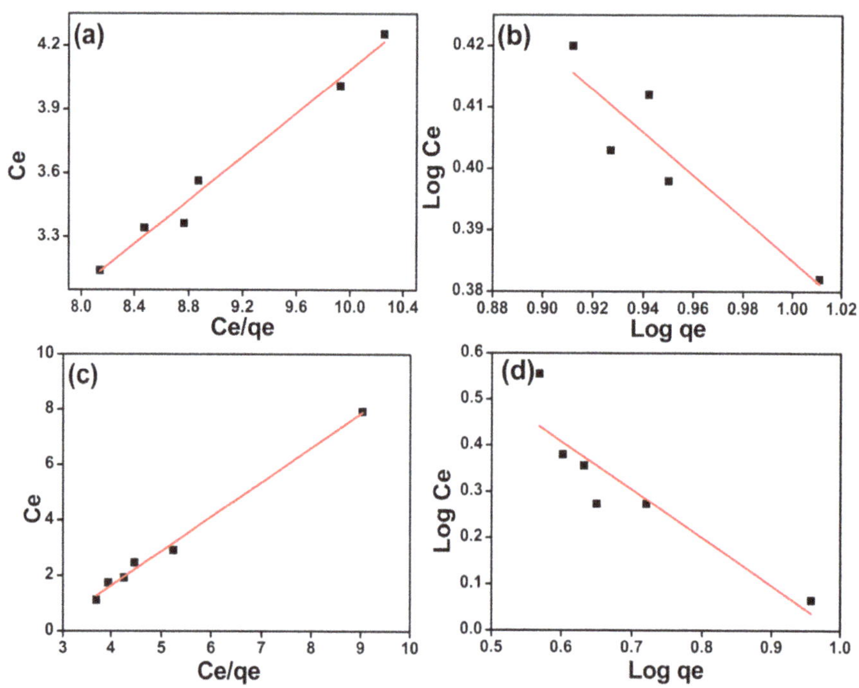

Figure 2. (**a**,**c**) Langmuir and (**b**,**d**) Fredundlich isotherms for Cd^{2+} removal by *T. fasciculatum* and *T. longibrachiatum*, respectively.

Table 2. Langmuir and Freundlich isotherm constants for removal of Cd^{2+} by *T. fasciculatum* and *T. longibrachiatum*.

Isotherms	Parameters	*T. fasciculatum*	*T. longibrachiatum*
Langmuir	q_{max} (mg/g)	1.90	0.80
	b (L/mg)	−2.23	−2.75
	R^2	0.99	0.99
Freundlich	K_f (L/g)	−0.15	0.02
	n	−3.11	−0.95
	R^2	0.84	0.84

3.4. SEM and FTIR Analysis

The morphological changes before and after the Cd^{2+} ions accumulation in the fungal isolates, i.e., *T. fasciculatum* and *T. longibrachiatum*, were further analyzed by scanning electron microscopy (SEM), and the results are presented in Figure 3.

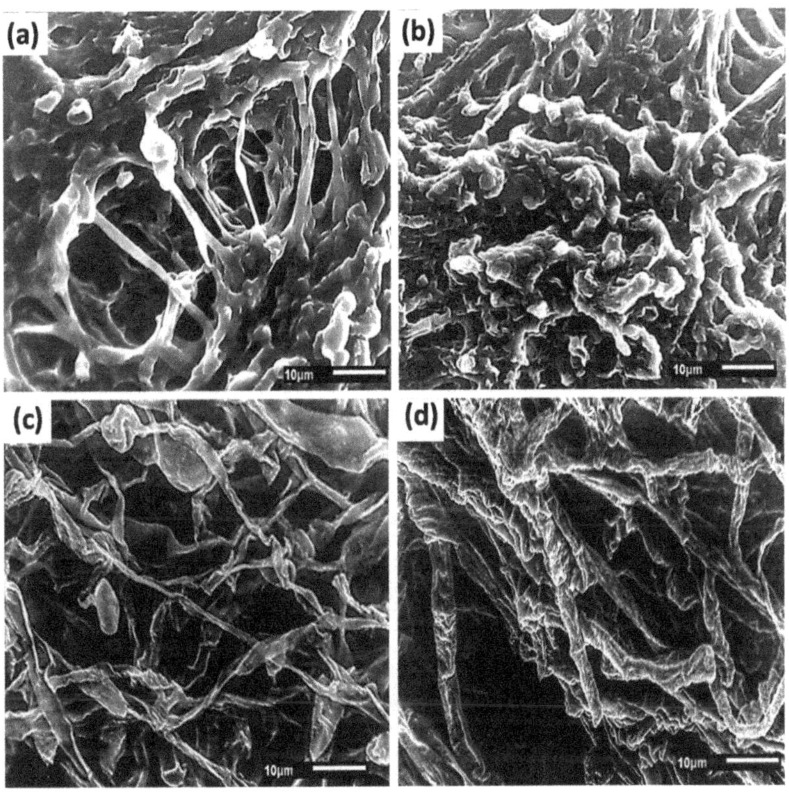

Figure 3. SEM images of (**a**) *T. fasciculatum* control and (**b**) after treatment with Cd^{2+}, (**c**) *T. longibrachiatum* control and (**d**) after treatment with Cd^{2+}.

The observation revealed that after 120 h, the hyphae of both the fungi was tube shaped, septate, and diverged with no metal treatment. However, after treatment with 20 mg/L of Cd^{2+} after 120 h, there was a thorough disruption and dissolution of mycelium of *T. fasciculatum* and *T. longibrachiatum*.

These metals were consistently intact to the fungal mycelium, and a higher absorption of Cd^{2+} ions along with flocculation in mycelium was seen. The formation of such complexes was most probably because of the detoxification mechanism, in which normally fungal isolates are used in order to manage the lethal concentrations of heavy metal ions [48].

To understand the metal ions bindings, the attachment of various functional groups on the fungal surface are important to study. To examine the functional groups available on the surfaces of fungal isolates, FTIR studies were performed for both the fungal biomass, i.e., *T. fasciculatum* and *T. longibrachiatum* before and after Cd^{2+} ions treatment. Figure 4 exhibits the typical FTIR spectra in the range of 450–4000 cm^{-1} for both the fungal biomass, i.e., *T. fasciculatum* and *T. longibrachiatum* before and after Cd^{2+} ions treatment. The Cd^{2+} free biomass of *T. fasciculatum* and *T. longibrachiatum* exhibited several absorption peaks, which show the complex nature of the biomass. The peak that appeared in the range of 3500–3200 cm^{-1} was due to stretching of the N–H bond of the amino groups and the O–H bond of the hydroxyl group. Upon Cd^{2+} loading to the biomass of *T. fasciculatum* and *T. longibrachiatum*, a significant change in the peak positions was seen, which confirmed the Cd^{2+} binding with hydroxyl and amino groups (Table 3). The peaks that appeared at 2853 cm^{-1} and 2922 cm^{-1} exhibited the presence of C–H methyl groups stretching. Furthermore, the peak shown at 1747 cm^{-1} represents the native carbonyl stretching, while the peak at 1373 cm^{-1} represented as CH symmetric/symmetric band. The peaks that appeared at 2853 cm^{-1}, 2922 cm^{-1}, and 1747 cm^{-1}did not show any significant change in samples exposed to Cd^{2+} except for a peak exhibited at 1373 cm^{-1}, which showed a slight change from its original position.

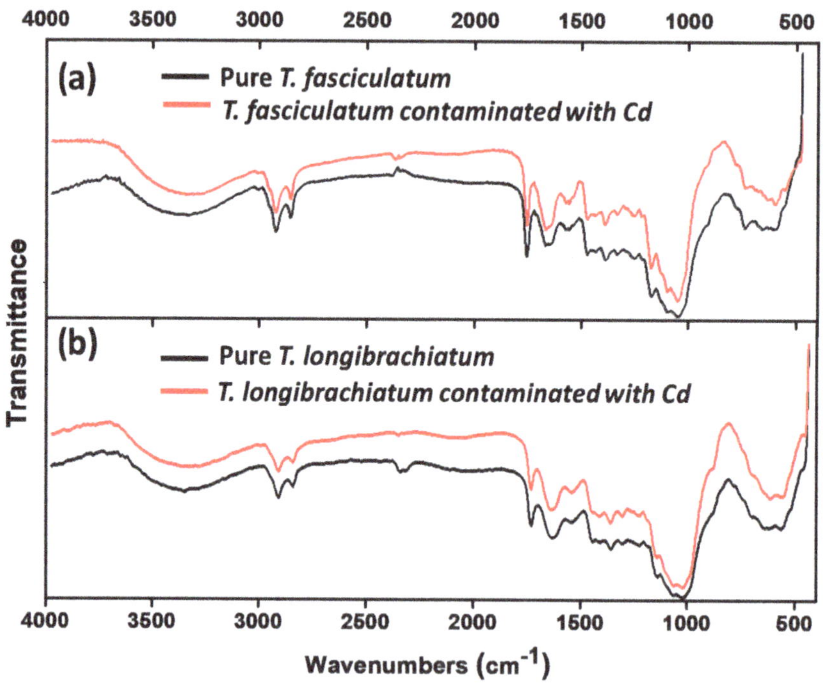

Figure 4. Fourier transform infrared spectroscopy (FTIR) spectra of (**a**) *T. fasciculatum* and (**b**) *T. longibrachiatum* fungal biomass before and after treatment with Cd^{2+}.

Table 3. FTIR peaks and their corresponding functional groups of *T. fasciculatum* and *T. longibrachiatum* before and after the addition of Cd^{2+}.

Sr. no.	*T. fasciculatum*		*T. longibrachiatum*		Assigned Groups
	Control	Cd^{2+} Addition	Control	Cd^{2+} Addition	
1	3332.6	3320.1	3363.5	3325.2	-OH, -NH
2	2923.6	2924.3	2923.3	2923.3	-CH, -OH
3	2853.3	2853.2	2854.6	2854.9	-CH
4	1745.4	1746.0	1744.9	1745.1	-C=O, of ester group
5	1654.8	1654.3	1645.5	1648.1	-C=O, COO
6	1560.2	1563.0	1554.5	1558.8	-NH
7	1373.8	1375.6	1372.3	1373.8	-CH

In case of Cd^{2+}-unloaded spectra of *T. Fasciculatum* and *T. longibrachiatum*, the peaks appeared at 1650 and 1544 cm^{-1} are related with the C=O of amide I and the NH/C=O blend of the amide II bond, respectively confirming the availability of the carboxyl groups [49,50]. It is interesting to see that the peak positioned at 1544 cm^{-1} expanded in the presence of Cd^{2+} ions, which confirms the interaction of Cd^{2+} with carboxyl groups. The above observations revealed the presence of various functional groups such as -CH, -OH, –C=O, and –NH in the binding of Cd^{2+}ions with the fungal biomass, i.e., *T. fasciculatum* and *T. longibrachiatum*. The observed results are in agreement with the reported literature [49–51].

3.5. Proposed Mechanism for Cd^{2+} Removal

Based on the findings of this work, we proposed the possible mechanism of *T. fasciculatum* and *T. longibrachiatum* response to Cd^{2+} stress (Figure 5). The bioaccumulation of Cd^{2+} in *T. fasciculatum* and *T. longibrachiatum* is a cumulative result of three pathways. First the Cd^{2+} get adsorbed over the fungal outer membrane [52]. It has been shown in the FTIR results that various functional groups available on the fungal surface showed significant changes in their peak positions, indicating the chemical adsorption of Cd^{2+} over the membrane. Then, the ions are engulfed in the fungal cytoplasm where they are assimilated and sequestrated by the cellular reactions [53]. In addition, the fungi can release exopolysaccharides (high-molecular-weight polymers composed of sugar residues) to entrap or adsorb the extracellular Cd^{2+} ions [54].

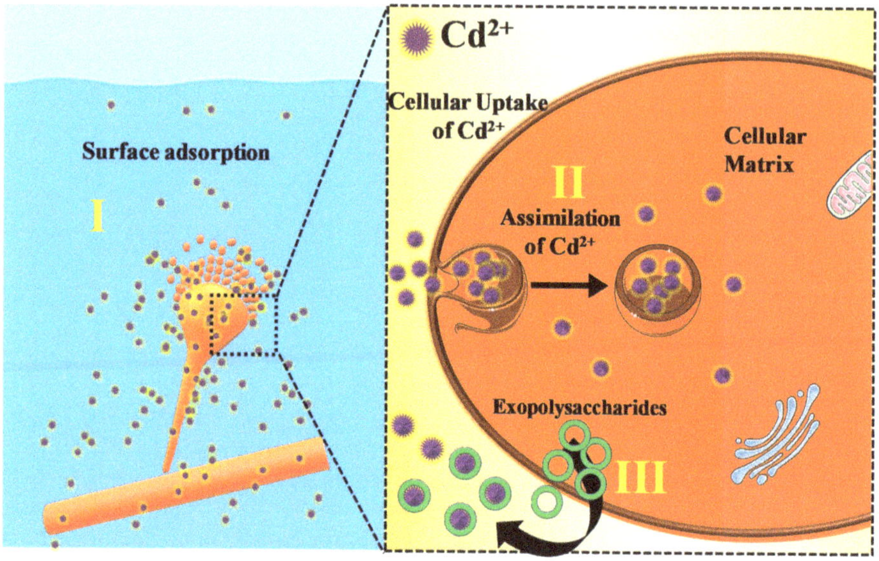

Figure 5. Pictorial illustration of Cd^{2+} removal by *T. fasciculatum* and *T. longibrachiatum*.

4. Conclusions

In summary, 25 sludge, industrial effluents and sewage samples were collected to isolate fungal isolates and screened to their tolerance to Cd^{2+} in PDA medium containing Cd^{2+} from 25 to 75 mg/L. There was a decline in the number of fungi for their tolerance to Cd^{2+} with the increment in concentration of Cd^{2+} from 25 to 75 mg/L. The removal of Cd^{2+} by highly effective fungi *T. fasciculatum* and *T. longibrachiatum* was dependent on various experimental parameters such as temperature, initial metal ion concentrations, pH, incubation time, inoculum size, etc. Under optimized conditions, i.e., pH 5.0, inoculum size (2.5% for *T. fasciculatum* and 2.0% for *T. longibrachiatum*), incubation time (120 h), initial Cd^{2+} concentration (20 mg/L), and temperature (30 °C), the maximum removal of Cd^{2+} was 67.10% for *T. fasciculatum* and 76.25% for *T. longibrachiatum*. The observed data were well fitted with the Langmuir and Freundlich isotherm models. The SEM examinations confirmed the morphological changes in response to the Cd^{2+} aggregation in *T. fasciculatum* and *T. longibrachiatum*. Both the fungal isolate exhibited Cd^{2+} binding on the cell wall surface as a mechanism of metal tolerance. The FTIR studies confirmed the presence of various functional groups such as –CH, –OH, -C=O and –NH in the binding of Cd^{2+} ions with the fungal biomass, i.e., *T. fasciculatum* and *T. longibrachiatum*. The observed results revealed the potential biosorption tendency of *T. fasciculatum* and *T. longibrachiatum* to remove Cd^{2+} from wastewater.

Author Contributions: Conceptualization, R.K. (Raman Kumar), N.S. and A.U.; methodology, R.K. (Raman Kumar), N.S. and A.U.; software, R.K. (Raman Kumar), N.S., P.S., A.U., R.K. (Rajeev Kumar), P.K.J., F.A.A., A.A.A. and N.A.-Z.; validation, R.K. (Raman Kumar), P.S., A.U., R.K. (Rajeev Kumar), P.K.J., F.A.A., A.A.A., and N.A.-Z.; formal analysis, R.K. (Raman Kumar), N.S. and A.U.; the investigation, R.K. (Raman Kumar), and A.U.; resources, R.K. (Raman Kumar), P.S., A.U., R.K. (Rajeev Kumar), N.S.,P.K.J., F.A.A., A.A.A., and N.A.-Z.; data curation, R.K. (Raman Kumar), P.S., A.U., R.K. (Rajeev Kumar), N.S., P.K.J., F.A.A., A.A.A., and N.A.-Z.; writing—original draft preparation, R.K. (Raman Kumar) and A.U.; writing—review and editing, R.K. (Raman Kumar) and A.U.; visualization, R.K. (Raman Kumar), P.S., A.U., R.K. (Rajeev Kumar), N.S., P.K.J., F.A.A., A.A.A. and N.A.-Z.; supervision, R.K. (Raman Kumar); N.S. project administration, R.K. (Raman Kumar), N.S. and A.U. All authors have read and agreed to the published version of the manuscript.

Funding: The authors extend their appreciation to the Deanship of Scientific Research at King Saud University for funding this work through the research group No. RGP-1441-305.

Acknowledgments: The authors are thankful to Haryana State Council of Science and Technology, Panchkula, Haryana for providing financial support for carrying out research work (Grant No. HSCST/R&D/2018/209R). The authors are thankful to Maharishi Markandeshwar (Deemed to be University), Mullana, India for providing the lab facilities. The authors extend their appreciation to the Deanship of Scientific Research at King Saud University for funding this work through the research group No. RGP-1441-305.

Conflicts of Interest: The authors declare no conflict of interest.

References

1. Wu, H.-Z.; Meng, L.-F.; Song, W.-G. Glutathione-Stabilized Silver Nanoparticles and Magnetic Nanoparticles Combination for Determination of Lead and Cadmium in Environmental Waters. *Sci. Adv. Mater.* **2019**, *11*, 1133–1139. [CrossRef]
2. Hoang, H.-G.; Lin, C.; Tran, H.-T.; Chiang, C.-F.; Bui, X.-T.; Cheruiyot, N.K.; Shern, C.-C.; Lee, C.-W. Heavy metal contamination trends in surface water and sediments of a river in a highly-industrialized region. *Environ. Technol. Innov.* **2020**, *20*, 101043. [CrossRef]
3. Song, T.; Li, R.; Li, N.; Gao, Y. Research Progress on the Application of Nanometer TiO_2 Photoelectrocatalysis Technology in Wastewater Treatment. *Sci. Adv. Mater.* **2019**, *11*, 158–165. [CrossRef]
4. Tan, K.; Ma, W.; Chen, L.; Wang, H.; Du, Q.; Du, P.; Yan, B.; Liu, R.; Li, H. Estimating the Distribution Trend of Soil Heavy Metals in Mining Area from HyMap Airborne Hyperspectral Imagery Based on Ensemble Learning. *J. Hazard. Mater.* **2020**, *401*, 123288. [CrossRef] [PubMed]
5. Zhou, X.-Y.; Wang, X.-R. Impact of industrial activities on heavy metal contamination in soils in three major urban agglomerations of China. *J. Clean. Prod.* **2019**, *230*, 1–10. [CrossRef]
6. Yao, S.; Li, Q.; Kong, N.; Li, C.; Wei, M.; Song, R.; Ren, X.; Pu, X.; Li, W. Influence of Initial Cu/(Zn+Sn) Concentration Ratio in Cu–Zn–Sn–S Composites on Their Microstructures, Adsorption and Visible-Light-Sensitive Photocatalytic Activities. *Sci. Adv. Mater.* **2018**, *10*, 1381–1388. [CrossRef]
7. Gulan, L.; Milenkovic, B.; Zeremski, T.; Milic, G.; Vuckovic, B. Persistent organic pollutants, heavy metals and radioactivity in the urban soil of Priština City, Kosovo and Metohija. *Chemosphere* **2017**, *171*, 415–426. [CrossRef]
8. Hassen, A.; Saidi, N.; Cherif, M.; Boudabous, A. Effects of heavy metals on *Pseudomonas aeruginosa* and *Bacillus thuringiensis*. *Bioresour. Technol.* **1998**, *65*, 73–82. [CrossRef]
9. Di Cesare, A.; Pjevac, P.; Eckert, E.; Curkov, N.; Šparica, M.M.; Corno, G.; Orlić, S.; Andrea, D.C.; Petra, P.; Ester, M.E.; et al. The role of metal contamination in shaping microbial communities in heavily polluted marine sediments. *Environ. Pollut.* **2020**, *265*, 114823. [CrossRef]
10. Bost, M.; Houdart, S.; Oberli, M.; Kalonji, E.; Huneau, J.-F.; Margaritis, I. Dietary copper and human health: Current evidence and unresolved issues. *J. Trace Elements Med. Boil.* **2016**, *35*, 107–115. [CrossRef]
11. Rayman, M.P. The importance of selenium to human health. *Lancet* **2000**, *356*, 233–241. [CrossRef]
12. Guo, J.; Xie, J.; Zhou, B.; Găman, M.-A.; Varkaneh, H.K.; Clark, C.C.; Salehi-Sahlabadi, A.; Li, Y.; Han, X.; Hao, Y.; et al. The influence of zinc supplementation on IGF-1 levels in humans: A systematic review and meta-analysis. *J. King Saud Univ. Sci.* **2020**, *32*, 1824–1830. [CrossRef]
13. Jaishankar, M.; Tseten, T.; Anbalagan, N.; Mathew, B.B.; Beeregowda, K.N. Toxicity, mechanism and health effects of some heavy metals. *Interdiscip. Toxicol.* **2014**, *7*, 60–72. [CrossRef]
14. Piñón-Gimate, A.; Jakes-Cota, U.; Tripp-Valdez, A.; Casas-Valdez, M.; Almendarez-Hernández, L.C.; Alejandra, P.-G.; Ulianov, J.-C.; Arturo, T.-V.; Margarita, C.-V.; Cesar, A.-H.L. Assessment of human health risk: Copper and lead concentrations in Stone Scorpionfish (*Scorpaena mystes*) from the coastal region of Santa Rosalia in the Gulf of California, Mexico. *Reg. Stud. Mar. Sci.* **2020**, *34*, 101003. [CrossRef]
15. Bi, Y.; Liu, D.; Huang, M. Preparation and Characterization of Nano Hydroxyapatite and Its Adsorption Behavior toward Lead Ions. *Sci. Adv. Mater.* **2018**, *10*, 896–903. [CrossRef]
16. Hosseini, S.; Alibakhshi, H.; Jashni, E.; Parvizian, F.; Shen, J.; Taheri, M.; Ebrahimi, M.; Rafiei, N. A novel layer-by-layer heterogeneous cation exchange membrane for heavy metal ions removal from water. *J. Hazard. Mater.* **2020**, *381*, 120884. [CrossRef]
17. Dabrowski, A.; Hubicki, Z.; Podkościelny, P.; Robens, E. Selective removal of the heavy metal ions from waters and industrial wastewaters by ion-exchange method. *Chemosphere* **2004**, *56*, 91–106. [CrossRef] [PubMed]

18. Tian, D.; Lv, H.; Mao, H.; Wang, D.; Zheng, Y.; Peng, C.; Hou, S. Synthesis, Characterization and Adsorption Behavior of Cd(II) Ion-Imprinted Mesoporous Materials. *Sci. Adv. Mater.* **2018**, *10*, 324–330. [CrossRef]
19. Ya, V.; Martin, N.; Chou, Y.-H.; Chen, Y.; Choo, K.-H.; Chen, S.-S.; Li, C.-W. Electrochemical treatment for simultaneous removal of heavy metals and organics from surface finishing wastewater using sacrificial iron anode. *J. Taiwan Inst. Chem. Eng.* **2018**, *83*, 107–114. [CrossRef]
20. Azarian, G.; Rahmani, A.R.; Khoram, M.M.; Atashzaban, Z.; Nematollahi, D. New batch electro-coagulation process for treatment and recovery of high organic load and low volume egg processing industry wastewater. *Process. Saf. Environ. Prot.* **2018**, *119*, 96–103. [CrossRef]
21. Shih, Y.-J.; Chien, S.-K.; Jhang, S.-R.; Lin, Y.-C. Chemical leaching, precipitation and solvent extraction for sequential separation of valuable metals in cathode material of spent lithium ion batteries. *J. Taiwan Inst. Chem. Eng.* **2019**, *100*, 151–159. [CrossRef]
22. Jantunen, N.; Virolainen, S.; Latostenmaa, P.; Salminen, J.; Haapalainen, M.; Sainio, T. Removal and recovery of arsenic from concentrated sulfuric acid by solvent extraction. *Hydrometallurgy* **2019**, *187*, 101–112. [CrossRef]
23. Xia, Y.; Tsai, F.-C.; Ma, N.; Zhang, K.-D.; Liu, H.-L.; Jiang, T.; Chiang, T.-C. Green Synthesis of H3O40PW12@MIL-100(Fe): Enhanced and Selective Adsorption of Methylene Blue from Aqueous Solution. *Sci. Adv. Mater.* **2018**, *10*, 172–180. [CrossRef]
24. Hao, J.; Ji, L.; Li, C.; Hu, C.; Wu, K. Rapid, efficient and economic removal of organic dyes and heavy metals from wastewater by zinc-induced in-situ reduction and precipitation of graphene oxide. *J. Taiwan Inst. Chem. Eng.* **2018**, *88*, 137–145. [CrossRef]
25. Kim, T.; Kim, T.-K.; Zoh, K.-D. Removal mechanism of heavy metal (Cu, Ni, Zn, and Cr) in the presence of cyanide during electrocoagulation using Fe and Al electrodes. *J. Water Process. Eng.* **2020**, *33*, 101109. [CrossRef]
26. Fan, Y.; Wang, Y.; Kang, D.; Deng, L.; Li, C. Oil-Tea Shell Derived N-Doped Porous Carbon for Selective Separation of CO_2, CH_4, and N_2. *Sci. Adv. Mater.* **2019**, *11*, 1146–1155. [CrossRef]
27. Zhao, X.; Liu, C. Efficient removal of heavy metal ions based on the optimized dissolution-diffusion-flow forward osmosis process. *Chem. Eng. J.* **2018**, *334*, 1128–1134. [CrossRef]
28. Sadeghi, M.H.; Tofighy, M.A.; Mohammadi, T. One-dimensional graphene for efficient aqueous heavy metal adsorption: Rapid removal of arsenic and mercury ions by graphene oxide nanoribbons (GONRs). *Chemosphere* **2020**, *253*, 126647. [CrossRef]
29. Naseem, K.; Begum, R.; Wu, W.; Usman, M.; Irfan, A.; Al-Sehemi, A.G.; Farooqi, Z.H. Adsorptive removal of heavy metal ions using polystyrene-poly(N-isopropylmethacrylamide-acrylic acid) core/shell gel particles: Adsorption isotherms and kinetic study. *J. Mol. Liq.* **2019**, *277*, 522–531. [CrossRef]
30. Gola, D.; Dey, P.; Bhattacharya, A.; Mishra, A.; Malik, A.; Namburath, M.; Ahammad, S.Z. Multiple heavy metal removal using an entomopathogenic fungi *Beauveria bassiana*. *Bioresour. Technol.* **2016**, *218*, 388–396. [CrossRef]
31. Talukdar, D.; Jasrotia, T.; Sharma, R.; Jaglan, S.; Kumar, R.; Vats, R.; Kumar, R.; Mahnashi, M.H.; Umar, A. Evaluation of novel indigenous fungal consortium for enhanced bioremediation of heavy metals from contaminated sites. *Environ. Technol. Innov.* **2020**, *20*, 101050. [CrossRef]
32. Rangabhashiyam, S.; Balasubramanian, P. Characteristics, performances, equilibrium and kinetic modeling aspects of heavy metal removal using algae. *Bioresour. Technol. Rep.* **2019**, *5*, 261–279. [CrossRef]
33. Qin, H.; Hu, T.; Zhai, Y.; Lu, N.; Aliyeva, J. The improved methods of heavy metals removal by biosorbents: A review. *Environ. Pollut.* **2020**, *258*, 113777. [CrossRef] [PubMed]
34. Wei, Y.-Q.; Zhao, Y.; Zhao, X.; Gao, X.; Zheng, Y.; Zuo, H.; Wei, Z. Roles of different humin and heavy-metal resistant bacteria from composting on heavy metal removal. *Bioresour. Technol.* **2019**, *296*, 122375. [CrossRef]
35. Bano, A.; Hussain, J.; Akbar, A.; Mehmood, K.; Anwar, M.; Hasni, M.S.; Ullah, S.; Sajid, S.; Ali, I. Biosorption of heavy metals by obligate halophilic fungi. *Chemosphere* **2018**, *199*, 218–222. [CrossRef]
36. El Hameed, A.H.A.; Eweda, W.E.; Abou-Taleb, K.A.; Mira, H. Biosorption of uranium and heavy metals using some local fungi isolated from phosphatic fertilizers. *Ann. Agric. Sci.* **2015**, *60*, 345–351. [CrossRef]
37. Visoottiviseth, P.; Panviroj, N. Selection of fungi capable of removing toxic arsenic compounds from liquid medium. *Sci. Asia* **2011**, *27*, 83–92. [CrossRef]
38. Say, R.; Yılmaz, N.; Denizli, A. Biosorption of Cadmium, Lead, Mercury, and Arsenic Ions by the Fungus *Penicillium purpurogenum*. *Sep. Sci. Technol.* **2003**, *38*, 2039–2053. [CrossRef]

39. Goksungur, Y. Biosorption of cadmium and lead ions by ethanol treated waste baker's yeast biomass. *Bioresour. Technol.* **2005**, *96*, 103–109. [CrossRef]
40. Dönmez, G.; Aksu, Z. Bioaccumulation of copper(II) and nickel(II) by the non-adapted and adapted growing *Candida* SP. *Water Res.* **2001**, *35*, 1425–1434. [CrossRef]
41. Prakasham, R.; Merrie, J.; Sheela, R.; Saswathi, N.; Ramakrishna, S. Biosorption of chromium VI by free and immobilized *Rhizopus arrhizus*. *Environ. Pollut.* **1999**, *104*, 421–427. [CrossRef]
42. Kumar, R.; Negi, S.; Sharma, P.; Prasher, I.; Chaudhary, S.; Dhau, J.S.; Umar, A. Wastewater cleanup using *Phlebia acerina* fungi: An insight into mycoremediation. *J. Environ. Manag.* **2018**, *228*, 130–139. [CrossRef] [PubMed]
43. Chaudhary, S.; Sharma, P.; Kaur, A.; Kumar, R.; Mehta, S.K. Surfactant Coated Silica Nanoparticles as Smart Scavengers for Adsorptive Removal of Naphthalene. *J. Nanosci. Nanotechnol.* **2018**, *18*, 3218–3229. [CrossRef] [PubMed]
44. Chaudhary, S.; Sharma, P.; Singh, D.; Umar, A.; Kumar, R. Chemical and Pathogenic Cleanup of Wastewater Using Surface-Functionalized CeO_2 Nanoparticles. *ACS Sustain. Chem. Eng.* **2017**, *5*, 6803–6816. [CrossRef]
45. Sharma, Y. Cr(VI) removal from industrial effluents by adsorption on an indigenous low-cost material. *Colloids Surfaces A Physicochem. Eng. Asp.* **2003**, *215*, 155–162. [CrossRef]
46. Al-Asheh, S.; Duvnjak, Z. Adsorption of copper and chromium by *Aspergillus carbonarius*. *Biotechnol. Prog.* **1995**, *11*, 638–642. [CrossRef]
47. Juang, R.-S.; Wu, F.-C.; Tseng, R.-L. Adsorption Isotherms of Phenolic Compounds from Aqueous Solutions onto Activated Carbon Fibers. *J. Chem. Eng. Data* **1996**, *41*, 487–492. [CrossRef]
48. Manna, A.; Sundaram, E.; Amutha, C.; Vasantha, V.S. Efficient Removal of Cadmium Using Edible Fungus and Its Quantitative Fluorimetric Estimation Using (Z)-2-(4H-1,2,4-Triazol-4-yl)iminomethylphenol. *ACS Omega* **2018**, *3*, 6243–6250. [CrossRef]
49. Naumann, A.; Navarro-González, M.; Peddireddi, S.; Kües, U.; Polle, A. Fourier transform infrared microscopy and imaging: Detection of fungi in wood. *Fungal Genet. Biol.* **2005**, *42*, 829–835. [CrossRef]
50. Jilkine, K.; Gough, K.M.; Julian, R.; Kaminskyj, S.G. A sensitive method for examining whole-cell biochemical composition in single cells of filamentous fungi using synchrotron FTIR spectromicroscopy. *J. Inorg. Biochem.* **2008**, *102*, 540–546. [CrossRef]
51. Akar, T.; Tunali, S. Biosorption performance of *Botrytis cinerea* fungal by-products for removal of Cd(II) and Cu(II) ions from aqueous solutions. *Miner. Eng.* **2005**, *18*, 1099–1109. [CrossRef]
52. Morley, G.F.; Gadd, G.M. Sorption of toxic metals by fungi and clay minerals. *Mycol. Res.* **1995**, *99*, 1429–1438. [CrossRef]
53. Krämer, U.; Talke, I.N.; Hanikenne, M.; Krämer, U. Transition metal transport. *FEBS Lett.* **2007**, *581*, 2263–2272. [CrossRef] [PubMed]
54. Rana, S.; Upadhyay, L.S.B. Microbial exopolysaccharides: Synthesis pathways, types and their commercial applications. *Int. J. Biol. Macromol.* **2020**, *157*, 577–583. [CrossRef]

processes

MDPI

Article

Design, Cost Estimation and Sensitivity Analysis for a Production Process of Activated Carbon from Waste Nutshells by Physical Activation

Marcelo León [1,*], Javier Silva [1], Samuel Carrasco [1] and Nelson Barrientos [2]

1 Escuela de Ingeniería Química, Pontificia Universidad Católica de Valparaíso, Av. Brasil 2162, Valparaíso 2362854, Chile; javier.silva@pucv.cl (J.S.); samuel.carrasco@pucv.cl (S.C.)

2 Trazado Nuclear e Ingeniería Ltda., Francisco de Villagra 385, Santiago 7760016, Chile; n.barrientos@trazadonuclear.cl

* Correspondence: marcelo.leon@pucv.cl; Tel.: +56-32-237-2630

Received: 2 July 2020; Accepted: 3 August 2020; Published: 6 August 2020

Abstract: A conceptual design of an industrial production plant for activated carbon was developed to process 31.25 tons/day of industrial waste nutshells as the raw material and produce 6.6 ton/day of activated carbon using steam as an activation agent. The design considered the cost of the main equipment, the purchase price of the nutshells, basic services, and operation. A sensitivity analysis was developed, considering the price of the finished product and the volume of raw material processing varied up to ±25%. Furthermore, the total annual cost of the product was determined based on the production of 2100 tons/year of activated carbon. Two cash flows were developed and projected to periods of 10 years and 15 years of production, using a tax rate of 27%, a low discount rate (LDR) of 10% per year, and without external financing. For a 10-year production project, the net present value (NPV) was USD 2,785,624, the internal return rate (IRR) 21%, the return on investment (ROI) 25%, and the discounted payback period (DPP) after the fifth year. Considering a project with 15 years of production, the NPV was USD 4,519,482, the IRR at 23%, the ROI 24%, and the DPP after the fifth year of production.

Keywords: economic evaluation; production cost; nutshell waste; activated carbon

1. Introduction

Currently, there is a great interest in adopting more efficient and low-cost processes for the treatment of wastewater. The rapid growth of the world population is resulting in increased contamination of freshwater sources, generating conditions of water stress in the short term [1]. The adsorption technique using activated carbon is one of the most-used methods for the removal of organic pollutants and metal ions in wastewaters, with previous studies reporting effective removal of impurities [2,3].

Activated carbons are highly porous carbon materials with a high specific internal surface area and commonly serve as adsorbent material in various industrial separation and purification applications [4]. Activated carbons can be obtained from chemical activation processes or physical activation of organic precursors. Chemical activation consists of the impregnation of the raw material with chemical agents, such as KOH, $ZnCl_2$, and H_3PO_4, among others, and simultaneous carbonization (pyrolysis) of the impregnated biomass in an inert gaseous atmosphere, where the main role of the activating compound is the degradation of the cellulosic material [5,6]. On the other hand, two relevant processes are involved in physical activation. The first process involves the pyrolysis of the raw material in an inert atmosphere that usually uses a gaseous stream of nitrogen. The second stage of activation involves gases, such as water vapor, carbon dioxide, or a mixture of these gasses with nitrogen or air in different proportions at high temperatures for the final activation [5,6].

Generally, the raw material's pyrolysis process takes place at temperatures between 400 and 600 °C, while the activation stage takes place between 800 and 1100 °C [5,6]. In comparison with chemical activation, physical activation can be considered clean and environmentally friendly, which would avoid the incorporation of impurities from chemical activation agents. There are various organic materials that serve as raw materials for obtaining activated coals, such as coconut shells [7], rice shells [8], palm shells [9], peanut shells [10], and nutshells [11], among others [5,6]. Recent studies have shown that physically activating carbon with steam using nutshells as a precursor has presented interesting characteristics for applications in the removal of heavy metals from effluents of polluted waters [12,13]. In this sense, it could be interesting to develop a complete economic evaluation of a nutshell activated carbon production plant, using the parameters and conditions used in the laboratory.

There are few reports on the analysis of production costs of activated carbon production plants. Noticeably, the study of Ng et al. [14] in 2003 considers the production-cost analysis for an activated carbon plant from pecan shells. The study compares the physical activation with steam and the chemical activation with phosphoric acid and reaches a production cost of 2.72 USD/kg and 2.89 USD/kg, respectively. Choy et al. [15] in 2005 reported on the production of activated carbon from bamboo waste and the evaluation of two production plants: one independent and one integrated. The study presents a thorough economic investigation and sensitivity analysis, estimating values for the internal return rate of 13.0% and 20.1%, respectively, among other economic indicators of interest. Subsequently, Lima et al. [16] in 2008 reported the capital and operating costs for an activated carbon plant from poultry waste, obtaining a global production cost of 1.44 USD/kg. Next, Stavropoulos et al. [17] in 2009 reported different production costs and other economic indicators for physically and chemically activated carbon production processes based on precursors, such as used tires, wood, petroleum coke, carbon black, coal, and lignite. The authors obtained production costs of 2.23 USD/kg, 2.49 USD/kg, 1.08 USD/kg, 1.22 USD/kg, 1.25 USD/kg and 2.18 USD/kg, respectively. Moreover, Vanreppelen et al. [18] in 2011 reported results on the feasibility of a process to produce nitrogenous activated carbon by co-pyrolysis of a mixture of particle board (chipboard) and melamine formaldehyde resin, estimating different economic indicators and developing sensitivity analyses. Furthermore, Santadkha and Skolpap [19] in 2017 reported the results of the economic evaluation for three types of production processes: first, a process of generating activated carbon from coconut shells; second, a process of regenerating coal obtained from the petrochemical industry, and, third, an integrated process that combines the production and regeneration of activated carbon. Nandiyanto [20] in 2018 reported the economic feasibility of the production of activated carbon and silica particles from rice straw residues, obtaining various economic indicators, such as an internal return rate of 44.06% for a case study.

The present work updates, complements, and discusses new economic approaches on the implementation of an industrial plant that produces physically activated carbon from nutshells. Results can be extrapolated to various types of raw material, as long as the selected raw materials and process conditions are similar. Although the assumptions made for the selected production process provide an adequate initial technical basis for the economic evaluation of a production plant at the industrial scale, there are limitations related to the conditions provided in this particular work. In this sense, the precision of this study is highly dependent on the variation in the cost of the main equipment, the cost of the raw material, the sale price of the product, and the estimated operating conditions of the production process. Another weakness may be related to the use of the factorial method to estimate the total capital investment used to calculate the economic indicators of profitability. Future work should be directed towards the optimization of the selected production process, through the use of a suitable chemical process simulation software. This information would greatly improve the economic study.

2. Materials and Methods

2.1. Production Process

The production process and operational parameters were proposed based on recent bibliographical data, which reported the obtaining of activated carbons from nutshells, using an inert nitrogen-based atmosphere for the pyrolysis stage and a flow of steam as an activating agent for the activation stage [13]. The parameters and conditions reported by Nazem et al. [13] were considered for the conceptual plant design considering a specific area of 1248 m^2/g for the activated carbon. This report indicates the pyrolysis stage temperature of 600 °C and a residence time of 1 h. For the activation stage, the operational conditions considered were 950 °C for temperature and 1 h for residence time. In the present work, a conversion rate of 30% was considered for the pyrolysis stage, in which nutshells are transformed into natural carbon, and a conversion rate of 70% for the activation stage, in which previously obtained natural carbon is converted into activated carbon. An overall conversion rate of 21% was established; consequently, this percentage of the available raw material of nutshells is finally transformed into activated carbon. On the other hand, material and energy balances were developed to obtain all the input and output flows of the production process, adjusted to the operating conditions considering production for the base case of 6.6 ton/day of activated carbon with ±25% variation. A sufficiently wide range of variation was considered to adequately establish the effect of the parameters on the net present value. Finally, the requirements for basic water, nitrogen, steam, and fuel services necessary for the operation of the process were estimated, as well as the size parameters for the sizing of the main process equipment.

2.2. Economic Analysis

For the economic evaluation, the installation of the production process plant in Chile was considered, because it has an appropriate availability of the raw material, access to ports of shipment, stability, and economic integration at a global level. On the other hand, considering that the mass ratio between the internal fruit and the shell is close to one, it was possible to establish that the quantity of available nutshells in Chile is similar to the exports of shelled nuts considered under the International Trade Center (ITC) code 080232. In this way, it was estimated that the availability of raw material was close to 34 kton/year, sufficient to supply the 10 kton/year (±25%) needed for the proper development of the investment project.

A working year of 360 days was considered, with a typical working period of 320 days/year for production and 40 days/year for plant maintenance tasks, and with a density of 600 kg/m^3 in the case of nutshells and 500 kg/m^3 in the case of the activated carbon obtained to size the main process equipment. The cost of the main equipment was estimated from actual local commercial quotations and costs present in the available literature [14–16]. The costs were updated to present value using the chemical engineering plant cost index (CEPCI), applying Equation (1). The costs of the main equipment were adjusted for required production capacity through Equation (2), using a typical exponent of 0.6 (six-tenths-factor rule) [21,22]. The operational labor costs were estimated graphically for an average condition plant, yielding 27 employee-hours/day/processing step.

$$\text{Cost Item (2019)} = \text{Cost Item (20XX)} \times \left[\frac{\text{Cost Index 2019}}{\text{Cost Index 20XX}}\right] \quad (1)$$

$$\text{Cost New Capacity} = \text{Cost Old Capacity} \times \left[\frac{\text{New Capacity}}{\text{Old Capacity}}\right]^{0.6} \quad (2)$$

The total capital investment established from the total cost of the main process equipment installed at the plant—including the auxiliary equipment, the total product cost including the operational labor, the projected cash flows using the selling price of the product and the sensitivity analysis—were determined taking as a guide the standard procedures described in *Plant Design and Economics for*

Chemical Engineers by Peter & Timmerhaus [21,22]. In particular, the total capital investment was based on the percentage of the delivered equipment cost method for a solids and liquids processing plant, described in the previous reference. The depreciation of process equipment was calculated using the linear method, and the income tax rate used was 27% (case of Chile) [23]. The expected error in the estimation of factored costs is around ±25%. The sales price of the finished product (activated carbon) for the base case was estimated at 2.75 USD/kg. Finally, economic indicators, such as net present value (NPV), internal return rate (IRR), return on investment (ROI), and discounted payback period (DPP), were determined. The equations for calculating NPV (Equation (3)), IRR (Equation (4)), and ROI (Equation (5)) are shown below [15,17]. In the case of the DPP, it was estimated graphically using the discounted cumulative cash flow.

$$\mathrm{NPV}\ [\mathrm{USD}] = \sum_{t=1}^{N} \frac{F_t}{(1+d)^t} - I_0 \tag{3}$$

$$\mathrm{NPV} = \sum_{t=1}^{N} \frac{F_t}{(1+d^*)^t} - I_0 = 0 \rightarrow d^* = \mathrm{IRR} \tag{4}$$

$$\mathrm{ROI}\ [\%] = \frac{F_t}{I_0} \times 100 \tag{5}$$

where N = project duration [years], F_t = annual profit [USD], d = low discount rate (LDR = 10% average market value), I_0 = total capital investment [USD], and d* = internal return rate (it is the discount rate when NPV = 0).

3. Results

3.1. Production Process Description (Base Case)

A processing capacity of 31,250 kg/day of nutshells was considered for the production of 6563 kg/day of activated carbon from nutshells for 320 effective working days of operation. Figure 1 shows the proposed production process for obtaining activated carbon from nutshells. The first stage is to grind 31,250 kg/day of nutshells to reach a maximum granular particle size of 18 mesh (1.0 mm) in a mill capable of processing 1302 kg/h and then to collect the ground raw material in two storage silos with a capacity of 182 m^3 each, allowing the raw material stock to be maintained for one week of production.

Subsequently, in the second stage, the 31,250 kg/day of ground nutshells must be processed in six independent loading processes per day of 5208 kg/load. The nutshells are fed to a rotary kiln where the raw material's carbonization process is carried out at a temperature of 600 °C with a residence time of 1 h. To avoid the combustion of organic matter inside the furnace chamber, a nitrogen flow is incorporated to displace the air. In this way, 1563 kg/load of natural coal is obtained from nutshells as an intermediate product, with an estimated yield for this stage of 30%. The process of activation is carried out at a temperature of 900 °C in the presence of a steam current as an activating agent for a residence time of 1 h. 1094 kg/load of activated carbon from nutshells is obtained as the final product of this stage, with a yield of 70%. The rotary reactor was designed with a volumetric load ratio of 15% and an internal volume of approximately 58 m^3. In the third stage, the activated carbon produced passes to cooling equipment where the temperature decreases to 50 °C using a nitrogen atmosphere and is stored in a silo of 92 m^3 of capacity that means the stock can be kept for a week of production. The fourth stage consists of passing the activated carbon through sieving equipment capable of processing 273 kg/h of product. Finally, in the fifth stage, the packaging process of the activated carbon produced is developed.

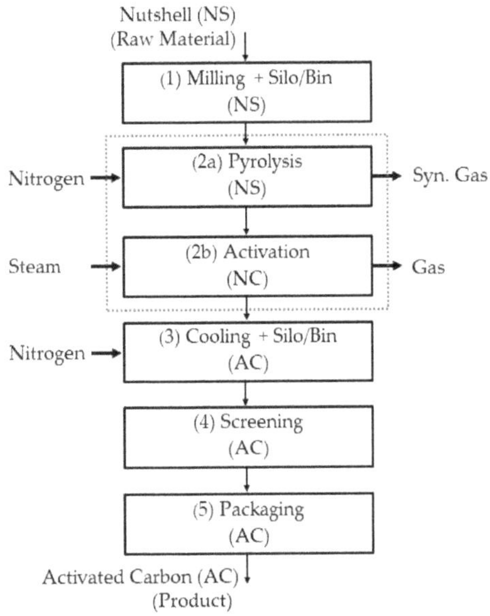

Figure 1. The proposed production process for obtaining activated carbon from industrial waste nutshells by physical activation with steam.

3.2. Cost Estimation (Base Case)

The cost of the main equipment of the production process placed in the plant for the base case considered is shown in Table 1. The total cost of the equipment, including auxiliary equipment, is USD 1,063,044. Furthermore, it was considered that the cost of the auxiliary equipment is 10% of the main process equipment. The construction material of the equipment considered was carbon steel.

Table 1. Main process equipment specifications and costs for the base case.

	MAIN PROCESS EQUIPMENTS	CAPACITY	UNITS	[USD]
1	Mill (Nutshell)	1302	kg/h	22,245
2	Silo/Bin (Nutshell Storage)	292	m^3	103,473
3	Rotary Kiln Pyrolysis/Activation (6 process per day)	5208	kg/process	515,624
4	Rotary Cooler Activated Carbon (6 process per day)	1094	kg/process	78,479
5	Silo/Bin (Activated Carbon Storage)	84	m^3	34,297
6	Vibrating Screen Sieve (Activated Carbon)	273	kg/h	7546
7	Packaging (Activated Carbon)	273	kg/h	45,816
8	Diesel Boiler	563	kg/h	27,896
9	Fuel Diesel Tank	109	m^3	68,488
10	Cooling Water Tank	174	m^3	62,540
C1	**TOTAL MAIN EQUIPMENT COST**			**966,404**
C2	**TOTAL MAIN + AUXILIARY EQUIPMENT COST (1.1 × C1)**			**1,063,044**

By using the total cost of the process equipment (including auxiliary equipment) as the 100% value, the total capital investment for the base case was estimated at USD 5,427,905, as shown in Table 2. In this sense, the total fixed-capital investment reached a value of USD 4,613,613 and a working capital value of USD 814,292. Concerning the base case, the annual cost of raw materials reached a value of USD 1,124,894 per year. This amount considered the purchase of 10 kton/year of nutshells at a unit price of 100 USD/ton, the supply of boiler water of 2700 m^3/year at a unit service price of 4 USD/m^3, and the supply of liquid nitrogen of 185 ton/year at a unit price of 618 USD/ton. On the other hand, the cost for basic fuel and cooling water services reached a value of USD 936,936 per year, considering a diesel oil supply of 1162 m^3/year at a unit price of 800 USD/m^3 and the supply of cooling water of 1866 m^3/year at a service price of 4 USD/m^3. Operational labor costs were estimated at USD 280,800, considering 27 employee-hours/day/processing step, five process stages as described in Figure 1, 320 days of annual operation, and a cost of 6.5 USD/employee-hours. All mentioned costs were estimated based on local values, considering a high rank to ensure an economic evaluation in the most extreme case (the case of Chile). Consequently, the total cost of the product for the base case was estimated at USD 4,523,987, as shown in Table 3.

Table 2. Total capital investment for the base case.

	ITEM		[USD]
A	**TOTAL FIXED-CAPITAL INVESTMENT**	**A1 + A2**	**4,613,613**
A1	TOTAL DIRECT PLANT COST	1 to 9	3,274,177
1	Delivered main equipment (includes auxiliary equipment)	100%	1,063,044
2	Purchased-equipment installation	39%	414,587
3	Instrumentation and controls (installed)	26%	276,392
4	Piping (installed)	31%	329,544
5	Electrical (installed)	10%	106,304
6	Buildings (including services)	29%	308,283
7	Yard improvements	12%	127,565
8	Service facilities (installed)	55%	584,674
9	Land (purchase is required)	6%	63,783
A2	TOTAL INDIRECT PLANT COST	10 to 14	1,339,436
10	Engineering and supervision	32%	340,174
11	Construction expenses	34%	361,435
12	Legal expenses	4%	42,522
13	Contractor's fee	19%	201,978
14	Contingency	37%	393,326
B	**WORKING CAPITAL**	**15 + 16**	**814,292**
15	About 15% of total capital investment	75%	707,988
16	Safety and hazard analyses	10%	106,304
	TOTAL CAPITAL INVESTMENT	**A + B**	**5,427,905**

Table 3. Total annual product cost for the base case.

	ITEM		[USD]
C	**MANUFACTURING COST**	**C1 + C2 + C3**	**3,821,529**
C1	DIRECT PRODUCTION COSTS	1 to 8	2,862,319
1	Raw materials (calculated)	-	1,124,894
2	Operating labor (calculated)	-	280,800
3	Direct supervisory and clerical labor (17.5% of operating labor)	17.5%	49,140
4	Utilities (calculated)	-	936,936
5	Maintenance and repairs (6% of fixed-capital investment)	6.0%	276,817
6	Operating supplies (15% of cost for maintenance and repairs)	15.0%	41,523
7	Laboratory charges (15% of operating labor)	15.0%	42,120
8	Patents and royalties (4% of C1.1 to C1.7)	4.0%	110,089
C2	INDIRECT PRODUCTION COSTS	9 to 11	595,156
9	Depreciation (10% of fixed-capital investment)	10.0%	461,361
10	Local taxes (2.5% of fixed-capital investment)	2.5%	115,340
11	Insurance (0.4% of fixed-capital investment)	0.4%	18,454
C3	PLANT-OVERHEAD COSTS (60% of 2 + 3 + 5)	60.0%	364,054
D	**GENERAL EXPENSES**	**14 to 16**	**702,458**
14	Administrative costs (15% of 2 + 3 + 5)	15.0%	91,014
15	Distribution and selling costs (11% of manufacturing cost)	11.0%	420,368
16	Research and development costs (5% of manufacturing cost)	5.0%	191,076
	TOTAL PRODUCT COST	**C + D**	**4,523,987**

4. Discussion

It is important to mention that the costs reported in the bibliography are linked to different time periods, which may affect the comparisons with respect to the values estimated in this work. In this sense, the cost comparison described below considers this as a limitation of the present work. As established in Table 1, the cost estimate of the main equipment and auxiliary equipment reached an updated value of USD 1,063,064, which is, in some cases, similar to the costs found in the literature for activated carbon production processes, using the physical route with steam as an activating agent. This value was estimated considering a plant size to process 6.6 ton/day of waste nutshells as raw materials. On the other hand, some of the most current costs are related to the estimate made by Lima et al. [16] in 2008, which reported a total cost of main equipment of USD 1,776,000, to process 20 tons/day of poultry litter. Subsequently, Stavropoulos et al. [17] in 2009 reported a total cost for process equipment of USD 1,154,416, for a production size of 4.5 ton/day considering various raw materials. More recently, Santadkha and Skolpap [19] in 2017 reported a total cost of main machinery and equipment of USD 1,301,429 for the generation of activated carbon from coconut shells and the regeneration of spent activated carbon obtained from the petrochemical industries, considering a plant capacity of 12 ton/day and 10 ton/day, respectively.

As previously mentioned, Table 3 shows that the total manufacturing cost for the base case reached a value of USD 3,821,529, and the total cost of the product reached a value of USD 4,523,987. If we divide these values by the total annual activated carbon production quantity of 2,100,000 kg/year, we obtain a manufacturing cost of USD 1.82 per kg of product and USD 2.15 per kg of product, respectively. For the calculation of these values, a non-zero cost for the raw material of USD 1,124,894 per year was considered. This makes a difference with other published studies that consider a marginal cost or do not consider costs in this important item, which can decrease the total cost of the product and lead to unrealistic conclusions about production costs, since any waste material used in industrial processes acquires economic value [17].

The values mentioned above can be compared with some costs reported in similar studies, being lower, for example, than the cost of the product obtained by Ng et al. [14] in 2003 who reached a production cost of activated carbon from pecan shells of USD 2.72 per kg and USD 2.89 per kg when the process was carried out by physical activation with steam and chemical activation with phosphoric acid, respectively, considering a minimum cost of USD 35,000 for raw materials. The costs obtained in the present work compare favorably with the costs reported by Stavropoulos et al. [17] in 2009 who reported production costs of USD 2.23 per kg, USD 2.49 per kg, and USD 2.18 per kg for the production of physically activated carbon with steam, using worn tires, wood, and lignite as raw materials, respectively. It is important to highlight that these last values were obtained considering a zero cost for raw materials. On the other hand, in the same study, lower production costs of USD 1.92 per kg, USD 1.84 per kg, and USD 1.72 per kg were reported for the production of chemically activated carbon with KOH, using worn tires, wood, and lignite as raw materials, respectively, considering a zero cost for raw materials. However, when a non-zero value was considered for raw materials, higher costs were calculated as USD 11.4 per kg, USD 6.39 per kg, and USD 5.38 per kg for the same precursors, respectively. In another study developed by Lima et al. [16] a lower production cost of USD 1.44 per kg was reported for an activated carbon plant from poultry waste, using physical activation with steam and a subsequent washing step with hydrochloric acid, followed by a step rinse with water. However, the cost was obtained considering a cost for raw materials of USD 269,537 per year, including the transportation of poultry waste, which is below the value considered here.

4.1. Economic Evaluation (10-Year Production Project)

Figure 2 shows the cumulative discounted cash-flow diagrams (CDCF) at different applied discount rates and without external financing, to determine the payback period for the capital investment. The payback period was after the fourth year of production, giving a net present value of USD 7,939,235 for a zero discount rate. For a 10% discount rate, the payback period was after the fifth year of production, giving a net present value of USD 2,785,624. Finally, for a discount rate of 20%, the investment recovery period was after the ninth year of production, giving a net present value of USD 176,231. In all cases the return on investment was 25%. On the other hand, Figure 3 shows the variation curve of the net present value at different discount rates. The internal rate of return (IRR) was estimated at 21%, which is higher than the minimum discount rate (LDR) of 10%, which is generally used by companies for the evaluation of investment projects in chemical plants; therefore, the project is viable for this particular condition.

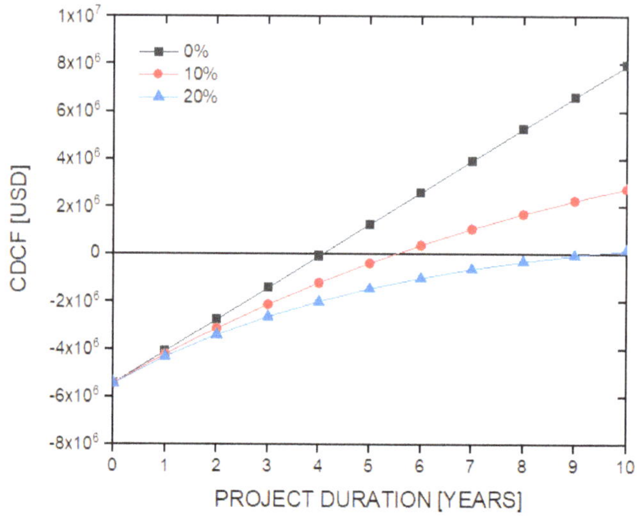

Figure 2. Cumulative discounted cash-flow (CDCF) diagrams at different discount rates of 0%, 10%, and 20%, for the 10-year production investment project.

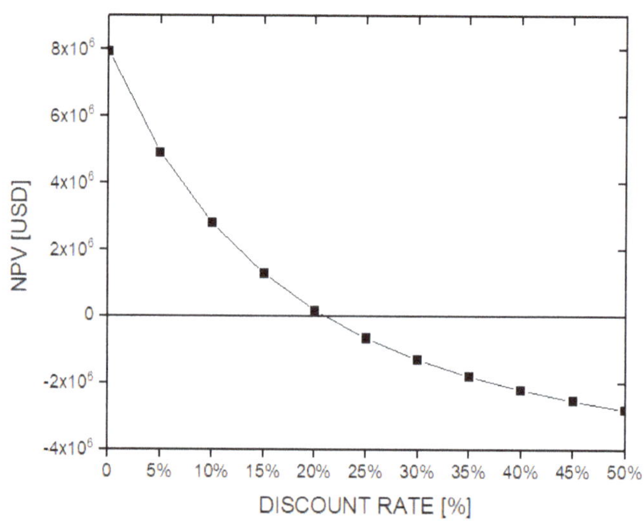

Figure 3. Net present value (NPV) of the investment project at different discount rates for the 10-year production investment project.

4.2. Sensitivity Analysis (Base Case of 10 Years of Production)

Figure 4 shows the effect on the NPV of the project's cash flow at ten years of production as a base case (using a tax rate of 27% and an LDR of 10% per year) due to the variation of ±25% in the cost of the equipment delivered to the plant, including the auxiliary equipment, the purchase price of the raw material (nutshells), the cost of basic services, and the cost of operational labor. The NPV varied by up to ±78% when the cost of equipment delivered was modified from the base case (USD 1,063,044). When the equipment delivered cost changed by +25%, a minimum NPV of USD 611,708 was obtained, along with an IRR of 12% and an ROI of 18% for this case. On the other hand, the NPV had a maximum variation of ±49% when the nutshells' raw material price changed for the base case (100 USD/ton), this variation being lower than the previous example. The NPV reached a minimum value of USD 1,432,787, when the price of the raw material was increased by 25%, obtaining an IRR of 16% and an ROI of 21%.

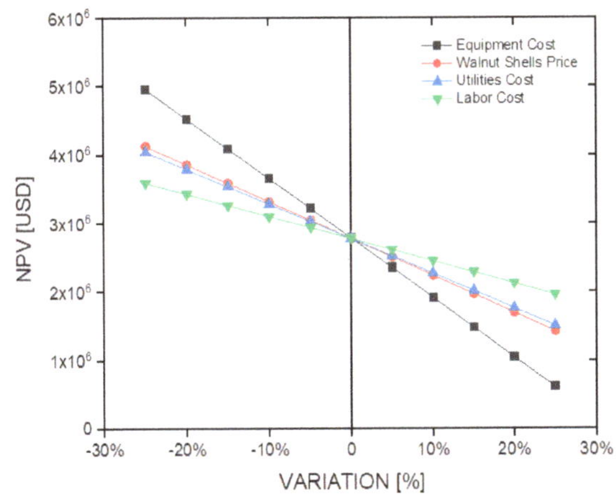

Figure 4. Effect by variation in ±25% of the cost of equipment, the purchase price of nutshells, the cost of services, and the cost in operating labor on the NPV considering an investment project of 10 years of production.

When the costs of the basic services were modified, the NPV had a ±46% variation lower than in the previous cases. Thus, the NPV reached a minimum value of USD 1,518,102 when the cost of basic services increased by 25%, obtaining an IRR and ROI value of 16% and 21%, respectively. Finally, the NPV had an ±29% variation when the cost of operational labor was modified for the base case (USD 280,800). It was the smallest variation within the cases studied in this particular item. In this case, the NPV minimum was USD 1,969,277, when the operational labor cost increased by 25%, obtaining an 18% IRR and an ROI of 22%.

Figure 5 shows the variation in net present value for the ±25% change in the volume of raw material processing (nutshells) and in the sales price of the finished product (activated carbon), which for the base case under consideration were 10 kton/year and 2.75 USD/kg, respectively. The NPV varied by up to ±178% when the processing volume of the raw material was modified to the base case. When the processing volume decreased by 25%, a minimum NPV with a negative value of USD 2,170,121 was obtained.

The NPV had a ±234% variation when the sales price of the finished product was modified to the base case. When this sales price decreased by 25%, a minimum NPV with a negative value of USD 3,731,774 was obtained. Consequently, the final product's sales price is the parameter with the highest sensitivity of the series studied, as it delivers the most upper range of variation in the net present value of the net cash flow of the investment project at 10 years of production.

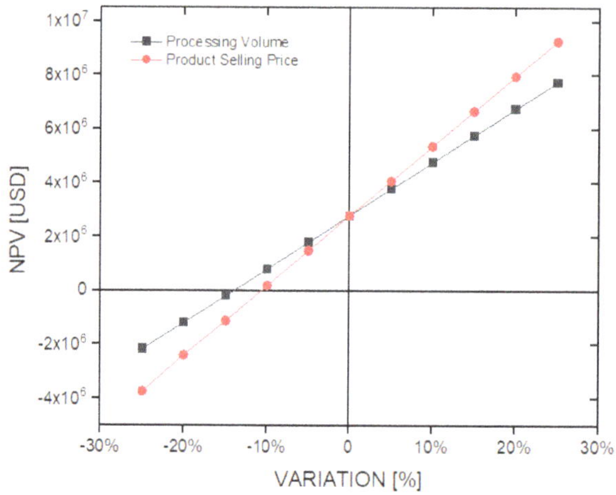

Figure 5. Effect on the NPV by variation in ±25% in the volume of processing of raw material (nutshells) and the selling price of the finished product (activated carbon) for the investment project of 10 years of production.

4.3. Economic Evaluation (15-Year Production Project)

Similar to the previous case, Figure 6 shows the cumulative discounted cash-flow diagrams at different discount rates to determine the payback period. For a zero discount rate, the payback period was after the fourth year of production, giving a net present value of USD 14,189,402. For a 10% discount rate, the payback period was after the fifth year of production, giving a net present value of USD 4,519,482. For a discount rate of 20%, the payback period was after the ninth year of production, giving a net present value of USD 686,744. In all cases the return on investment was 24%. Figure 7 shows the variation curve of the net present value at different discount rates. From this figure, the cost of the internal rate of return (IRR) can be graphically estimated at 23%. Therefore, the project is feasible for this particular condition, as discussed above.

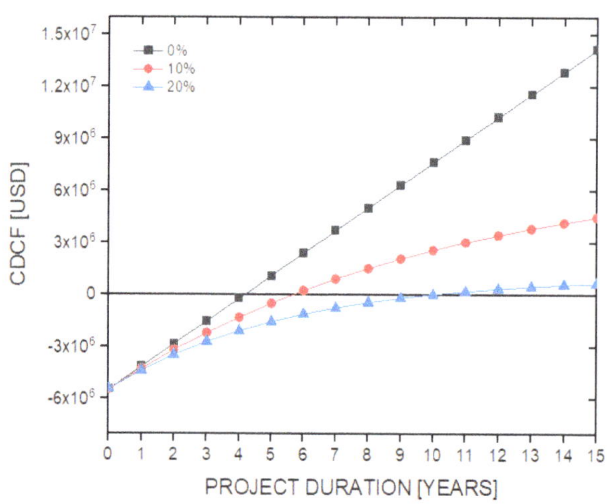

Figure 6. Cumulative discounted cash-flow diagrams (CDCF) at different discount rates of 0%, 10%, and 20% for the 15-year production investment project.

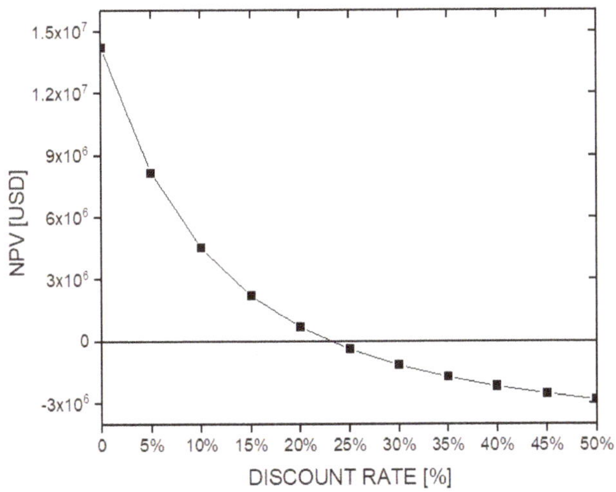

Figure 7. Net present value (NPV) of the investment project, at different discount rates for the 15-year production investment project.

4.4. Sensitivity Analysis (Base Case of 15 Years of Production)

Figure 8 shows the change in the NPV of the project cash flow at 15 years of production as a base case (using a tax rate of 27% per year and an LDR of 10% per year) due to the variation in ±25% in the cost of the equipment delivered (including auxiliary equipment), the price of the raw material (nutshells), the cost of basic services, and the cost of operational labor. This analysis is similar to the case seen above, which involves a 10-year production project. The NPV varied by up to ±54% when the cost of the equipment delivered was modified from the base case (USD 1,063,044). When the delivered equipment cost was increased by 25%, a minimum NPV was obtained of USD 2,096,311, along with an IRR of 15% and an ROI of 17%.

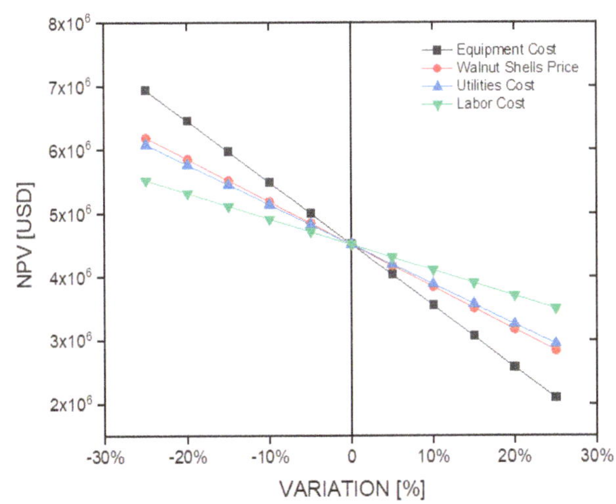

Figure 8. Effect by variation in ±25% of the cost of equipment, the purchase price of nutshells, the cost of services, and the cost in operating labor on the NPV considering an investment project of 15 years of production.

The NPV had a maximum variation of ±37% when the price of the nutshells' raw material was modified for the base case (100 USD/ton), this variation being lower than the previous case. When the price of the raw material increased by 25%, a minimum NPV for this case of UDS 2,844,866 was obtained, along with an IRR of 18% and an ROI of 20%. When the costs of the basic services were modified, the NPV had a ±35% variation lower than in the previous cases. Thus, NPV reached a minimum value of USD 2,950,474 for a +25% variation in the cost of basic services, obtaining a 19% IRR and a 20% ROI. Finally, the NPV had a maximum difference of ±22% when the cost of operational labor was modified to the base case (USD 280,800), the lowest variation being within the cases studied in this particular item. In this case, a minimum NPV of USD 3,508,962, an IRR of 20%, and an ROI of 22% were obtained if the operational labor cost is increased by 25%.

Figure 9 shows the change in net present value for the ±25% change in the volume of processing of the raw material and the selling price of the finished product for their base cases discussed above. The NPV had a variation of up to ±136% when the processing volume of the raw material was changed from the base case. When the processing volume decreased by 25%, a minimum NPV with a negative value of USD 1,615,009 was obtained. On the other hand, the NPV had a ±177% variation when the sales price of the finished product was modified for the base case. When this sales price decreased by 25%, a minimum NPV with a negative value of USD 3,496,851 was obtained. Consequently, the sales price of the finished product is the most sensitive parameter of the series studied for the 15-year production investment project, in a similar way to the 10-year operation project.

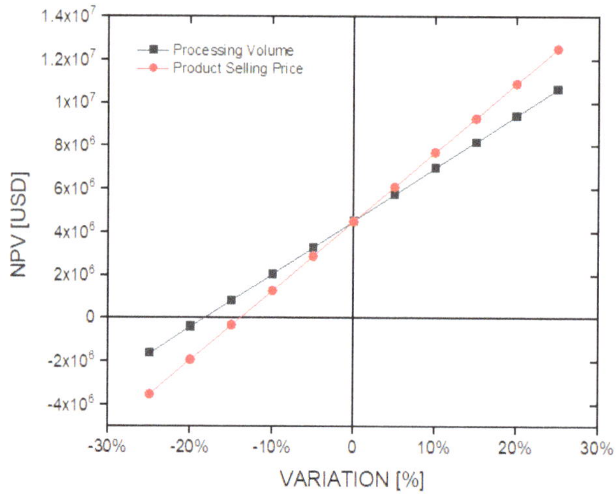

Figure 9. Effect on the NPV by variation in ±25% in the volume of processing of raw material (nutshells) and the selling price of the finished product (activated carbon) for the investment project of 15 years of production.

5. Conclusions

Based on the economic analysis developed, it is estimated that the generation of activated carbon from industrial waste nutshells by physical methods through the use of water vapor as an activating agent is economically profitable under the particular characteristics of the cases studied in this work. The activated carbon from nutshells would have a total cost of approximately USD 2.15 per kilogram of final product. In addition, the total manufacturing cost reached a value of USD 1.82 per kilogram of final product. For the calculation of these values, a non-zero cost was considered for the purchase of the raw material. This consideration makes a difference with other published studies that consider a marginal cost or do not consider the costs of purchasing raw materials. The estimated internal rate of return was 21% and 23% for 10 and 15 years of operating time, respectively. Such results are higher than the typical minimum discount rate of 10%, which is generally used by companies to evaluate investment projects in chemical plants, thus making the investment project viable. It was estimated that for the projected case of 10 years of operation (tax rate of 27% and LDR of 10% per year), the return on investment reaches 25%, and that for the planned case of 15 years of service (tax rate of 27% and LDR of 10% per year) the return on investment reaches 24%. In both cases, the payment period was made after the fifth year of production. The analysis of various sensitivity factors showed the limits to ensure the profitability of the project. Among these factors, the selling price of the finished product is the most sensitive parameter. On the other hand, one of the most important contributions of this work is to try to reduce the degree of uncertainty in the estimation of production costs and in the analysis of profitability indicators to provide a better approximation of the real costs involved in the economic analysis of an activated carbon production plant and provide a greater amount of information for potential industrial investors. In summary, this study can support the academic, research and financial analysis of investment projects and provide valuable information to industrial investors who could identify a good return on their investment capital.

Author Contributions: Conceptualization, M.L. and J.S.; Methodology, M.L. and J.S.; Validation, All Authors; Formal Analysis, M.L. and J.S.; Investigation, M.L. and J.S.; Resources, M.L. and J.S.; Data Curation, All Authors; Writing—Original Draft Preparation, M.L., J.S. and S.C.; Writing—Review & Editing, All Authors; Visualization, All Authors; Supervision, M.L. and J.S.; Project Administration, M.L. and J.S. All authors have read and agreed to the published version of the manuscript.

Funding: This research received no external funding.

Conflicts of Interest: The authors declare no conflict of interest.

References

1. Wong, S.; Ngadi, N.; Inuwa, I.; Hassan, O. Recent advances in applications of activated carbon from biowaste for wastewater treatment: A short review. *J. Clean. Prod.* **2018**, *175*, 361–375. [CrossRef]
2. Guillossou, R.; Le Roux, J.; Mailler, R.; Vulliet, E.; Morlay, C.; Nauleau, F.; Gasperi, J.; Rocher, V. Organic micropollutants in a large wastewater treatment plant: What are the benefits of an advanced treatment by activated carbon adsorption in comparison to conventional treatment? *Chemosphere* **2019**, *218*, 1050–1060. [CrossRef] [PubMed]
3. Azimi, A.; Azari, A.; Rezakazemi, M.; Ansarpour, M. Removal of Heavy Metals from Industrial Wastewaters: A Review. *ChemBioEng Rev.* **2017**, *4*, 37–59. [CrossRef]
4. Danish, M.; Tanweer, A. A review on utilization of wood biomass as a sustainable precursor for activated carbon production and application. *Renew. Sust. Energ. Rev.* **2018**, *87*, 1–21. [CrossRef]
5. Yahya, M.; Al-Qodah, Z.; Ngah, C. Agricultural bio-waste materials as potential sustainable precursors used for activated carbon production: A review. *Renew. Sustain. Energy Rev.* **2015**, *46*, 218–235. [CrossRef]
6. Cha, J.; Park, S.; Jung, S.-C.; Ryu, C.; Jeon, J.-K.; Shin, M.-C.; Park, Y.-K. Production and utilization of biochar: A review. *J. Indust. Eng. Chem.* **2016**, *40*, 1–15. [CrossRef]
7. Njoku, V.; Islam, M.; Asif, M.; Hameed, B. Preparation of mesoporous activated carbon from coconut frond for the adsorption of carbofuran insecticide. *J. Anal. Appl. Pyrol.* **2014**, *110*, 172–180. [CrossRef]
8. Alvarez, J.; Lopez, G.; Amutio, M.; Bilbao, J.; Olazar, M. Upgrading the rice husk char obtained by flash pyrolysis for the production of amorphous silica and high quality activated carbon. *Bioresour. Technol.* **2014**, *170*, 132–137. [CrossRef] [PubMed]
9. Adinata, D.; Wandaud, W.; Aroua, M. Preparation and characterization of activated carbon from palm shell by chemical activation with K_2CO_3. *Bioresour. Technol.* **2007**, *98*, 145–149. [CrossRef] [PubMed]
10. Wu, M.; Guo, Q.; Fu, G. Preparation and characteristics of medicinal activated carbon powders by CO_2 activation of peanut shells. *Powder Technol.* **2013**, *247*, 188–196. [CrossRef]
11. Kambarova, G.; Sarymsakov, S. Preparation of activated charcoal from walnut shells. *Solid Fuel Chem.* **2008**, *42*, 183–186. [CrossRef]

12. Bae, W.; Kim, J.; Chung, J. Production of granular activated carbon from food-processing wastes (walnut shells and jujube seeds) and its adsorptive properties. *J. Air Waste Manag. Assoc.* **2014**, *64*, 879–886. [CrossRef] [PubMed]
13. Nazem, M.; Zare, M.; Shirazian, S. Preparation and optimization of activated nano-carbon production using physical activation by water steam from agricultural wastes. *RSC Adv.* **2020**, *10*, 1463–1475. [CrossRef]
14. Ng, C.; Marshall, W.; Rao, R.; Bansode, R.; Losso, J. Activated carbon from pecan shell: Process description and economic analysis. *Ind. Crop. Prod.* **2003**, *17*, 209–217. [CrossRef]
15. Choy, K.; Barford, J.; McKay, G. Production of activated carbon from bamboo scaffolding waste-process design, evaluation and sensitivity analysis. *Chem. Eng. J.* **2005**, *109*, 147–165. [CrossRef]
16. Lima, I.; McAloon, A.; Boateng, A. Activated carbon from broiler litter: Process description and cost of production. *Biomass Bioenergy* **2008**, *32*, 568–572. [CrossRef]
17. Stavropoulos, G.; Zabaniotou, A. Minimizing activated carbons production cost. *Fuel Process. Technol.* **2009**, *90*, 952–957. [CrossRef]
18. Vanreppelen, K.; Kuppens, T.; Thewys, T.; Carleer, R.; Yperman, J.; Schreurs, S. Activated carbon from co-pyrolysis of particle board and melamine (urea) formaldehyde resin: A techno-economic evaluation. *Chem. Eng. J.* **2011**, *172*, 2–3. [CrossRef]
19. Santadkha, T.; Skolpap, W. Economic comparative evaluation of combination of activated carbon generation and spent activated carbon regeneration plants. *JESTEC* **2017**, *12*, 3329–3343.
20. Nandiyanto, A. Cost analysis and economic evaluation for the fabrication of activated carbon and silica particles from rice straw waste. *JESTEC* **2018**, *13*, 1523–1539.
21. Peter, M.; Timmerhaus, K. *Plant Design and Economics for Chemical Engineers*, 4th ed.; McGraw-Hill: New York, NY, USA, 1991.
22. Peter, M.; Timmerhaus, K. *Plant Design and Economics for Chemical Engineers*, 5th ed.; McGraw-Hill: New York, NY, USA, 2003.
23. Internal Revenue Service. Available online: http://homer.sii.cl. (accessed on 7 June 2020).

processes

Article

Exploration of a Molecularly Imprinted Polymer (MIPs) as an Adsorbent for the Enrichment of Trenbolone in Water

Anele Mpupa [1,2], Mehmet Dinc [3], Boris Mizaikoff [1,4,*] and Philiswa Nosizo Nomngongo [1,2,5,*]

1 Department of Chemical Sciences, University of Johannesburg, Doornfontein Campus, P.O. Box 17011, Johannesburg 2028, South Africa; 216051138@student.uj.ac.za
2 DSI/NRF SARChI Nanotechnology for Water, University of Johannesburg, Doornfontein 2028, South Africa
3 Hahn-Schickard, Sedanstrasse 14, 89077 Ulm, Germany; mehmet.dinc@hahn-schickard.de
4 Institute of Analytical and Bioanalytical Chemistry, Ulm University, 89081 Ulm, Germany
5 DSI/Mintek Nanotechnology Innovation Centre, University of Johannesburg, Doornfontein 2028, South Africa
* Correspondence: boris.mizaikoff@uni-ulm.de (B.M.); pnnomngongo@uj.ac.za (P.N.N.); Tel.: +27-11-559-6187 (P.N.N.)

Abstract: The presence of endocrine disruptors in surface waters can have negative implications on wildlife and humans both directly and indirectly. A molecularly imprinted polymer (MIP) was explored for its potential to enhance the UV-Vis determination of trenbolone in water using solid-phase extraction (SPE). The synthesized MIP was studied using Fourier transform infrared spectra (FTIR) and scanning electron microscopy (SEM). Using the MIP resulted in a preconcentration and enrichment factor of 14 and 8, respectively. Trenbolone binding on the MIP was shown to follow a Langmuir adsorption and had a maximum adsorption capacity of 27.5 mg g^{-1}. Interference studies showed that the MIP selectivity was not compromised by interferences in the sample. The MIP could be recycled three times before significant loss in analyte recovery.

Keywords: androgenic hormones; solid-phase extraction; molecularly imprinted polymers; trenbolone; endocrine disruptors

Citation: Mpupa, A.; Dinc, M.; Mizaikoff, B.; Nomngongo, P.N. Exploration of a Molecularly Imprinted Polymer (MIPs) as an Adsorbent for the Enrichment of Trenbolone in Water. *Processes* **2021**, *9*, 186. https://doi.org/10.3390/pr9020186

Received: 2 November 2020
Accepted: 7 January 2021
Published: 20 January 2021

Publisher's Note: MDPI stays neutral with regard to jurisdictional claims in published maps and institutional affiliations.

1. Introduction

Water is the most important natural resource for the survival of all living organisms [1]. Due to the rise of urbanization, climate change, industrial production and population, the quality of water has been negatively affected [2–4]. As a result, many cities are facing water shortages [1]. Moreover, pollution to surface waters not only contributes to water shortages but can also lead to the degradation of grain quality. Pollutants entering the soil or water pose a varying number of threats to the functioning of the ecosystem as well as human health [5]. Soil is usually polluted with polycyclic aromatic hydrocarbons (PAHs), petroleum-related compounds, chlorophenols, heavy metals, organic pollutants and pesticides [5]. There is an overlap of the types of pollutants present in water and soil. This is because of water runoff from contaminated soils, wastewater and septic tanks [6]. The United States and European Union, in 2002 and 2009, reported the presence of organic pollutants, especially ones with endocrine-disrupting properties, in at least 80% of samples [7]. Since 2013, there have been approximately 1000 synthetic and natural compounds with endocrine-disrupting properties [7]. The most investigated endocrine disruptors include some personal care products, pharmaceuticals, poly brominated diphenyl ethers (PBDEs) and hormones [8].

Androgenic hormones such as testosterone are C_{19} steroids and are responsible for the development of masculine characteristics [9]. Thus, they are normally used as growth promoters in human and veterinary therapies [10]. Anabolic androgenic steroids such as testosterone are widely used by athletes to improve athletic performance because of their mytrophic action [11]. Trenbolone and zeranol are hormonally active substances that can be used as growth promoters [12]. The use of these anabolic growth agents is banned by the European Union, Russia and China [13]. This is because a number of studies have shown that residues of these substances can be found in meat and other food stuffs [14]. Whereas this is the case in some countries such as South Africa, there is no information on whether these substances are used legally or illegally. These substances have been detected in the liver, bile, plasma, kidney and urine of the cattle [15]. This means that hormonal residues can be found in animal excreta, which can directly pollute the soil and indirectly affects water quality through leaching [16]. In the environment, these substances are persistent [17]. More concerning is their endocrine-disrupting properties, even at extremely low concentrations [18]. It is suspected that trenbolone can result in neurodegenerative disorders such as Alzheimer's disease when people are exposed to it for extended periods [19]. Therefore, analytical methods with the ability to detect these substances in water are required.

With modern analytical chemistry, it is possible to determine an analyte of interest in different kinds of sample matrices [20]. The analysis and quantification of hormones in water matrices has predominantly been done using chromatographic techniques coupled with different detectors [21–23] and electrochemical [24,25] and spectrophotometric [26,27] techniques. Among these, gas and liquid chromatography (GC and LC) are the most prevalent due to efficient separation and identification success [28]. However, the biggest limitation of GC determination is the need for derivatisation and conversion [29]. These manipulations can result in loss of analyte [29]. This has driven a surge in liquid chromatography (LC) methods, which do not require chemical pre-treatment for hormonal analysis and quantification [30]. LC methods often lack the GC specificity in complex matrices and thus require sample preparation in order to ensure accurate quantification [31]. Spectrophotometric determination offers short analysis run times. It is also readily available and simple, which makes it an attractive technique for routine analysis [32,33]. The major drawback of UV-Vis spectrophotometry is low sensitivity and poor selectivity in comparison to chromatographic and electrochemical methods [34,35]. A sample preparation step can be used to combat this disadvantage [34,35]. Hence, the goals of sample preparation include the preconcentration of analytes to concentrations higher than the instrument's limits of detection, removal of interferences and isolation of the analytes from matrix effects [36].

Solid phase extraction (SPE) is the most widely used sample preparation technique for the preconcentration and analysis of organic pollutants [37,38]. This is due to its advantages, such as its simplicity, potential for automation, high enrichment factors and high selectivity [39]. In addition, the most attractive advantage for analytical chemists is the possibility of using a reduced amount of organic solvents and the fact that SPE is highly tuneable with regard to the adsorbents used [35]. Previously, silica-bonded phases such as C_{18} have been used for as adsorbents for SPE of wide range of organic pollutants [33,40–44]. However, recently, modified and tuned solid phase materials have been used to achieve more specificity [45]. One of the most promising adsorbents with regard to selectivity are molecularly imprinted polymers. Molecularly imprinted polymers (MIPs) have been demonstrated to be useful for the specific recognition, binding and isolation of chemically active target molecules [46]. MIPs can be prepared by moulding polymeric structures around a target molecule through polymerisation in the presence of a cross-linking agent and initiator [47]. The formation of a MIP follows a two-step process, where the first step includes the formation of a template-monomer complex followed by polymerisation around the template. Then, the template is removed by washing resulting in a polymer matrix with specific analyte-binding sites [47,48]. MIPs provide high selectivity combined with easy preparation and application. As such, they can be used for a variety of applications, such as sensors [49–51], catalysis [52,53], capillary electrophoresis and separation [54] and solid phase extraction [55,56].

The aim of this work was to prepare a molecularly imprinted polymer using trenbolone as a template. The synthesised MIP was then evaluated for its binding and adsorption characteristics. The effects of the sample pH, mass of the adsorbent, extraction time and eluent volume on the preconcentration of trenbolone using the prepared MIP were investigated using multivariate optimisation. Quantification and method performance evaluation were carried out using simple UV spectrophotometry. In our knowledge, molecularly imprinted polymers have not been explored for the quantification of trenbolone using UV-Vis spectrophotometry.

2. Materials and Methods

2.1. Materials

Ethanol, methanol, acetic acid, sodium hydroxide, methylacrylic acid, trenbolone (analytical grade), anhydrous toluene, ethyleneglycoldimethacryate (EGDMA) and azobisisobutyronitrile (AIBN) were all purchased from Sigma-Aldrich (München, Germany). A 1000 mg L^{-1} trenbolone stock solution was prepared by dissolving the accurate appropriate amount with methanol. The stock solution was stored at 2 °C in a fridge until it was required for use. Working solutions were prepared by diluting the stock solution with ultra-pure water (Thermo Scientific, Deionizer Nanopure Diamond UV TOC Water System, Waltham, MA, USA).

2.2. Instrumentation

A Heraeus Megafuge 16 centrifuge (Thermo Scientific) was used for all separations during the study. A Heraeus vacuum oven (Thermo Scientific) was used for drying, and a Thermomixer C (Eppendorf, Hamburg, Germany) was used for agitation. The polymer surface was studied using a Quanta 3DFEG scanning electron microscope. The Carl Ziess Axio light microscope was used to evaluate the surface of the particles using the Axiovision software. Analyte quantification was done using an Analytik Jena Specord S600 UV-Vis spectrophotometer in the wavelength range 180–400 nm with a 10 mm UV cuvette.

2.3. Synthesis of Molecularly Imprinted Polymer

The method for the synthesis of core-shell molecularly imprinted polymers (MIP) was adopted from the authors of [57]. First, we added 50 mL of anhydrous toluene to a 2-neck round bottom flask, followed by the template molecule (0.264 mmol), methacrylic acid (1.06 mmol), EGDMA (2.11 mmol) and 20 mg of AIBN. The solution was mixed for 10 min under a gentle stream of argon. The solution was heated at 50 °C for 6 h, then at 60 °C for a further 24 h. The products were then matured at a temperature of 85 °C for 6 h. The template was then removed by washing the resulting polymer using methanol/acetic acid mixture (9:1, *v/v*). The polymer was then dried under vacuum at 40 °C for 24 h before use. Nonimprinted polymers (NIPs) were synthesised in the same way in the absence of the template molecule.

2.4. Binding Characteristics and Adsorption Capacity

The binding studies were performed on solutions made using ultra-pure water at room temperature by appropriately modifying the method described by Fernandes et al., 2017 [58]. To investigate adsorption capacity, the MIP and NIP were weighed into centrifuge tubes containing 10 mL of the trenbolone solution with concentrations varying from 10–50 mg L^{-1} and shaken for 180 min at room temperature. The adsorbents were then separated by centrifuging at 3500 rpm for 10 min and the supernatant absorbance was measured at 350 nm. The difference between the initial and equilibrium trenbolone concentration, Equation (1), can be considered as the amount of trenbolone bound to the adsorbent. The adsorption capacity (Q_e) was calculated using Equation (1).

$$Q_e = \frac{(C_0 - C_e) \times V}{w} \quad (1)$$

where C_0 and C_e represent the initial and equilibrium concentrations (mg L^{-1}), respectively, V is the volume of the trenbolone solution used (L) and w is the amount of polymer used for the adsorption process (g). The specific MIP recognition properties were evaluated using the imprinting factor (IF), which is the ratio between the MIP and NIP adsorption capacity ratio, Equation (2):

$$\text{IF} = \frac{Q_{e(MIP)}}{Q_{e(NIP)}} \quad (2)$$

where $Q_{e(MIP)}$ and $Q_{e(NIP)}$ are the MIP and NIP adsorption capacities, respectively.

2.5. Solid Phase Extraction of Trenbolone Using MIP

The extraction of trenbolone was done using solid-phase extraction in which an appropriate amount of polymer was weighed into a centrifuge tube containing 10 mL (3 mg L^{-1}) of the trenbolone solution. The mixture was then agitated using a thermo mixer. After extraction, the polymer and bulk solution were separated by means of centrifugation at 3500 rpm for 10 min. After which, the analyte was eluted with a mixture of methanol and acetic acid (9:1, *v/v*) by agitation and separated by centrifugation. UV-Vis spectrophotometry was used to analyse the bulk and elute solution.

2.6. Optimisation of SPE Method

The optimisation of the factors (Table 1) that could affect the extraction of trenbolone was done multivariately using a 2-level central composite design with mass of adsorbent (MA), sample pH, extraction time (the time the sample and polymer were agitated, ET) and eluent volume (EV) as the factors of interest.

Table 1. Parameter and levels used in central composite experimental design.

Factors	Low Level (−1)	Central Point (0)	High Level (+1)
Mass of adsorbent (MA) (mg)	10	30	50
pH	4	6.5	9
Extraction time (ET) (min)	10	20	30
Eluent volume (μL)	400	700	1000

2.7. Analytical Performance and Validation of MIP Extraction Method

The selectivity of the MIP was tested using both adsorption and signal enhancement. Briefly, 30 mg of polymers (NIP and MIP) were weighed into 15 mL centrifuge tubes. Then, 10 mL of 3 mg L^{-1} spiked lake water was added to the tube. The tubes were then agitated for 20 min using a thermo mixer, after which centrifugation for 10 min at 3500 rpm was used to separate the polymers from the bulk solution. The analyte was then eluted using a methanol/acetic acid mixture (9:1, *v/v*) (700 μL). The elute and bulk solutions were both analysed using a UV-Vis spectrophotometer. The analytical performance of the method was evaluated using the limit of detection (LOD), limit of quantification (LOQ), linear range (LR) and preconcentration and enrichment factors (PF and EF).

2.8. Reusability Studies

The reusability and regeneration of the polymer was studied by weighing 30 mg of the polymer and 10 mL of a synthetic sample. The mixer was then agitated using a thermo mixer for minutes. The polymer and supernatant were separated using a centrifuge at 3500 rpm for 10 min. The analyte was then eluted with a mixture of methanol and acetic acid (9:1, *v/v*) by shaking the eluent with polymer in a thermo mixer for 10 min. Subsequently, centrifugation was used to separate the polymer and eluent. The elute was quantified using UV-Vis spectrophotometry at 350 nm.

2.9. Selectivity Studies

The selectivity of the MIP was assessed using β-estradiol, which is one of the most commonly detected hormones in the environment. In order to test the real sample selectivity of the MIP in comparison to the NIP, lake water was spiked with 3 mg L^{-1} of trenbolone and β-estradiol. After, the adsorption and preconcentration methods were applied on the samples. The distribution ratio was calculated according to Azodi-Deilami et al., 2014 [59] using the equation:

$$K_D = \frac{\left(C_i - C_f\right)V}{C_f m} \tag{3}$$

where C_i, C_f, V and m are the initial and equilibrium concentrations, solution volume and polymer mass, respectively. The K_D was then used to calculate the selectivity coefficient (α) defined by the equation:

$$\alpha = \frac{K_D(Trenbolone)}{K_D(\beta - estradiol)} \tag{4}$$

3. Results and Discussion

3.1. Characterisation

The characterisation of the molecularly imprinted polymer (MIP) was done after the removal of the template. The synthesized polymers were characterised to understand their most prominent functional groups using Fourier transform infrared spectra (FTIR) (Figure 1) and their surface characteristics were studied using scanning electron microscopy (SEM) equipped with an energy dispersive X-ray spectrometer (EDS) in order to investigate the chemical compositions of the polymers (Figure 2A–D).

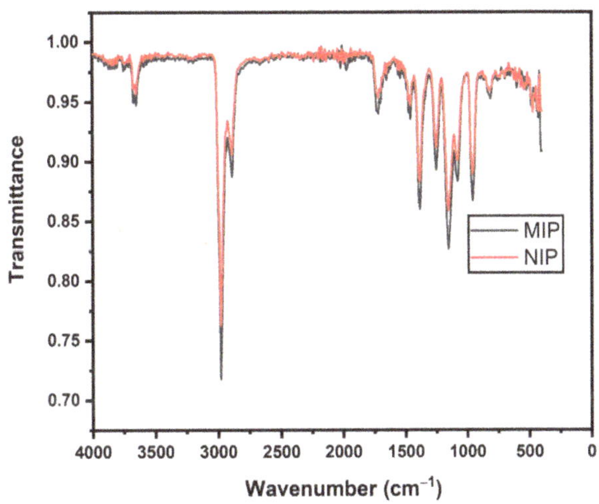

Figure 1. Fourier transform infrared spectra (FTIR) for the molecularly imprinted polymer (MIP) and nonimprinted polymer (NIP).

As seen in Figure 1, the MIP and NIP spectra shared a similar FTIR profile. This was due to the fact that the MIP and NIP synthesis were similar except for the absence of the template molecule in the synthesis of the NIP. The band at 1720 cm^{-1} was a characteristic peak attributed to the COOH functional group. The second characteristic peak for poly methacrylic acid was found at 1456 cm^{-1}, which was as a result of the C-C double bond, whereas the bands at 2900 cm^{-1} and 3000 cm^{-1} were due to the C-H stretching. A similar profile observed by Gupta and colleagues [60].

The SEM images of the MIP and NIP (Figure 2A,B) showed the presence of a bulk polymer. There were no significant differences between the MIP and NIP surfaces, as expected. This was because the synthesis routes of the two polymers was similar except for the presence of the template in the preparation of the MIP. The EDS spectra (Figure 2C,D) for both the MIP and NIP showed that both polymers had similar elemental compositions. This was due to the removal of the template from the MIP, resulting in the MIP and NIP having relatively similar elemental compositions [61].

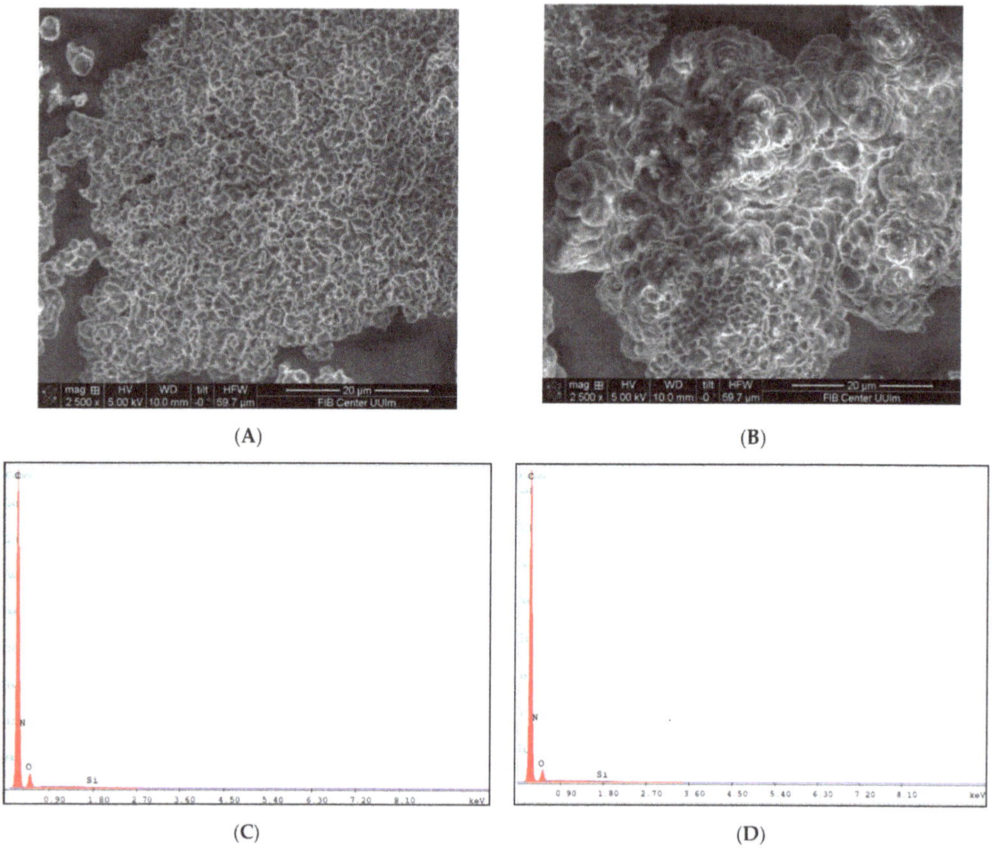

Figure 2. Scanning electronic microscopy (SEM) images for (**A**) MIP and (**B**) NIP and X-ray spectrometer (EDS) spectra for (**C**) MIP and (**D**) NIP.

3.2. Binding Characteristics and Adsorption Capacity

The binding characteristics and adsorption capacity studies were carried out to evaluate the differences in analyte binding between the MIP and NIP. From the sorption studies, there was a clear difference between the adsorption of trenbolone on the MIP and NIP (Figure 3 shows the UV spectra for the analyte solution before incubation (sample) and the solutions after incubation on the MIP and NIP). This was shown by the difference in the absorbance, which can be directly correlated to the trenbolone concentration on the respective solutions. The imprinting factor was used to evaluate the MIP recognition properties. According to the literature, a MIP with an IF of >1.2 can be classified as suitable for the recognition of the analyte [55]. The calculated imprinting factor for the synthesized polymer was 2.41, which is greater than 1.2, meaning that the MIP had suitable recognition properties [62,63].

The concentration difference was further confirmed by actual concentrations in studying the adsorption capacities of the respective polymers, as shown in Figure 4. The MIP generally had higher adsorption capacity in comparison with the NIP. This was expected as the MIP has specific imprinted sites on the polymer for the analyte [56]. Thus, this informed the choice of using the MIP only for the optimisation of the extraction procedure. Adsorption studies were done to investigate the adsorption behaviour of the MIP and NIP.

The equilibrium adsorption data for trenbolone onto the polymers was investigated using Langmuir [64] and Freundlich [64] isotherms with the nonlinear Equations (5) and (6).

$$\text{Langmuir}: \; q_e = \frac{q_m k_L C_e}{1 + k_L C_e} \tag{5}$$

where q_e, q_m, k_l and C_e are amount of trenbolone adsorbed, maximum adsorption capacity, Langmuir constant and concentration of trenbolone at equilibrium respectively.

$$\text{Freundlich}: \; q_e = K_F C_e^{\frac{1}{n}} \tag{6}$$

where K_F, C_e and n are the measure of adsorption capacity, equilibrium concentration and indicator of adsorption effectiveness, respectively.

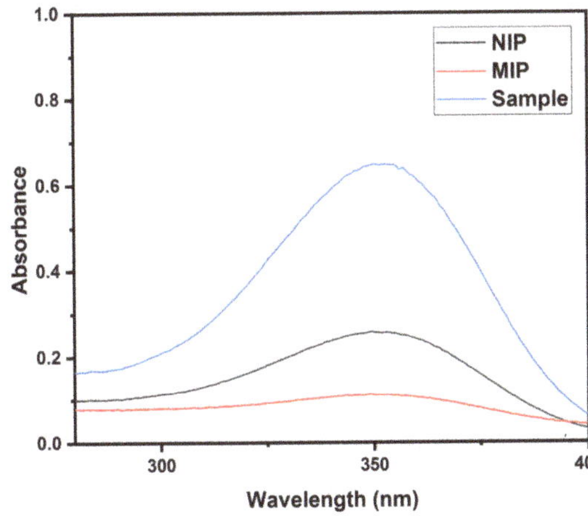

Figure 3. Graph showing the effect of the MIP for the removal of trenbolone in a 10 mg/L solution. NIP = 10 mg/L solution after adsorption onto the nonimprinted polymer, MIP = 10 mg/L solution after adsorption onto the imprinted polymer and Sample = original 10 mg/L solution before adsorption.

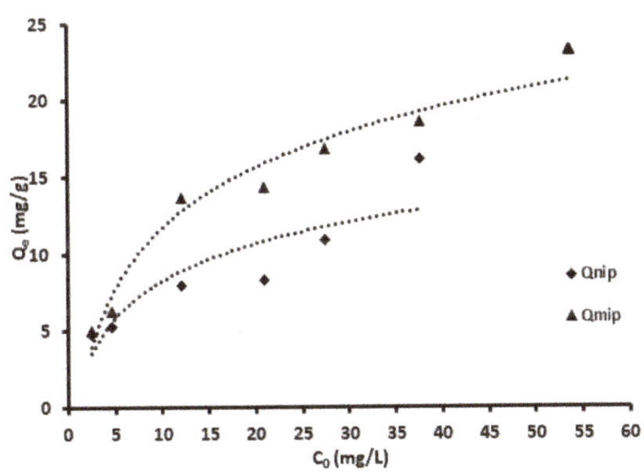

Figure 4. Adsorption capacity studies for the NIP and MIP.

The adsorption data obtained for trenbolone adsorption onto the MIP were shown to fit the Langmuir isotherm with a correlation coefficient of 0.9263, which was greater than the

R^2 of 0.8100 for the Freundlich isotherm. This meant that the adsorption of trenbolone onto the MIP was homogenous and thus assumed monolayer adsorption [65–67]. The maximum adsorption capacity calculated using the Langmuir isotherm was found to be 27.5 mg g^{-1}. Trenbolone adsorption onto the NIP was characterized by the Freundlich adsorption isotherm with an adsorption capacity of 11.4 mg g^{-1}.

3.3. *Multivariate Optimisation*

The optimisation results were analysed using the analysis of variance (ANOVA), response surface methodology and desirability function in order to determine the significant factors, interaction effects and desirable conditions for optimum extraction [68]. The analysis of variance is reported in the form of a Pareto chart (Figure 5), where the redline is indicative of the 95% confidence interval that a factor has significant effect on the extraction of trenbolone using the MIP [69]. From the Pareto chart, there was no factor that was significant for the extraction at the 95% confidence level. The reason for this could be the dependency of the analyte adsorption on the sites created during the imprinting, resulting in the extraction being a result of the analyte removal from the MIP.

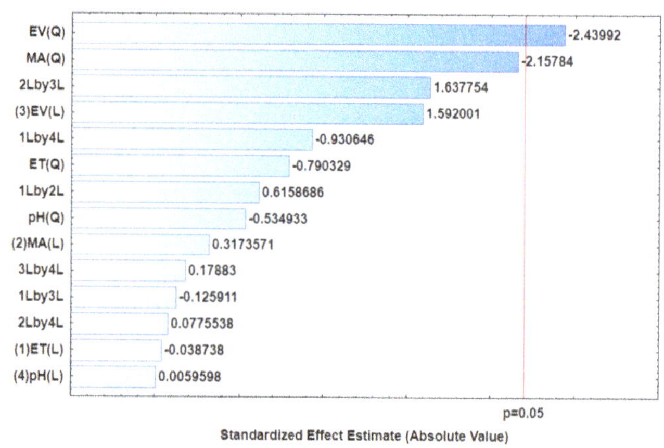

Figure 5. Pareto chart of standardized factors for the extraction of trenbolone. Factor (Q): Factor (quadratic); factor (L): Factor (linear); 2Lby3L: MA-EV (mass of adsorbent and eluent volume) interaction; 1Lby4L: ET-pH (extraction time—solution pH) interaction; 1Lby2L: ET-MA (extraction time—mass of adsorbent) interaction; 3Lby4L: EV-pH interaction; 1Lby3L: ET-EV interaction; 2Lby4L: MA-pH interaction.

The response surface methodology was used to investigate the interactive effects of the independent factors on the analytical response (extracted trenbolone concentration). The 3D response surface plots are presented in Figure 6. The combined effected of MA and other factors (ET, pH and EV) revealed that better adsorption was obtained as the mass increased from 10–50 mg (Figure 6A,D,E). This suggests that an increased mass of a polymer increases the available template binding sites, thus leading to quantitative trenbolone extraction. The interactive effect between MA and EV shows that high recoveries were obtained when the eluent volume was between 700 µL and 800 µL (Figure 6D). This implies that higher eluent volumes (>500 µL) are required for the complete removal of the template (trenbolone) from the polymer. A similar trend was observed in Figure 6B,F. The combined effect of ET with pH, MA and EV shows that quantitative extraction of trenbolone took place at any point where ET > 0. These findings correspond to the Pareto chart results, which show that this factor was not significant at the 95% confidence level. The effect of pH showed that, at highly acidic pH < 4, the extraction efficiency of the adsorbent was low (Figure 6C,E,F). The reason for this could be that an acidic environment causes the unbinding of the analyte from the MIP.

The profile of desirability and predicted values (Figure 7) was used to estimate the optimum conditions based on the desirability of each factor [70]. The highest extraction was assigned a desirability of 1, while the central and lowest were assigned 0.5 and 0, respectively. A desirability value of 1 was selected as the target and used to estimate the conditions required for optimum signal enhancement. These conditions included a sample pH of 6.5, MA of 30 mg, ET of 20 min and EV of 700 µL. The predicted conditions were then experimentally confirmed and resulted in an average concentration of 19.57 ± 0.15 and enhancement factor of 7.8.

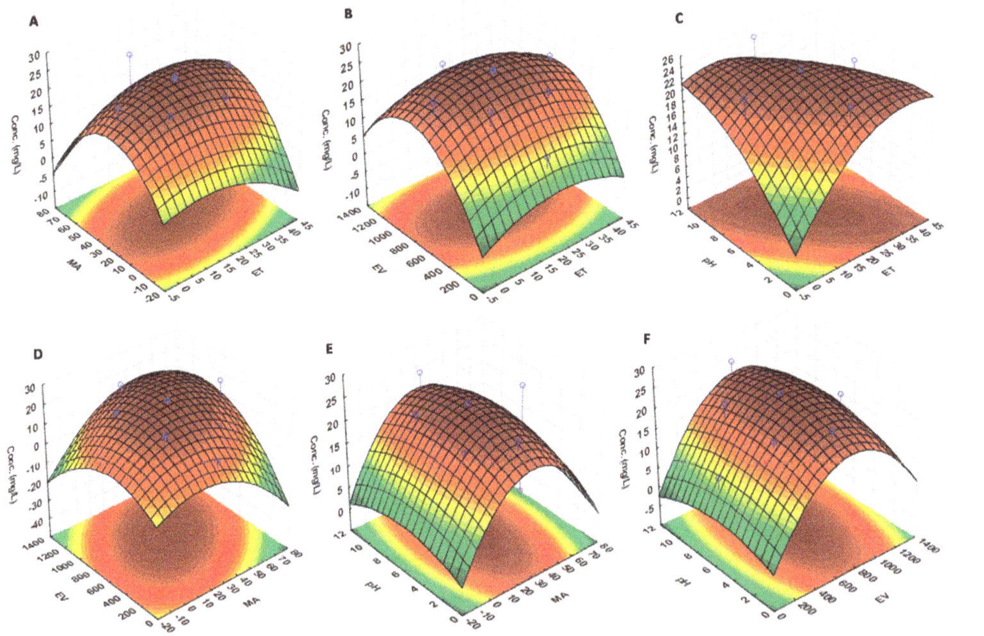

Figure 6. Response surface methodology plots showing the interaction between the independent factors (**A**) MA and ET, (**B**) EV and ET, (**C**) pH and MA, (**D**) pH and ET, (**E**) EV and MA and (**F**) pH and EV.

3.4. Analytical Performance and Validation of MIP Extraction Method

The analytical performance of the method was evaluated using the limits of detection (*LOD*) and quantification (*LOQ*), the linear range (LR), %RSD, enrichment factor (EF) and preconcentration factor (PF). The *LOD* and *LOQ* are described by the expressions $LOD = \frac{3 \times sd}{m}$ and $LOQ = \frac{10 \times sd}{m}$, respectively, where *sd* is the standard deviation of seven replicates of the lowest concentration standard sample prepared in ultra-pure water and *m* is the slope of the calibration curve [71]. From this method, the *LOD*, *LOQ*, LR, %RSD, EF and PF were found to be 0.07, 0.24 mg L^{-1}, LOQ—10 mg L^{-1}, 0.75, 7.2 and 14, respectively.

The adsorption and recovery % of trenbolone was then used to determine the effectiveness of the polymers in selectively binding trenbolone. The experimental results are presented in Figure 8, where it can be noted that the MIP had superior adsorption and enhancement in comparison to the NIP. This means that the MIP could selectively bind trenbolone in the presence of interferences better than the NIP. The selectivity coefficient (α) for the MIP was found to be 4.96, with distribution ratio (K_D) values of 1.14 and 0.23 for trenbolone and b-estradiol, respectively. The α for the NIP was found to be 1.17, with K_D values of 0.15 and 0.13, respectively. The higher K_D for the trenbolone confirmed better MIP selectivity when compared to the NIP.

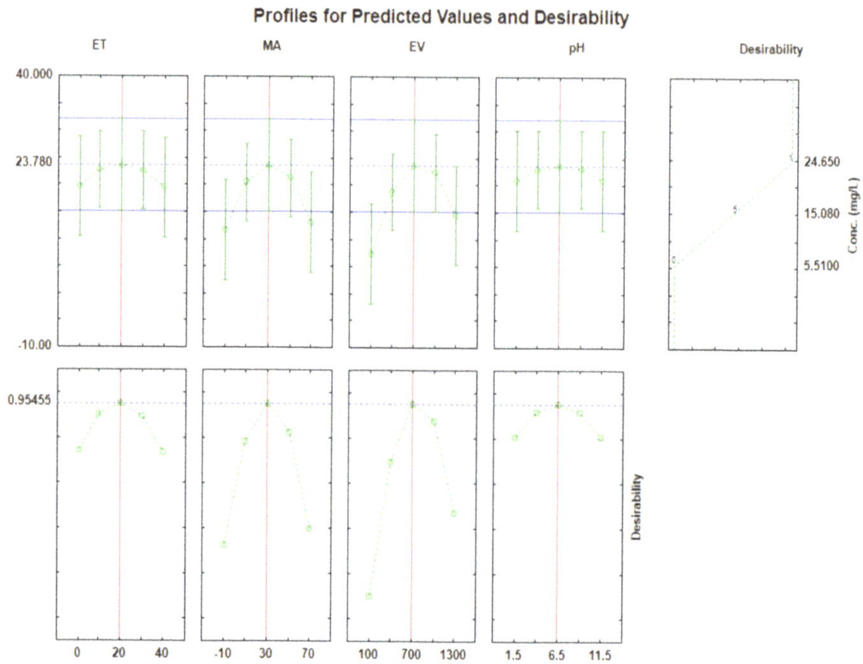

Figure 7. Desirability profiles with predicted values for the investigated factors affecting the extraction of trenbolone.

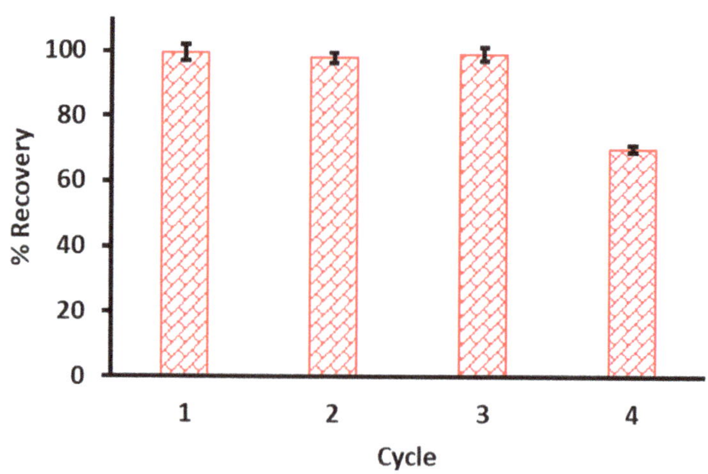

Figure 8. Adsorption and extraction performance of the MIP in comparison to the NIP.

3.5. Polymer Reusability

The reusability studies of the polymer were carried out using the polymer in the preconcentration procedure, eluting the analyte, washing the polymer with the eluent and vacuum-drying the polymer for reuse after each cycle. The reusability results are presented in Figure 9, where it can be noted that, in cycles 1–3, the trenbolone recovery was above 95%. A significant decrease in recovery was observed in the fourth cycle, where the recovery was 70.7 ± 0.32%.

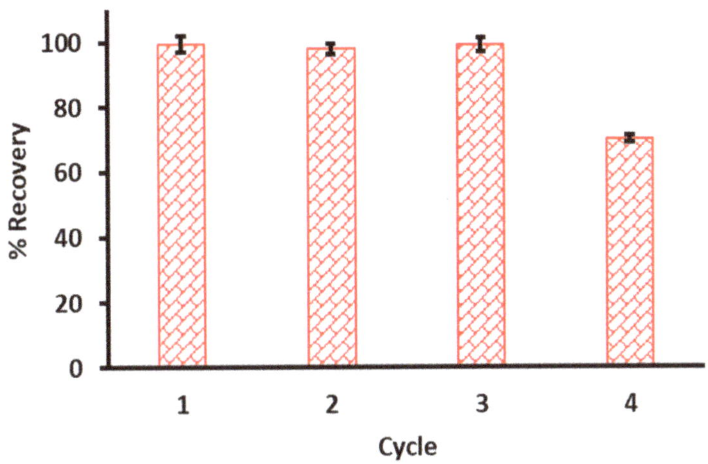

Figure 9. Trenbolone percent recovery during different cycles of reusing the MIP (n = 3).

4. Conclusions

A molecularly imprinted polymer based on a trenbolone template was prepared. The FTIR spectra of MIP and NIP showed that the chemical compositions of the polymers were similar, while the SEM image of MIP was more ordered in comparison to that of NIP. Using UV-Vis spectrophotometry quantification and adsorption data, the imprinting factor and isotherms showed that the MIP was favourable. The MIP was then used in solid-phase extraction of trenbolone. The extraction procedure was optimised using a multivariate experimental design based on central composite, and the SPE optimum conditions were found to be pH of 6.5, MA of 30 mg, ET of 20 min and EV of 700 µL. The adsorption of trenbolone was found to favour Langmuir, which meant the adsorption followed a monolayer mechanism with an imprinting factor (IF) of 2.41. The MIP was found to be recyclable for up to three times with recoveries above 95%.

Author Contributions: A.M., M.D., B.M. and P.N.N., Conceptualization; A.M. and P.N.N., method development; P.N.N., chemometrics; A.M. and M.D., sampling; A.M. and M.D., investigation; A.M., M.D., synthesis and characterization, A.M. and M.D., data collection; A.M., writing—original draft; B.M. and P.N.N., writing—review and editing; B.M., resources; B.M. and P.N.N., supervision. All authors have read and agreed to the published version of the manuscript.

Funding: This research was funded by Deutscher Akademischer Austauschdienst (DAAD, Germany grant no. 91752008).

Institutional Review Board Statement: Not applicable.

Informed Consent Statement: Not applicable.

Data Availability Statement: The data presented in this study are available on request from the corresponding author.

Acknowledgments: National Research Foundation (NRF, South Africa, grant nos. 113010 and 91230).

Conflicts of Interest: The authors declare no conflict of interest.

References

1. Halder, J.N.; Islam, M.N. Water Pollution and its Impact on the Human Health. *J. Environ. Hum.* **2015**, *2*, 36–46. [CrossRef]
2. Kumar, P.; Masago, Y.; Mishra, B.K.; Fukushi, K. Evaluating future stress due to combined effect of climate change and rapid urbanization for Pasig-Marikina River, Manila. *Groundw. Sustain. Dev.* **2018**, *6*, 227–234. [CrossRef]
3. Salerno, F.; Gaetano, V.; Gianni, T. Urbanization and climate change impacts on surface water quality: Enhancing the resilience by reducing impervious surfaces. *Water Res.* **2018**, *144*, 491–502. [CrossRef]
4. Miller, J.D.; Hutchins, M. The impacts of urbanisation and climate change on urban flooding and urban water quality: A review of the evidence concerning the United Kingdom. *J. Hydrol. Reg. Stud.* **2017**, *12*, 345–362. [CrossRef]
5. Chen, M.; Zhang, J. Bioremediation of soils contaminated with polycyclic aromatic hydrocarbons, petroleum, pesticides, chlorophenols and heavy metals by composting: Applications, microbes and future research needs. *Biotechnol. Adv.* **2015**, *33*, 745–755. [CrossRef]
6. Jacobsen, C.S.; Hjelmsø, M.H. Agricultural soils, pesticides and microbial diversity. *Curr. Opin. Biotechnol.* **2014**, *27*, 15–20. [CrossRef]
7. Kassotis, C.D.; Alvarez, D.A.; Taylor, J.A.; Frederick, S.; Nagel, S.C.; Tillitt, D.E. Characterization of Missouri surface waters near point sources of pollution reveals potential novel atmospheric route of exposure for bisphenol A and wastewater hormonal activity pattern. *Sci. Total Environ.* **2015**, *524*, 384–393. [CrossRef] [PubMed]
8. Harley, K.G.; Gunier, R.B.; Kogut, K.; Johnson, C.; Bradman, A.; Calafat, A.M.; Eskenazi, B. Prenatal and early childhood bisphenol A concentrations and behavior in school-aged children. *Environ. Res.* **2013**, *126*, 43–50. [CrossRef] [PubMed]
9. Palanza, P.; Nagel, S.C.; Parmigiani, S.; Saal, F.S. Perinatal exposure to endocrine disruptors: Sex, timing and behavioral endpoints. *Curr. Opin. Behav. Sci.* **2015**, *7*, 69–75. [CrossRef]
10. Peng, H.; Luo, M.; Xiong, H.; Yu, N.; Ning, F.; Fan, J.; Zeng, Z.; Li, J.; Chen, L. Preparation of photonic-magnetic responsive molecularly imprinted microspheres and their application to fast and selective extraction of. *J. Chromatogr. A* **2016**, *1442*, 1–11. [CrossRef]
11. López-García, M.; Romero-González, R.; Frenich, A.G. Determination of steroid hormones and their metabolite in several types of meat samples by ultra high performance liquid chromatography—Orbitrap high resolution mass spectrometry. *J. Chromatogr. A* **2015**, *1540*, 21–30. [CrossRef]
12. Horie, M.; Nakazawa, H. Determination of trenbolone and zeranol in bovine muscle and liver by liquid chromatography—electrospray mass spectrometry. *J. Chromatogr. A* **2000**, *882*, 53–62. [CrossRef]
13. Tomkins, P.; Saaristo, M.; Bertram, M.G.; Michelangeli, M.; Tomkins, R.B.; Wong, B.B.M. An endocrine-disrupting agricultural contaminant impacts sequential female mate choice in fish. *Environ. Pollut.* **2018**, *237*, 103–110. [CrossRef] [PubMed]
14. Kaklamanos, G.; Theodoridis, G. Rapid multi-method for the determination of growth promoters in bovine milk by liquid chromatography—tandem mass spectrometry. *J. Chromatogr. B* **2013**, *930*, 22–29. [CrossRef]
15. Lega, F.; Angeletti, R.; Stella, R.; Rigoni, L.; Biancotto, G.; Giusepponi, D.; Moretti, S.; Saluti, G.; Galarini, R. Abuse of anabolic agents in beef cattle: Could bile be a possible alternative matrix? *Food Chem.* **2017**, *229*, 188–197. [CrossRef]
16. Zhang, F.; Xie, Y.; Li, X.; Wang, D.; Yang, L.; Nie, Z. Science of the Total Environment Accumulation of steroid hormones in soil and its adjacent aquatic environment from a typical intensive vegetable cultivation of North China. *Sci. Total Environ.* **2015**, *538*, 423–430. [CrossRef]
17. Lagesson, A.; Saaristo, M.; Brodin, T.; Fick, J.; Klaminder, J.; Martin, J.M.; Wong, B.B.M. Fish on steroids: Temperature-dependent effects of 17 b-trenbolone on predator escape, boldness, and exploratory behaviors. *Environ. Pollut.* **2019**, *245*, 243–252. [CrossRef]
18. Feng, R.; Zhang, Y.; Li, H.; Wu, D.; Xin, X.; Zhang, S.; Yu, H. Ultrasensitive electrochemical immunosensor for zeranol detection based on signal amplification strategy of nanoporous gold films and nano-montmorillonite as labels. *Anal. Chim. Acta* **2013**, *758*, 72–79. [CrossRef]
19. Ma, F.; Liu, D. 17β-trenbolone, an anabolic–androgenic steroid as well as an environmental hormone, contributes to neurodegeneration. *Toxicol. Appl. Pharmacol.* **2015**, *282*, 68–76. [CrossRef]
20. Płotka-wasylka, J.; Szczepan, N. Modern trends in solid phase extraction: New sorbent media. *Trends. Anal. Chem.* **2016**, *77*, 23–43. [CrossRef]
21. Liu, K.; Kang, K.; Li, N.; An, J.; Lian, K.; Kang, W. Simultaneous Determination of Five Hormones in Milk by Automated Online Solid-Phase Extraction Coupled to High-Performance Liquid Chromatography. *J. AOAC Int.* **2020**, *103*, 265–271. [CrossRef] [PubMed]
22. Sampaio, N.M.F.M.; Castilhos, D.B.N.; da Silva, B.C.; Riegel-Vidotti, I.C.; Silva, B.J.G. Evaluation of Polyvinyl Alcohol/Pectin-Based Hydrogel Disks as Extraction Phase for Determination of Steroidal Hormones in Aqueous Samples by GC-MS/MS. *Molecules* **2019**, *24*, 40. [CrossRef] [PubMed]
23. Zheng, J.; Xi, C.; Wang, G.; Cao, S.; Tang, B.; Mu, Z. Rapid Screening and Determination of the Residues of Hormones and Sedatives in Milk Powder Using the UHPLC-MS/MS and SPE. *Food Anal. Methods* **2018**, *11*, 3435–3451. [CrossRef]
24. Jodar, L.V.; Santos, F.A.; Zucolotto, V.; Janegitz, B.C. Electrochemical sensor for estriol hormone detection in biological and environmental samples. *J. Solid State Electrochem.* **2018**, *22*, 1431–1438. [CrossRef]

25. Serafín, V.; Arévalo, B.; Martínez-García, G.; Aznar-Poveda, J.; Lopez-Pastor, J.A.; Beltrán-Sánchez, J.F.; Garcia-Sanchez, A.J.; Garcia-Haro, J.; Campuzano, S.; Yáñez-Sedeño, P. Enhanced determination of fertility hormones in saliva at disposable immunosensing platforms using a custom designed field-portable dual potentiostat. *Sens. Actuators B Chem.* **2019**, *299*, 126934. [CrossRef]
26. Savaris, D.L.; Alberton, M.B.; de Matos, R.; Fernanda, M.; Zalazar, D.J.R.; Lindino, C.A. Development of Spectrophotometric Method for the Determination of 17α-Methyltestosterone. *Br. J. Anal. Chem.* **2019**, *6*, 12–19.
27. Najim, S.S. Spectrophotometric determination of progesterone and dopamine in breast cancer serum. *Univ. Thi Qar. J.* **2017**, *12*, 1_21. [CrossRef]
28. Pérez, R.L.; Escandar, G.M. Multivariate calibration-assisted high-performance liquid chromatography with dual UV and fluorimetric detection for the analysis of natural and synthetic sex hormones in environmental waters and sediments. *Environ. Pollut.* **2016**, *209*, 114–122. [CrossRef]
29. Naldi, A.C.; Fayad, P.B.; Prévost, M.; Sauvé, S. Analysis of steroid hormones and their conjugated forms in water and urine by on-line solid-phase extraction coupled to liquid chromatography tandem mass spectrometry. *Chem. Cent. J.* **2016**, *10*, 30. [CrossRef]
30. Tomšíková, H.; Aufartová, J.; Solich, P.; Nováková, L.; Sosa-Ferrera, Z.; Santana-Rodríguez, J.J. High-sensitivity analysis of female-steroid hormones in environmental samples. *TrAC Trends Anal. Chem.* **2012**, *34*, 35–58. [CrossRef]
31. Migowska, N.; Caban, M.; Stepnowski, P.; Kumirska, J. Simultaneous analysis of non-steroidal anti-inflammatory drugs and estrogenic hormones in water and wastewater samples using gas chromatography–mass spectrometry and gas chromatography with electron capture detection. *Sci. Total Environ.* **2012**, *441*, 77–88. [CrossRef] [PubMed]
32. Kazemi, E.; Dadfarnia, S.; Shabani, A.M.H.; Abbasi, A.; Vaziri, M.R.R.; Behjat, A. Iron oxide functionalized graphene oxide as an efficient sorbent for dispersive micro-solid phase extraction of sulfadiazine followed by spectrophotometric and mode-mismatched thermal lens spectrometric determination. *Talanta* **2016**, *147*, 561–568. [CrossRef] [PubMed]
33. Mpupa, A.; Nomngongo, P.N. Multivariate-Assisted Solid Phase Extraction Procedure for Simultaneous Preconcentration and Assessment of UV-Filters in Wastewater Prior to UV-Vis Spectrophotometric Determination. In *Emerging Pollutants: Some Strategies for the Quality Preservation of Our Environment*; IntechOpen: London, UK, 2018; pp. 45–59.
34. Mpupa, A.; Mashile, G.P.; Nomngongo, P.N. Vortex assisted-supramolecular solvent based microextraction coupled with spectrophotometric determination of triclosan in environmental water samples. *Open Chem.* **2017**, *15*, 255–262. [CrossRef]
35. Sobańska, A.W.; Kałębasiak, K.; Pyzowski, J.; Brzezińska, E. Quantification of sunscreen benzophenone-4 in hair shampoos by hydrophilic interactions thin-layer chromatography/densitometry or derivative UV spectrophotometry. *J. Anal. Methods Chem.* **2015**, 1–7. [CrossRef]
36. Zygler, A.; Wasik, A.; Namie, J. Retention behaviour of some high-intensity sweeteners on different SPE sorbents. *Talanta* **2010**, *82*, 1742–1748. [CrossRef]
37. Dimpe, K.M.; Nomngongo, P.N. Current sample preparation methodologies for analysis of emerging pollutants in different environmental matrices. *TrAC Trends Anal. Chem.* **2016**, *82*, 199–207. [CrossRef]
38. Chigome, S.; Darko, G.; Torto, N. Electrospun nanofibers as sorbent material for solid phase extraction. *Analyst* **2011**, *136*, 2879–2889. [CrossRef]
39. Wierucka, M.; Biziuk, M. Application of magnetic nanoparticles for magnetic solid-phase extraction in preparing biological, environmental and food samples. *Trends Anal. Chem.* **2014**, *59*, 50–58. [CrossRef]
40. Sun, X.; Wang, M.; Peng, J.; Yang, L.; Wang, X.; Wang, F.; Zhang, X.; Wu, Q.; Chen, R.; Chen, J. Dummy molecularly imprinted solid phase extraction of climbazole from environmental water samples. *Talanta* **2019**, *196*, 47–53. [CrossRef]
41. Zhang, M.; Mao, Q.; Feng, J.; Yuan, S.; Wang, Q.; Huang, D.; Zhang, J. Validation and application of an analytical method for the determination of selected acidic pharmaceuticals and estrogenic hormones in wastewater and sludge. *J. Environ. Sci. Heal Part A* **2016**, *51*, 914–920. [CrossRef]
42. He, S.; Wang, R.; Wei, W.; Liu, H.; Ma, Y. Simultaneous determination of 22 residual steroid hormones in milk by liquid chromatography–tandem mass spectrometry. *Int. J. Dairy Technol.* **2020**, *73*, 357–365. [CrossRef]
43. Medina, G.S.; Acquaviva, A.; Reta, M. Development of monolithic sorbent cartridges (m-SPE) for the extraction of non-steroidal anti-inflammatory drugs from surface waters and their determination by HPLC. *Microchem. J.* **2020**, *159*, 105447. [CrossRef]
44. Raeke, J.; Lechtenfeld, O.J.; Wagner, M.; Herzsprung, P.; Reemtsma, T. Selectivity of solid phase extraction of freshwater dissolved organic matter and its effect on ultrahigh resolution mass spectra. *Environ. Sci. Process Impacts* **2016**, *18*, 918–927. [CrossRef] [PubMed]
45. Bizkarguenaga, E.; Ros, O.; Iparraguirre, A.; Navarro, P.; Vallejo, A.; Usobiaga, A.; Zuloaga, O. Solid-phase extraction combined with large volume injection-programmable temperature vaporization—gas chromatography—mass spectrometry for the multiresidue determination of priority and emerging organic pollutants in wastewater. *J. Chromatogr. A* **2012**, *1247*, 104–117. [CrossRef] [PubMed]
46. Gast, M.; Sobek, H.; Mizaikoff, B. Advances in Imprinting Strategies for Selective Virus Recognition A Review. *Trends Anal. Chem.* **2019**, *114*, 218–232. [CrossRef]
47. Chen, L.; Wang, X.; Lu, W.; Wu, X.; Li, J. Molecular imprinting: Perspectives and applications. *Chem. Soc. Rev.* **2016**, *45*, 2137–2211. [CrossRef]

48. Ming, W.; Wang, X.; Lu, W.; Zhang, Z.; Song, X.; Li, J.; Chen, L. Magnetic molecularly imprinted polymers for the fluorescent detection of trace 17β-estradiol in environmental water. *Sens. Actuators B Chem.* **2017**, *238*, 1309–1315. [CrossRef]
49. Ahmad, O.S.; Bedwell, T.S.; Esen, C.; Garcia-Cruz, A.; Piletsky, S.A. Molecularly imprinted polymers in electrochemical and optical sensors. *Trends Biotechnol.* **2019**, *37*, 294–309. [CrossRef]
50. Ansari, S.; Masoum, S. Molecularly imprinted polymers for capturing and sensing proteins: Current progress and future implications. *TrAC Trends Anal. Chem.* **2019**, *114*, 29–47. [CrossRef]
51. Karaseva, N.; Ermolaeva, T.; Mizaikoff, B. Piezoelectric sensors using molecularly imprinted nanospheres for the detection of antibiotics. *Sens. Actuators B Chem.* **2016**, *225*, 199–208. [CrossRef]
52. Zhang, Z.; Li, Y.; Zhang, X.; Liu, J. Molecularly imprinted nanozymes with faster catalytic activity and better specificity. *Nanoscale* **2019**, *11*, 4854–4863. [CrossRef] [PubMed]
53. Muratsugu, S.; Shirai, S.; Tada, M. Recent progress in molecularly imprinted approach for catalysis. *Tetrahedron. Lett.* **2020**, *61*, 151603. [CrossRef]
54. Turiel, E.; Martin-Esteban, A. Molecularly imprinted polymers: Towards highly selective stationary phases in liquid chromatography and capillary electrophoresis. *Anal. Bioanal. Chem.* **2004**, *378*, 1876–1886. [CrossRef] [PubMed]
55. Zink, S.; Moura, F.A.; da Silva Autreto, P.A.; Galvao, D.S.; Mizaikoff, B. Efficient prediction of suitable functional monomers for molecular imprinting via local density of states calculations. *Phys. Chem. Chem. Phys.* **2018**, *20*, 13153–13158. [CrossRef]
56. Fernandes, R.S.; Dinc, M.; Raimundo, I.M., Jr.; Mizaikoff, B. Synthesis and characterization of porous surface molecularly imprinted silica microsphere for selective extraction of ascorbic acid. *Microporous Mesoporous Mater.* **2018**, *264*, 28–34. [CrossRef]
57. Ma, J.; Yuan, L.; Ding, M.; Wang, S.; Ren, F.; Zhang, J.; Du, S.; Li, F.; Zhou, X. The study of core-shell molecularly imprinted polymers of 17β-estradiol on the surface of silica nanoparticles. *Biosens. Bioelectron.* **2011**, *26*, 2791–2795. [CrossRef]
58. Fernandes, R.S.; Dinc, M.; Raimundo, M.R., Jr.; Mizaikoff, B. Analytical Methods microspheres for the selective extraction of vanillin. *Anal. Methods* **2017**, *9*, 2883–2889. [CrossRef]
59. Azodi-Deilami, S.; Najafabadi, A.H.; Asadi, E.; Abdouss, M.; Kordestani, D. Magnetic molecularly imprinted polymer nanoparticles for the solid-phase extraction of paracetamol from plasma samples, followed its determination by HPLC. *Microchim. Acta* **2014**, *181*, 1823–1832. [CrossRef]
60. Gupta, N.V.; Shivakumar, H.G. Investigation of swelling behavior and mechanical properties of a pH-sensitive superporous hydrogel composite. *Iran J. Pharm. Res. IJPR* **2012**, *11*, 481.
61. Arabi, M.; Ghaedi, M.; Ostovan, A. Synthesis and application of in-situ molecularly imprinted silica monolithic in pipette-tip solid-phase microextraction for the separation and determination of gallic acid in orange juice samples. *J. Chromatogr. B* **2017**, *1048*, 102–110. [CrossRef]
62. Nantasenamat, C.; Naenna, T.; Ayudhya, C.I.N.; Prachayasittikul, V. Quantitative prediction of imprinting factor of molecularly imprinted polymers by artificial neural network. *J. Comput. Aided Mol. Des.* **2005**, *19*, 509–524. [CrossRef] [PubMed]
63. Kidakova, A.; Boroznjak, R.; Reut, J.; Öpik, A.; Saarma, M.; Syritski, V. Molecularly imprinted polymer-based SAW sensor for label-free detection of cerebral dopamine neurotrophic factor protein. *Sens. Actuators B Chem.* **2020**, *308*, 127708. [CrossRef]
64. Langmuir, I. The constitution and fundamental properties of solids and liquids. II. Liquids. *J. Am. Chem. Soc.* **1917**, *39*, 1848–1906. [CrossRef]
65. Nqombolo, A.; Mpupa, A.; Gugushe, A.S.; Moutloali, R.M.; Nomngongo, P.N. Adsorptive removal of lead from acid mine drainage using cobalt-methylimidazolate framework as an adsorbent: Kinetics, isotherm, and regeneration. *Environ. Sci. Pollut. Res.* **2019**, *26*, 3330–3339. [CrossRef]
66. Mashile, P.P.; Mpupa, A.; Nomngongo, P.N. Adsorptive removal of microcystin-LR from surface and wastewater using tyre-based powdered activated carbon: Kinetics and isotherms. *Toxicon* **2018**, *145*, 25–31. [CrossRef]
67. Gugushe, A.S.; Nqombolo, A.; Nomngongo, P.N. Application of Response Surface Methodology and Desirability Function in the Optimization of Adsorptive Remediation of Arsenic from Acid Mine Drainage Using Magnetic Nanocomposite: Equilibrium Studies and Application to Real Samples. *Molecules* **2019**, *24*, 1792. [CrossRef]
68. Mashile, G.P.; Mpupa, A.; Nomngongo, P.N. In-syringe micro solid-phase extraction method for the separation and preconcentration of parabens in environmental water samples. *Molecules* **2018**, *23*, 1450. [CrossRef]
69. Mpupa, A.; Mashile, G.P.; Nomngongo, P.N. Ultrasound-assisted dispersive solid phase nanoextraction of selected personal care products in wastewater followed by their determination using high performance liquid chromatography-diode array detector. *J. Hazard. Mater.* **2019**, *370*, 33–41. [CrossRef]
70. Khor, C.P.; Jaafar, M.B.; Ramakrishnan, S. Optimization of Conductive Thin Film Epoxy Composites Properties Using Desirability Optimization Methodology. *J. Optim.* **2016**, *2016*, 1–8. [CrossRef]
71. Manzo, V.; Miró, M.; Richter, P. Programmable flow-based dynamic sorptive microextraction exploiting an octadecyl chemically modified rotating disk extraction system for the determination of acidic drugs in urine. *J. Chromatogr. A* **2014**, *1368*, 64–69. [CrossRef]

processes

MDPI

Article

Enhanced Adsorptive Removal of β-Estradiol from Aqueous and Wastewater Samples by Magnetic Nano-Akaganeite: Adsorption Isotherms, Kinetics, and Mechanism

Anele Mpupa [1,2], Azile Nqombolo [1,2,3], Boris Mizaikoff [1,4] and Philiswa Nosizo Nomngongo [1,2,3,*]

1 Department of Chemical Sciences, University of Johannesburg, Doornfontein Campus, P.O. Box 17011, Johannesburg 2028, South Africa; 216051138@student.uj.ac.za (A.M.); azilenqombolo@gmail.com (A.N.); boris.mizaikoff@uni-ulm.de (B.M.)
2 DSI/NRF SARChI Chair: Nanotechnology for Water, University of Johannesburg, Doornfontein 2028, South Africa
3 DSI/Mintek Nanotechnology Innovation Centre, University of Johannesburg, Doornfontein 2028, South Africa
4 Institute of Analytical and Bioanalytical Chemistry, Ulm University, Albert-Einstein-Allee 11, 89081 Ulm, Germany
* Correspondence: pnnomngongo@uj.ac.za; Tel.: +27-11-559-6187

Received: 15 August 2020; Accepted: 14 September 2020; Published: 22 September 2020

Abstract: A surfactant-free method was used to synthesize iron oxyhydroxide (akaganeite, β-FeOOH) nanorods and characterized using Fourier transform infrared spectroscopy (FTIR), X-ray diffraction (XRD), scanning electron microscopy combined with energy-dispersive X-ray spectroscopy (SEM-EDS), and transmission electron microscopy (TEM). The synthesized nanoadsorbent was applied for the adsorptive removal of β-estradiol from aqueous solutions. The parameters affecting the adsorption were optimized using a multivariate approach based on the Box–Behnken design with the desirability function. Under the optimum conditions, the equilibrium data were investigated using two and three parameter isotherms, such as the Langmuir, Freundlich, Dubinin–Radushkevich, Redlich–Peterson, and Sips models. The adsorption data were described as Langmuir and Sips isotherm models and the maximum adsorption capacities in Langmuir and Sips of the β-FeOOH nanorods were 97.0 and 103 mg g^{-1}, respectively. The adjusted non-linear adsorption capacities were 102 and 104 mg g^{-1} for Langmuir and Sips, respectively. The kinetics data were analyzed by five different kinetic models, such as the pseudo-first order, pseudo-second order, intraparticle, as well as Boyd and Elovich models. The method was applied for the removal β-estradiol in spiked recoveries of wastewater, river, and tap water samples, and the removal efficiency ranged from 93–100%. The adsorbent could be reused up to six times after regeneration with acetonitrile without an obvious loss in the removal efficiency (%RE = 95.4 ± 1.9%). Based on the results obtained, it was concluded that the β-FeOOH nanorods proved to be suitable for the efficient removal of β-estradiol from environmental matrices.

Keywords: β-estradiol; akaganeite nanorods; adsorptive removal; endocrine disruptors; desirability function

---

1. Introduction

Water is the most important natural resource for the survival of all organisms [1]. Due to the increase in urbanization, climate change, industrial production, and population growth, the quality of water can be negatively affected. Pollutants entering soil or water pose a variety of threats to the functioning of ecosystems and to human health [2]. Soil is frequently polluted with polycyclic aromatic hydrocarbons (PAHs), petroleum-related compounds, heavy metals, chlorophenols, organic pollutants, and pesticides [2]. There is an overlap of the types of pollutants present in water and soil, because of water run-off from contaminated soils, wastewater, and septic tanks [3]. The United States and the European Union in 2002 and 2009 reported that at least 80% of all collected samples were contaminated with organic pollutants, which frequently have endocrine-disrupting properties [4]. The most widely investigated endocrine disruptors include molecules or metabolites derived from personal care products, pharmaceuticals, polybrominated diphenyl ethers (PBDEs), and hormones [5].

β-estradiol is a natural hormone excreted by mammals and is primarily produced by the reproductive organs [6]. Naturally, β-estradiol plays a role in physiological processes, such as reproduction and sexual function [7]. Estrogens from animal-derived food (for example, but not limited to, eggs, milk, and fish) can be introduced to people via their usage as growth-promoting agents, thus resulting in unnaturally high concentrations [8]. Elevated β-estradiol levels may interfere with the function of the thyroid in birds and fish [9], may reduce fertility, and may lead to sexual disorders. It is also known to generate congenital malformations in children [10]. The contamination route is via animal and human excretions, as well as subsequent transport by sewage effluent into the water systems [11]. Therefore, the decontamination of water systems used for later water consumption is crucial for human and environmental wellbeing.

Numerous decontamination techniques, such as adsorption [12], catalytic degradation [13], oxidation [14], and biodegradation [15], have been used to remove β-estradiol from aqueous systems. Though seldom used because of difficulties in adsorbent removal, adsorption has great potential for decontamination owing to its high efficiency and tunability. Adsorption can be tailored by the wide variety of sorbets available, including carbon nanotubes [7], membranes [16], molecularly imprinted polymers [17], and carbon-based materials [18]. Akaganeite is an iron oxyhydroxide (β-FeOOH) with double chains of octahedral corners forming a tunnel-like framework of mesopores [19,20]. Akaganeite occurs naturally in a variety of soil types. Iron oxyhydroxide has also been prepared in the laboratory and has been shown to remove both anions and cations [21]. It is also paramagnetic at room temperature, which makes it easy to remove from any matrix it has been dispersed in using an external magnet [22]. Consequently, it is useful as both a catalyst and an adsorbent material in adsorption/preconcentration [19].

The main objective of this study was to explore the adsorption properties of β-estradiol onto a synthesized akaganeite nanomaterial serving as the adsorbent material. Multivariate optimization strategies were used to explore the effect of sample pH, adsorbent dosage, and extraction time for the adsorptive removal of β-estradiol as an exemplary yet highly relevant analyte. The adsorption process was studied in detail via adsorption isotherms, while the rate-limiting steps were investigated using adsorption kinetics. Finally, the adsorption performance of the synthesized akaganeite was evaluated using spiked wastewater samples.

2. Materials and Methods

2.1. Materials

The ferric (III) chloride, sodium hydroxide, and ethanol were all of analytical reagent grade and β-estradiol the (98%) purchased from Sigma Aldrich (St. Loius, MO, USA) were used in the study. A stock solution (20 mg L^{-1}) was prepared by dissolving an appropriate amount of the β-estradiol in methanol and refrigerated until required for use. Standard solutions were prepared daily by diluting the stock solution with ultra-pure water (Direct-Q® 3UV-R purifier system, Millipore, Merck, KGaA, Darmstadt, Germany).

2.2. Instrumentation

All pH measurements were done using an OHAUS ST series pen pH meter (Nanikon, Switzerland). The adsorption studies were carried out using the Branson 5800 Ultrasonic Cleaner (Danbury, CT, USA). The FTIR measurements were taken using the KBr pellet technique and recorded in the region 400–4000 cm^{-1} using a Perkin Elmer Spectrum 100 spectrometer. Scanning electron microscopy (SEM, TESCAN VEGA 3 XMU, LMH instrument, Czech Republic) coupled with energy-dispersive X-ray spectroscopy (EDS) was used to study the morphology and elemental composition of the adsorbent at an accelerating voltage of 20 kV. The transmission electron microscopic image was captured using TEM (JEM-2100, JEOL, Tokyo, Japan). An Agilent high-performance liquid chromatography (HPLC) 1200 Infinity series, equipped with a photodiode array detector (Agilent Technologies, Waldbronn, Germany), was used for all analysis. The separation was done using an Agilent Zorbax Eclipse Plus C18 column (3.5 μm × 150 mm × 4.6 mm) (Agilent, Newport, CA, USA) operated at an oven temperature of 25 °C. The chromatograms were recorded using a 1.00 mL min^{-1} flow rate, a solvent mixture of 55% mobile phase A (water) and 45% mobile phase B (acetonitrile), and an adsorption wavelength of 260 nm using an isocratic elution system.

2.3. Sampling and Storage

Influent and effluent wastewater samples were collected from a local wastewater treatment plant (WWTP, Gauteng, South Africa) in winter 2019. River water samples were collected from Apies River, which is upstream from the wastewater treatment plant. Tap water samples were collected from the University of Johannesburg Doornforntein campus. All samples were collected in 1 L glass bottles and refrigerated until use.

2.4. Preparation of Akaganeite Nanorods

Akaganeite nanoparticles were prepared according to the method described by [23]. Briefly, a 5 mol L^{-1} solution of NaOH was prepared in 30 mL deionized water and stirred at room temperature in a beaker equipped with a magnetic stirrer. A 25 mL 2 mol L^{-1} $FeCl_3$ was rapidly added to the NaOH solution using a syringe; the stirring was continued for an hour. Ethanol was then added, and the resulting precipitate was washed several times with a 1:1 ethanol–water solution. The precipitate was then dried in an oven at 50 °C for 5 h.

2.5. Batch Adsorption of β-Estradiol

The adsorption was done using the batch incubation method described in [24]. The method was optimized using the Box–Behnken design. The levels of independent factors, such as sample pH, contact time (CT), and mass adsorbent (MA), are presented in Table S1. The appropriate amount of adsorbent was then weighed and added to a glass bottle. Aliquots of a 10 mL sample solution containing 6 mg L^{-1} β-estradiol were added into the sample bottle. The samples were then agitated using an ultrasound water bath; thereafter, the adsorbent and supernatant were separated using an external magnet. Thereafter, 1 mL of sample was filtered via a 0.22 μm PVDF filter before analysis with HPLC. All the experiments were carried out in triplicates, after which the adsorption removal efficiency was calculated using Equation (1):

$$\%RE = \left(\frac{C_0 - C_e}{C_0}\right) \times 100 \quad (1)$$

where C_0 and C_e are the initial and equilibrium concentrations, respectively.

An optimization process is a critical stage in analytical chemistry. This is because of regulations due to the high cost of reagents and quantity of variables that affect the development and validation of analytical methods [25]. Multivariate optimization is an important technique because it takes less time, resources, and effort than the univariate approach. The design of experiments is important in analytical applications [25]. Multivariate optimization strategies allow to efficiently minimize the parameter space-saving time and resources when determining the optimum conditions, here for the deployment of an adsorbent material [25]. For the adsorption of β-estradiol onto akaganeite, the (1) pH of the sample, (2) mass of adsorbent (MA), and (3) contact time (CT) were optimized using a multivariate approach, as presented in Table S1.

Under optimum conditions, the equilibrium isotherms and kinetics of adsorption were studied. The β-estradiol solutions had concentrations ranging from 1 to 9 mg L^{-1}. The equilibrium isotherm studies were done in triplicates and the β-estradiol concentration adsorbed onto akaganeite nanorods was calculated using Equation (2):

$$q_e = \left(\frac{C_0 - C_e}{m}\right)V \tag{2}$$

where C_0 and C_e are the initial and equilibrium concentrations (mg. L^{-1}) of β-estradiol, respectively, and m (g) and V (L) represent the mass of the adsorbent and volume of the sample used. The adsorption kinetics was investigated by the addition of 10 mL aliquots (6 mg. L^{-1}) into 7 glass bottles. The bottles were then sonicated for 1–60 min for each sample before magnetic separation and HPLC analysis.

2.6. Method Validation and Real Sample Analysis

The collected wastewater (influent and effluent), river, and tap water samples were analyzed before adsorption and β-estradiol could not be detected in the sample. As a result, the method was validated using spiked sample recoveries; this was due to the absence of certified reference materials for emerging pollutants. To validate the method, wastewater effluent samples were spiked at two levels (4 and 8 mg. L^{-1}) with β-estradiol by appropriate dilution with the effluent water. The validation experiments were carried out by adding 10 mL of the sample (adjusted to pH 5.25) onto 42.5 mg of akaganeite in a capped glass bottle. The mixture was then agitated in an ultrasonic bath for 60 min, after which the supernatant and adsorbent were separated using an external magnet before filtering and HPLC analysis of the supernatant.

3. Results and Discussion

3.1. Characterisation of the Akaganeite Nanoparticles

Figure 1 shows the Fourier transform infrared spectrum of the as-synthesized akaganeite. The peaks at 3437 and 3135 cm^{-1} were ascribed for OH stretching and the bending at 1620 cm^{-1} and 1009 cm^{-1} were assigned to the absorption bands of the iron oxyhydroxides arising from the Fe–OH and Fe–O vibration of the akaganeite [19,26]. The bending peak at 650 cm^{-1} is reported to be the characteristic vibrations of Fe–O in β-FeOOH [26]. Lastly, the OH bending band around the 1431 cm^{-1} region corresponded to the β-FeOOH FTIR spectrum reported in the literature [19,26].

Akaganeite nanoparticles are known to be low-crystallinity materials, which during XRD characterization are indicated by rather low-intensity peaks [27]. In Figure 2, an exemplary XRD data set for the synthesized akaganeite nanoparticles are presented. From the XRD pattern, the characteristic low-intensity peaks were observed, confirming low crystallinity. According to [26], decreasing the pH of the synthesis media results in more crystalline akageneite, while increasing the pH of the synthesis media results in less crystalline akaganeite nanoparticles, which was a similar case with the present study (i.e., sodium hydroxide was used). According to [27], the diffraction peaks at around 2θ = 20° and 40° are characteristic nano-akaganeite peaks and are evident in Figure 2.

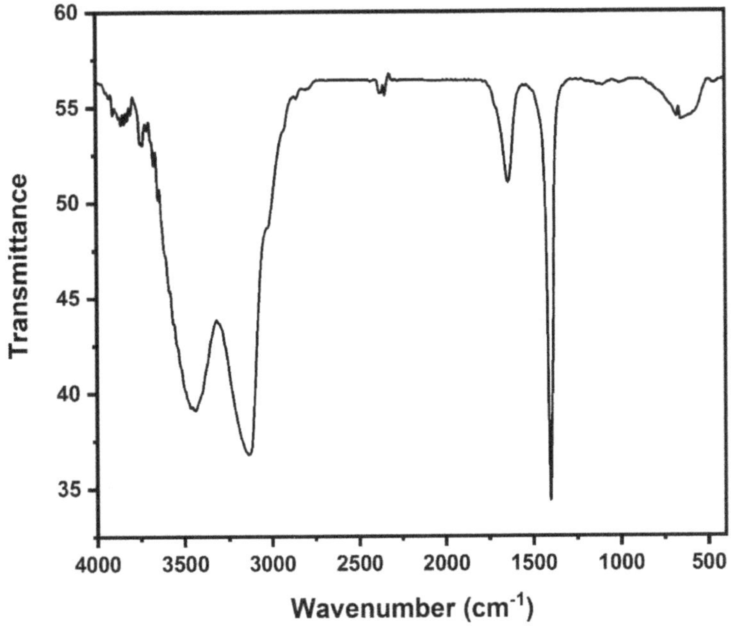

Figure 1. FTIR spectra of akaganeite.

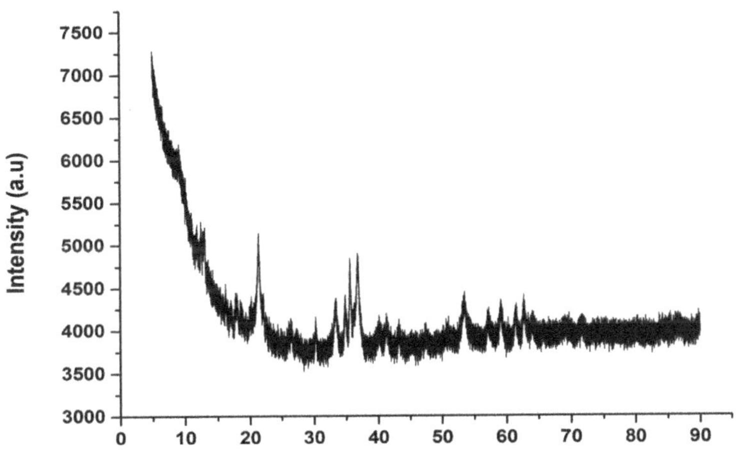

Figure 2. X-ray diffraction pattern of the akaganeite nanorods.

Scanning electron microscopy (SEM) was used to investigate the surface morphology of the akaganeite nanomaterial combined with energy-dispersive X-ray (EDX) for elemental analysis. The major components of the akaganeite nanomaterial were determined by EDX and given in the spectrum represented in Figure 3B. SEM images (Figure 3A,C) revealed aggregated akaganeite nanoparticles, which resulted from the absence of a surfactant during the synthesis, as was reported in [19]. The yellow frame in Figure 3C indicates the area used for evaluation by EDX.

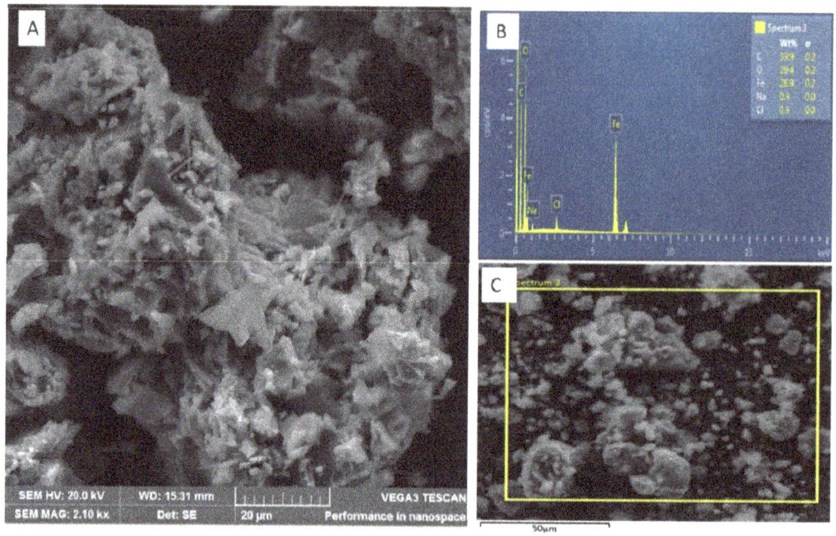

Figure 3. An exemplary (**A**) SEM image and (**B**) EDX spectrum and (**C**) section analysed using EDX of the akaganeite nanomaterial synthesized in absence of a surfactant.

Transmission electron microscopy (TEM) studies, as shown in Figure 4, indicate the characteristic rod-shaped akaganeite structures next to more spherical/octagonal particles, as described by [19]. While no surfactant was used herein, to better control the shape of the obtained akageneite material, a surfactant is needed, ensuring more ordered mesostructures. However, one aim of the present study was to ensure a straightforward and possibly simple synthesis route. Hence, a more uniform particle shape/size was traded against the simplicity of the synthesis strategy, which in absence of surfactants does not require further processing steps, including autoclaved heating and time-consuming Soxhlet extraction [19].

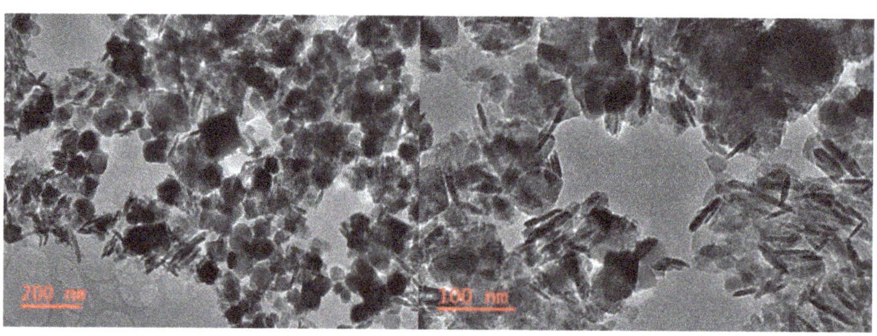

Figure 4. Exemplary TEM images of the synthesized akaganeite nanomaterial in absence of any surfactant.

3.2. *Optimisation of Removal Procedure*

A Pareto chart (see Figure S1) was used to identify critical factors during the adsorption process by evaluating the variance within the experimental design matrix [28]. The Pareto chart is based on analysis of variance (ANOVA), which is a linear model. As a result, during the analysis of the Pareto chart, only the linear factors (i.e., factors with the index L) were included, while quadratic factors (with the index Q) were not considered [29]. In the analysis of a Pareto chart, for a factor to be considered statistically significant, its bar should pass the redline that indicates a p-value of 0.05

(95% confidence level) [30,31]. The ultimately dominating factors were determined to include the (1) pH of the sample, (2) the mass of adsorbent (MA), and (3) the contact time (CT). Of these three factors, the only one beyond the 95% confidence line (red line in Figure S1) was the pH.

The response surface methodology (RSM) was used to investigate (a) the interaction and (b) the quadratic effects of the main parameters pH, MA, and CT, respectively, using the data generated via a Box–Behnken experimental design, yielding 3D surface plots (Figure S2) for these three parameters. As it can be seen from Figure S2A, increasing the mass of the adsorbent increases the percentage removal (% RE) with any pH below 7, and at a pH of around 7 the maximum % RE could be observed. The same could be observed from the interaction of pH and contact time. Figure S2C shows the interaction between CT and MA; masses between 25 and 10 together with a CT of 20 min gives a minimum 5 RE while a CT and MA above 30 min and 30 mg results in higher percentage removal (% RE). The plots were used in conjunction with the profiles of desirability to determine the optimum conditions.

Using the desirability function (DF, Figure S3) allows the simultaneous estimation of the optimal values for all the investigated factors. Desirability always takes values within the 0–1 range where 0 indicates the least desired result (33.4% removal), 0.5 being the central point 69.4%, and 1 being the most desirable value assigned a % removal of 105 [32,33]. According to the desirability profiles, the optimal conditions for the adsorption process were a pH of 5.25, MA of 42.5 mg, and CT of 60 min. The DF predicted values were then used to confirm the optimum conditions of the adsorption method. To understand the effect of pH on the adsorption, the point of zero charge of akaganeite was considered. At a pH lower than the point of zero charge, the surface of the material is positively charged, while at a pH higher than the point of zero charge, the surface of the adsorbent is negatively charged [34–36]. The point of zero charge for the akaganeite nanorods as described in the literature is around 7–7.5 [21]. The optimum pH for the adsorption was found to be 5.25, which is lower than the point of zero charge of akageneite. This means that the surface of the material was positively charged. β-estradiol also has a pKa of 10.71, meaning that at the optimum pH, the analyte is in its molecular state. This translates to the presence of the OH groups on the analyte to be available for ionic interaction between the β-FeOOH and β-estradiol.

3.3. *Adsorption Isotherms*

The ratio between the concentration of adsorbed analyte and the amount of analyte remaining in the supernatant solution at equilibrium conditions provides the absolute removal efficiency [24]. The interaction of the analyte species with the adsorbent material reaches a dynamic equilibrium after a certain incubation time, which may be described via so-called isotherm models [37]. In the present study, the isotherms for β-estradiol adsorbing at the surface of β-FeOOH nanorods were studied at a pH of 5.25 with a β-estradiol concentration of 1–10 mgL^{-1}, an adsorbent mass (MA) of 42.5 mg, and contact time (CT) of 60 min at 25 °C. It was observed that the adsorption capacity increased with an increase in initial β-estradiol concentration (Figure 5A). In this study, five isotherm models were used to describe the equilibrium. These include the Langmuir, Freundlich, Dubinin–Radushkevich, Redlich–Peterson, and Sips isotherm models, and the resulting plots are shown in Figure 5B–F. The R^2 values for the different models were used to determine which model best fits the adsorption process. The Langmuir and Freundlich models were used to determine the primary adsorption mechanism. As seen in Table 1 the adsorption process fitted the Langmuir model with an R^2 of 0.9996, which was higher than the 0.8811 for the Freundlich model.

The β-estradiol adsorption data were fitted with Langmuir, Freundlich, Dubinin–Radushkevich, and Redlich–Peterson isotherms models, as summarized in Table 1. Comparing the obtained goodness-of-fit values in Table 1, the experimental data followed a Langmuir isotherm (R^2 = 0.9996), which was higher than for any other isotherm. This implies that adsorption occurs homogeneously at the surface sites of the β-FeOOH nanomaterial and that the adsorption of β-estradiol may be assumed occurring in a monolayer fashion. Thus, a Langmuir constant (K_L) of 16 L g^{-1} and the maximum adsorption capacity of 97.0 mg g^{-1} was obtained. The calculated separation factor (R_L) of 0.10–0.25 for the β-estradiol adsorption on the β-FeOOH nanomaterial, according to the literature, demonstrates a favorable adsorption [38,39].

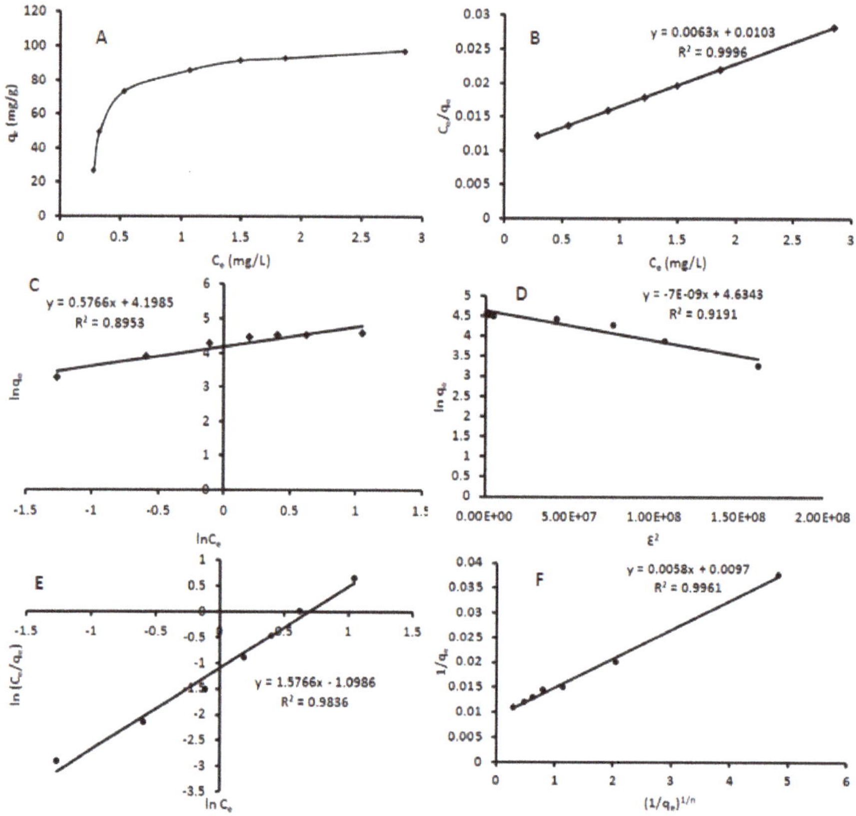

Figure 5. Adsorption isotherm plots for the sorption of β-estradiol using akaganeite (**A**)—Langmuir isotherm model; (**B**)—Langmuir linear model; (**C**)—Freundlich linear model; (**D**)—Dubinin–Radushkevich linear model; (**E**)—Redlich–Peterson linear model; (**F**)—Sips linear model.

The Dubinin–Radushkevich isotherm is conventionally used to differentiate between chemical and physical adsorption by evaluating the mean energy (E). It is said that an E value that is lower than 8 kJ mol indicates physisorption dominance while an E value above 8 indicates chemisorption dominance [40,41]. The calculated E value was 8.5 kJ mol^{-1}, which indicates a chemisorption process. The Redlich–Peterson model combines both the Freundlich and Langmuir isotherms, whereby B is the exponent from the linear plot ranging between 0 and 1. If B equals 1, the model reduces to the Langmuir equation, and if B = 0, it reduces to the Freundlich equation [42]. From the obtained results, B has a value of 0.88, which again favors the Langmuir model to describe the interaction between β-estradiol and the β-FeOOH nanorods. To further investigate monolayer adsorption, the Sips model was used. When the Sips model exponent $n_S = 1$, the adsorption process follows a Langmuir isotherm, and when $n_S > 1$, the data is considered to fit the Freundlich model [43–45]. In this work, the Sips model exponent was equal to 1, further confirming monolayer adsorption.

Table 1. Adsorption models and parameters investigated in the present study.

Model	Parameter	R^2
Langmuir $\frac{C_e}{q_e} = \frac{1}{q_{max} \cdot K_L} \cdot C_e + \frac{1}{q_{max}}, R_L = \frac{1}{1+K_L C_0}$	q_{max} (mg g^{-1}) = 97.0 K_L (L g^{-1}) = 16 R_L = 0.10–0.25	0.9996
Freundlich $\ln q_e = \ln k_f + \frac{1}{n} \ln C_e$	K_F (L/mg) = 66.6 n = 1.7	0.8853
Dubinin–Radushkevich $\ln q_e = \ln q_m - \beta E^2$ $\varepsilon = RT \ln\left(1 + \frac{1}{C_\rho}\right)$ $E = \frac{1}{\sqrt{2B}}$	$q_{D\text{-}R}$ (mg/g) = 103 E (kJ/mol) = 8.5	0.9191
Redlich–Peterson $\ln\left(k_R \cdot \frac{C_e}{q_e} - 1\right) = b_R \ln C_e + \ln \alpha_R$	α = 0.37 β = 1.6	0.9836
Sips $q_e = \frac{q_{ms} K_S C_e^{nS}}{1+K_S C_e^{nS}}$ $\frac{1}{q_e} = \frac{1}{q_{mS} K_S}\left(\frac{1}{C_e}\right)^{\frac{1}{nS}} + \frac{1}{q_{mS}}$	q_{ms} (mg L^{-1}) = 103 K_S (L g^{-1}) = 1.7 n_S = 0.97	0.9961

q_e: amount adsorbed; q_{max}: maximum monolayer adsorption; K_L: Langmuir constant; C_e; concentration of adsorbate at equilibrium; R_L: separation factor; K_F: adsorption capacity; 1/n: adsorption intensity; β: Dubin constant: E: mean adsorption energy (kJ/mol) ε: Polanyi potential; R: gas constant; T: temperature; K_R and α_R: Redlich-Peterson constants: B: slope qmS (mg L^{-1}): Sips maximum adsorption capacity; KS (L g^{-1}): Sips isotherm model constant; nS: Sips isotherm model exponent.

The adsorption data were also fitted to the nonlinear isotherm models. This because the major problem encountered when using linearized models is the violation of theories behind each model using the modification of the original equation [46–48]. In such cases, the linearized model may give best-fitting parameters as opposed to the original nonlinear model [46–48]. Therefore, to compare the linearised and nonlinearized regression models, the data were fitted to the original nonlinear isotherm equation (Table 2). The better-fitting kinetic of the nonlinear equations was investigated using residual standard error (RSE) and R^2. According to the RSE and R^2, the best model is Redlich–Peterson, followed by the Sips, Langmuir, Dubinin–Radushkevich, and Freundlich isotherms. In this study, the isotherm parameters obtained for both linear and nonlinear models were comparable, indicating that the two regression models were not significantly different at the 95% confidence level. Besides, these findings proved that the linearized model did not violate the theory behind the original models.

3.4. Adsorption Kinetics

Batch experiments were carried out to investigate the adsorption kinetics of β-estradiol (Figure 6A) using 42.5 mg of β-FeOOH nanorods, a concentration of 5 mg L^{-1} β-estradiol, and a pH of 5.25 at room temperature (25 °C). As seen, the rapid uptake of β-estradiol by the adsorbent happened from 5–20 min and the equilibrium was attained between 20 and 60 min. The kinetics data were fitted into various kinetics equations, as seen in Table 3. The plots for the pseudo-first order (ln (q_e-q_t) vs. t), pseudo second order (t/q_t vs. t), and Elovich (q_t vs. ln t) kinetic models are shown in Figure 6B–D. The derived parameters together with the correlation coefficient are summarized in Table 3. As seen in Figure 6B and Table 3, the correlation coefficients for the pseudo-first-order model was 0.6555. The experimental adsorption capacity (97.2 mg g^{-1}) was almost four times higher than the calculated (26.8 mg g^{-1}) adsorption capacity. These finding suggested that the adsorption of β-estradiol onto β-FeOOH nanorods was not an ideal pseudo-first-order reaction. The R^2 value for the pseudo-second-order model is higher than that of the first-order model. This suggests that the kinetic equilibrium data were best described by a pseudo-second-order model. The calculated q_e value of the pseudo-second-order model is also 100 mg g^{-1}, which is in agreement with the experimentally obtained value of q_e at 97.2 mg g^{-1}. This agreement suggested that the adsorption process was driven by chemisorption involving the electrostatic interaction between the positively charged adsorbent and lone pairs of electrons of β-estradiol. The Elovich kinetic model was used to describe the chemisorption process on the surface of the β-FeOOH nanorods. The kinetics data were fitted to the Elovich equation and gave a relatively good R^2 value (0.9595), which shows that the uptake of β-estradiol onto the β-FeOOH nanorods involved chemisorption [49].

Table 2. Non-linear adjusted adsorption isotherms.

Model	Parameter	Adjusted R^2	R^2
Langmuir $q_e = \frac{q_{max} K_L C_e}{1+K_L C_e}$	q_{max} (mg g^{-1}) = 102 K_L (L g^{-1}) = 0.44 RSE = 6.2	0.9442	0.9764
Freundlich $q_q = K_L C_e^{1/n}$	K_F (L/mg) = 74.4 n = 2.87 RSE = 13.2	0.7518	0.8906
D-R $q_e = q_{D-R} \exp^{-(K_{D-R}\,\varepsilon^2)}$ $\varepsilon = RT \ln\left(1+\frac{1}{C_e}\right)$	q_{D-R} (mg/g) = 105 E (kJ/mol) = 8.5 RSE: 8.3	0.8993	0.9191
Redlich–Peterson $q_e = \frac{K_R C_e}{\alpha_R C_e^{\beta_r}}$	K_R α = 1.68 β = 0.35 RSE = 2.21	0.9930	0.9976
Sips $q_e = \frac{q_{ms} K_S C_e^{ns}}{1+K_S C_e^{nS}}$	q_{ms} (mg L^{-1}) = 104 K_S (L g^{-1}) = 1.35 n_s = 1.28 RSE = 2.59	0.9903	0.9967

Table 3. Kinetic models and parameters investigated in the present study.

Model	Parameter	R^2
Pseudo-First order $\ln(q_e - q_t) = \ln q_e - K_2 t$	q_e (mg/g) = 26.8 k_1 (mg/g $min^{1/2}$) = 0.063	0.6555
Pseudo-Second order $\frac{1}{q_t} = \frac{1}{K_2 q_e^2} + \frac{1}{q_e} t$	q_e = 100 k_t (mg/g $min^{1/2}$) = 0.004	0.9989
Elovich $q_t = \frac{1}{\beta}\ln(\alpha\beta) + \frac{1}{\beta}\ln t$ $q_t = \frac{1}{b}\ln(1+\alpha B t)$	α (mg/g min) = 104 β (g/mg) = 0.058	0.9595
Intraparticle $q_t = K_d t^{\frac{1}{2}} + C$	K_{id1} (mg/g $min^{1/2}$) = 19.1 C_1 (mg/g) 12.2 K_{id2} (mg/g $min^{1/2}$) = 0.073 C_2 (mg/g) = 94.2	R_1^2 = 0.9589 R_2^2 = 0.9139
Boyd $B_t = -0.4977 - \ln(1-F)$ $F = \frac{q_t}{q_0}$	α = 0.056 β = 0.57	0.8122

q_t: amount adsorbed at time t; q_e: sorption capacity; k_1: rate constant; k_2: second-order constant; C: intercept; k_{id}: intraparticle diffusion rate constant; α is the initial rate constant and β is the desorption constant; B: chemisorption; B_t: represents F; F: the fraction of the solute adsorbed at any given time.

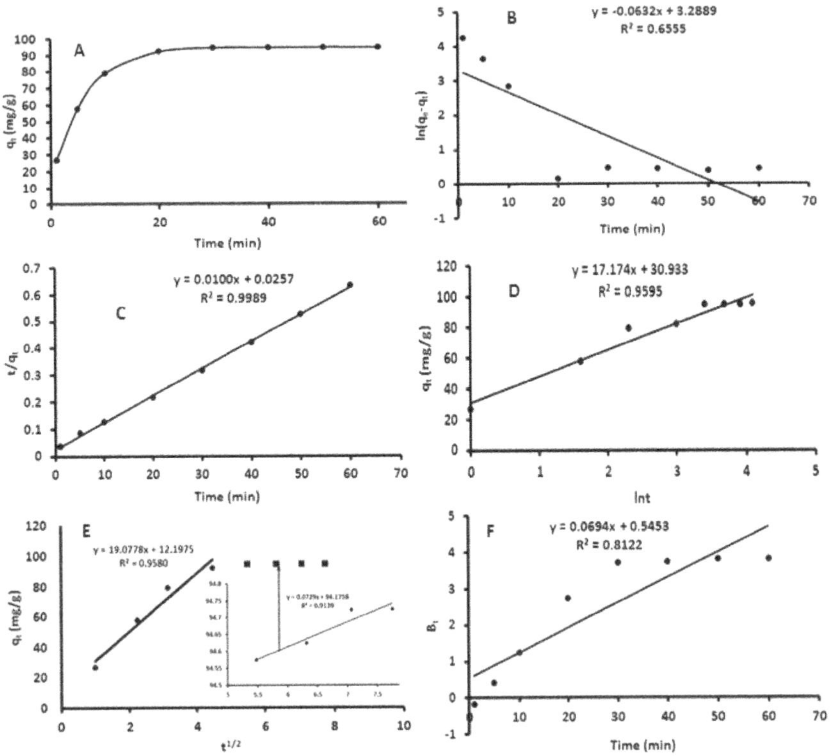

Figure 6. Adsorption kinetics plots for the sorption of β-estradiol using akaganeite: (**A**)—adsorption kinetics model; (**B**)—pseudo-first order linear model; (**C**)—pseudo-second order linear model; (**D**)—Elovich linear model; (**E**)—intraparticle diffusion linear model; (**F**)—Boyd linear model.

To understand the rate-determining step, various intraparticle diffusion models were applied, as summarized in Table 3. From the plots of the intraparticle diffusion (Figure 6E), it is evident a two-step adsorption process occurred during the uptake of β-estradiol by the akageneite nanorods. The first step is based on the so-called film diffusion or surface adsorption, i.e., diffusion of β-estradiol from the solution to the external surface of the adsorbent [24]. The plots also do not go through the origin, which indicates that the intraparticle diffusion is not the rate-determining step. The second step suggested that the adsorption process proceeds by intraparticle diffusion and the intraparticle diffusion happened at a higher concentration. Kid_1 derived from the film diffusion is exceeding the Kid_2 of the intraparticle diffusion, which indicates that the intraparticle diffusion is a slow process. The value of the intercept (C) is also larger for the intraparticle diffusion, thus providing a more relevant contribution to the adsorption process [50]. Because the intraparticle diffusion plot shows a non-linear nature, the Boyd model (Figure 6F) was used to investigate the definite rate-controlling step. According to Hu et al. [49], when the Boyd plot passes through the origin it means that particle diffusion is the dominant mechanism that governs the adsorption process. However, as seen in Figure 6F, the plot did not pass through the origin, suggesting that the adsorption process might be governed by a boundary layer diffusion mechanism [51]. However, the high correlation coefficient of the Elovich model meant that between the two steps of the rate-determining steps, intraparticle diffusion was more prominent when compared to film diffusion [52,53]. This validates the results obtained during the optimization, which essentially results in the fact that electrostatic interactions are driving the β-estradiol adsorption onto akageneite. This is also confirmed in the literature for the adsorption of β-estradiol onto carbon nanotubes [16].

Nonlinear regression of the kinetics data for each model was investigated using the adjusted R^2, R^2, and residual standard error. The calculated kinetics parameter and the correlation coefficients are illustrated in Table 4. Unlike in the linearized regression model where the trend was pseudo second order > Elovich > pseudo-first order, the data was best fitted by the pseudo-second order, followed by the pseudo-first-order and Elovich models. When comparing the kinetic parameters between the non-linear and linear models, it could be seen that the calculated q_e for the linearised pseudo-first order was the only model that violated the theory of the original non-linear model.

Table 4. Adjusted non-linear adsorption kinetics.

Model	Parameter	Adjusted R^2	R^2
Pseudo-First order $q_t = q_e\left(1 - e^{-k_1 t}\right)$	q_e (mg/g) = 94.1 k_1 (mg/g min$^{1/2}$) = 0.204 RSE = 4.106	0.9727	0.9882
Pseudo-Second order $q_t = \frac{q_e^2 k_2 t}{1+q_e k_2 t}$	q_e = 102 k_t (mg/g min$^{1/2}$) = 0.004 RSE = 3.083	0.9846	0.9934
Elovich $q_t = \frac{1}{\beta}\ln(1 + \alpha\beta t)$	α (mg/g min) = 100 β (g/mg) = 0.057 RSE = 6.953	0.9217	0.9659

3.5. Adsorption Thermodynamics Studies

Thermodynamic studies were investigated to ascertain the dominant adsorption mechanisms (that is, physisorption or chemisorption). The adsorption process was carried at different temperatures (298, 303, 308, and 313 K) and the thermodynamic parameters, such as enthalpy (ΔH°), entropy (ΔS°), and Gibbs energy (ΔG°), were calculated according to the expressions reported by References [46,48,50,52]. The graph of $\ln K_C$ against 1/T(K^{-1}) is presented in Figure 7. The values of ΔH° and ΔS° were estimated from the slope and intercept of the plot. The ΔH° and ΔS° were found to be 45.5 kJ mol^{-1} and 186 J mol–^{1} K^{-1} while the Gibbs energies were ranged from −9.86 to −12.6 kJ mol^{-1}. The positive ΔH° reveals that the adsorption process was endothermic in nature and the positive ΔS° suggested an increase in the randomness at the boundary of the solid/liquid phases during the adsorption process [46,48,50,52]. The negative ΔG° indicate the spontaneity of the adsorption process. Moreover, the ΔH° value was higher than 20.9 kJ/mol, confirming that the chemisorption was dominant [52].

3.6. Adsorption Mechanism

In order to gain further insight into the adsorption process of β-estradiol onto akaganeite, the adsorption was characterized by performing FTIR analysis (Figure 8a–c) on akaganeite before and after adsorption, as well as on the analyte. The adsorption of β-estradiol resulted in peaks in the 1500–1000 cm^{-1} region, which were ascribed to various C–O, C–H, and C–OH bonds. As evident in Figure 8c, the emergence of the prominent peaks at 2800–2900 cm^{-1}, corresponding to the stretching vibration of the C–H bonds of the β-estradiol, and the broadening of the OH vibration peak at 3135 cm^{-1} confirmed the presence β-estradiol on the surface of the adsorbent [54]. Figure 8a revealed that akageneite has distinctive vibration peaks of hydroxyl groups and according to the literature the pKa of -OH is around 9.5–13 [55,56]. This implies that at pH values below the pKa values, the hydroxyl groups of the adsorbent are protonated [55,56]. In addition, the pH_{pzc} of the adsorbent was reported to be between 7 and 7.5, suggesting that adsorbent is positively charged below the pH_{pzc}. Moreover, the pKa value of β-estradiol is reported to be 10.71, suggesting the analyte existed as a neutral species. There, the adsorption mechanism was dominated by the electrostatic attraction between the positively charged adsorbent and the lone pair of electrons of the oxygen atoms present in the analyte. Furthermore, the presence of the electron-rich aromatic ring in the structure of the analyte resulted in cation–π interactions between the positively charged adsorbent surface and the π-system of the β-estradiol molecule. These finding are in line with the observation in the optimization data where pH influenced the interactions between the positively charged adsorbent and the electron-rich β-estradiol molecule [6]. Moreover, the enhanced peak intensities, shifting of adsorption bands at 1700–1500 cm^{-1}, and disappearance of some peaks at 650–600 cm^{-1}, as seen in Figure 8b, confirmed that the adsorption mechanism was dominated by the chemisorption process, which is in line with the adsorption isotherms, kinetics data, and thermodynamics studies.

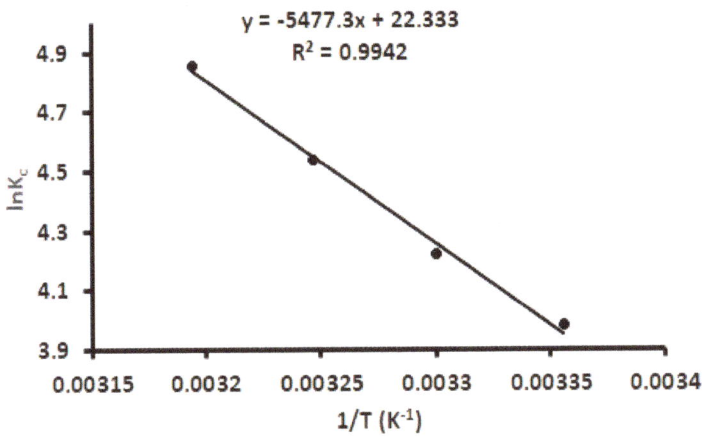

Figure 7. Van't Hoff plot calculation of for thermodynamic parameters.

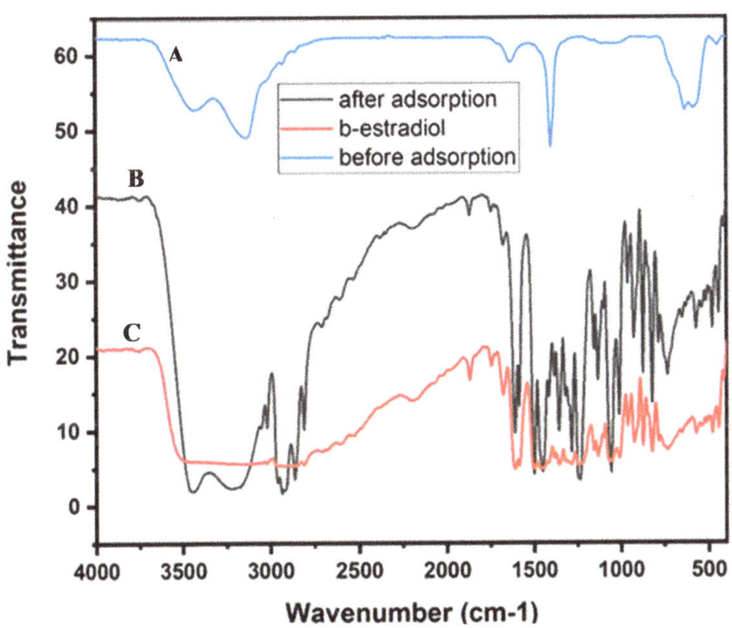

Figure 8. FTIR spectrum for (**A**) akageneite before adsorption, (**B**) akageneite before adsorption, and (**C**) β-estradiol.

3.7. Analysis of Real Samples

During the experimental studies, the actual removal of β-estradiol from the spiked aqueous samples was investigated. The two-level spiked effluent wastewater samples showed that the adsorption method could remove between 93 and 100%. This proved that no significant interferences on the adsorption properties were experienced due to the presence of other wastewater components. The performance parameters of the method developed in the present study were in fact comparable with other reported β-estradiol removal methods based on adsorption, except for References [18,57], as summarized in Table 5.

Table 5. Comparison of materials used in previous reports for the removal of β-estradiol from water samples.

Adsorbent	Analyte(s)	Adsorption Capacity (mg g^{-1})	Ref.
Nitrogen-doped porous carbon	β-estradiol	455	[18]
Multiwalled carbon nanotubes	β-estradiol	54	[58]
Carbon nanotubes	β-estradiol	21.1	[59]
Single-walled carbon nanotube membrane	β-estradiol	9.53	[16]
Cyclodextrin polymer	β-estradiol	210	[57]
Akaganeite nanorods	β-estradiol	97.0	Current work

3.8. Regeneration Studies

The reusability and regeneration of the prepared akaganeite nanorods were evaluated by performing a series of adsorption–desorption cycles (Figure S4). Desorption of the analyte was achieved using 100% acetonitrile. Thereafter, the adsorbent was washed with an ethanol–water solution and dried in an oven at 50 °C. The removal efficiency of the nanoadsorbent toward β-estradiol slightly decreased after the 6th usage/regeneration cycle; however, it remained at >90%. These findings suggested that the akaganeite nanoadsorbent could be regenerated and subjected to multiple usages with suitable robustness and reusability properties.

4. Conclusions

Akaganeite nanoparticles generated via a facile and simple synthesis route using a precipitation technique are highly useful for the adsorptive removal of β-estradiol. A variety of characterization techniques, including XRD, SEM, EDX, and TEM, confirmed the formation of akaganeite nanoparticles that are predominantly rod-shaped. The method was demonstrated to remove β-estradiol at concentrations up to 8 mg L^{-1} from a variety of spiked aqueous solutions with removal efficiencies of 93–100%. The adsorption process was characterized using both linear and adjusted non-linear models. The adsorption process followed pseudo-second order kinetics and a Langmuir isotherm model with a maximum adsorption capacity of 97.0 mg g^{-1} for the linearized isotherm model. The adjusted non-linear Langmuir isotherm resulted in a maximum adsorption capacity of 102 mg g^{-1}, which was comparable to the linear model. The akaganeite nanoparticles also could be reused during up to six adsorption/desorption cycles maintaining an approx. 90% removal efficiency. Hence, a low-cost and environmentally friendly adsorbent was demonstrated, which is a suitable candidate for water treatment via adsorption for hormone-like substances such as β-estradiol.

Supplementary Materials: The following are available online at http://www.mdpi.com/2227-9717/8/9/1197/s1, Figure S1: FTIR of akaganeite after adsorption; Figure S2: Pareto chart of the standardized relevance of the individual variables affecting the adsorbing performance of the synthesized materials. The red line indicates the 95% confidence interval. The linear interactions of these factors are: 1Lby2L (pH–MA interaction), 2Lby3L (MA–CT interaction) and 1Lby3L (pH–CT interaction); Figure S3: Response surface plots showing the interaction effects of the main parameters pH, adsorbent material mass (MA), and contact time (CT); Figure S4: Desirability function for the main parameters pH, adsorbent material mass (MA), and contact time (CT); Figure S5: Regeneration studies.

Author Contributions: A.M., B.M. and P.N.N., conceptualization; A.M. and P.N.N., method development; P.N.N., chemometrics; A.M. and A.N., sampling; A.M. and A.N., investigation; A.M., synthesis and characterization, A.M. and A.N., data collection; A.M., writing—original draft; A.M., A.N. and P.N.N., writing—review and editing; P.N.N., resources; B.M. and P.N.N., supervision. All authors have read and agreed to the published version of the manuscript.

Funding: This research was funded by the National Research Foundation (NRF, South Africa, grant nos. 113010 and 91230).

Acknowledgments: The authors would like to acknowledge the Department of Chemical Sciences, University of Johannesburg (Doornfontein Campus) for providing laboratory space and other resources required in this project

Conflicts of Interest: The authors declare no conflict of interest.

References

1. Halder, J.N.; Islam, M.N. Water Pollution and its Impact on the Human Health. *J. Environ. Hum.* **2015**, *2*, 36–46.
2. Chen, M.; Zhang, J. Bioremediation of soils contaminated with polycyclic aromatic hydrocarbons, petroleum, pesticides, chlorophenols and heavy metals by composting: Applications, microbes and future research needs. *Biotechnol. Adv.* **2015**, *33*, 745–755. [PubMed]
3. Jacobsen, C.S.; Hjelmsø, M.H. Agricultural soils, pesticides and microbial diversity. *Curr. Opin. Biotechnol.* **2014**, *27*, 15–20. [PubMed]
4. Kassotis, C.D.; Alvarez, D.A.; Taylor, J.A.; Frederick, S.; Nagel, S.C.; Tillitt, D.E. Characterization of Missouri surface waters near point sources of pollution reveals potential novel atmospheric route of exposure for bisphenol A and wastewater hormonal activity pattern. *Sci. Total Environ.* **2015**, *524–525*, 384–393.
5. Harley, K.G.; Gunier, R.B.; Kogut, K.; Johnson, C.; Bradman, A.; Calafat, A.M.; Eskenazi, B. Prenatal and early childhood bisphenol A concentrations and behavior in school-aged children. *Environ. Res.* **2013**, *126*, 43–50.
6. Jiang, L.H.; Liu, Y.G.; Zeng, G.M.; Xiao, F.Y.; Hu, X.J.; Hu, X.; Wang, H.; Li, T.T.; Zhou, L.; Tan, X.F. Removal of 17β-estradiol by few-layered graphene oxide nanosheets from aqueous solutions: External influence and adsorption mechanism. *Chem. Eng. J.* **2016**, *284*, 93–102.
7. Zaib, Q.; Khan, I.A.; Saleh, N.B.; Flora, J.R.V.; Park, Y.G.; Yoon, Y. Removal of bisphenol a and 17β-estradiol by single-walled carbon nanotubes in aqueous solution: Adsorption and molecular modeling. *Water. Air. Soil Pollut.* **2012**, *223*, 3281–3293.
8. Silva, C.P.; Lima, D.L.D.; Schneider, R.J.; Otero, M.; Esteves, V.I. Development of ELISA methodologies for the direct determination of 17β-estradiol and 17α-ethinylestradiol in complex aqueous matrices. *J. Environ. Manag.* **2013**, *124*, 121–127.
9. Fisher, A.M.; Thornton, B.J. Method for the Detection of 17-B-estradiol in Wastewater Facility Effluents Using HPLC. 2018. Available online: https://knowledge.e.southern.edu/research_bio/8/ (accessed on 15 August 2020).
10. Janegitz, B.C.; Dos Santos, F.A.; Faria, R.C.; Zucolotto, V. Electrochemical determination of estradiol using a thin film containing reduced graphene oxide and dihexadecylphosphate. *Mater. Sci. Eng. C* **2014**, *37*, 14–19.
11. Moraes, F.C.; Rossi, B.; Donatoni, M.C.; de Oliveira, K.T.; Pereira, E.C. Sensitive determination of 17β-estradiol in river water using a graphene based electrochemical sensor. *Anal. Chim. Acta* **2015**, *881*, 37–43.
12. Yin, Z.; Liu, Y.; Tan, X.; Jiang, L.; Zeng, G.; Liu, S.; Tian, S.; Liu, S.; Liu, N.; Li, M. Adsorption of 17B-estradiol by a novel attapulgite/biochar nanocomposite: Characteristics and influencing factors. *Process Saf. Environ. Prot.* **2019**, *121*, 155–164. [CrossRef]
13. Kawasaki, N.; Ogata, F.; Yamaguchi, I.; Tominaga, H. Degradation Characteristics of 17 b -Estradiol by Ozone Treatment with Activated Carbon. *J. Oleo Sci.* **2009**, *266*, 261–266. [CrossRef] [PubMed]
14. Naimi, I.; Bellakhal, N. Removal of 17 β -Estradiol by Electro-Fenton Process. *Mater. Sci. Appl.* **2012**, *3*, 880–886.
15. Hashimoto, T.; Murakami, T. Removal and degradation characteristics of natural and synthetic estrogens by activated sludge in batch experiments. *Water Res.* **2009**, *43*, 573–582. [CrossRef] [PubMed]
16. Heo, J.; Flora, J.R.V.; Her, N.; Park, Y.G.; Cho, J.; Son, A.; Yoon, Y. Removal of bisphenol A and 17β-estradiol in single walled carbon nanotubes-ultrafiltration (SWNTs-UF) membrane systems. *Sep. Purif. Technol.* **2012**, *90*, 39–52. [CrossRef]

17. Saifuddin, N.; Nur, Y.A.; Abdullah, S.F. Microwave enhanced synthesis of chitosan-graft-polyacrylamide molecular imprinting polymer for selective removal of 17 β-estradiol at trace concentration. *Asian J. Biochem.* **2011**, *6*, 38–54. [CrossRef]
18. Duan, Q.; Li, X.; Wu, Z.; Alsaedi, A.; Hayat, T.; Chen, C.; Li, J. Adsorption of 17 b -estradiol from aqueous solutions by a novel hierarchically nitrogen-doped porous carbon. *J. Colloid Interface Sci.* **2019**, *533*, 700–708. [CrossRef]
19. Yuan, Z.Y.; Su, B.L. Surfactant-assisted nanoparticle assembly of mesoporous β-FeOOH (akaganeite). *Chem. Phys. Lett.* **2003**, *381*, 710–714. [CrossRef]
20. Pepper, R.A.; Couperthwaite, S.J.; Millar, G.J. A novel akaganeite sorbent synthesised from waste red mud: Application for treatment of arsenate in aqueous solutions. *J. Environ. Chem. Eng.* **2018**, *6*, 6308–6316. [CrossRef]
21. Kyzas, G.Z.; Peleka, E.N.; Deliyanni, E.A. Nanocrystalline akaganeite as adsorbent for surfactant removal from aqueous solutions. *Materials* **2013**, *6*, 184–197. [CrossRef]
22. Kim, J.; Li, W.; Philips, B.L.; Grey, C.P. Phosphate adsorption on the iron oxyhydroxides goethite ([small alpha]-FeOOH), akaganeite ([small beta]-FeOOH), and lepidocrocite ([gamma]-FeOOH): A 31P NMR Study. *Energy Environ. Sci.* **2011**, *4*, 4298–4305. [CrossRef]
23. Chitrakar, R.; Tezuka, S.; Sonoda, A.; Sakane, K.; Ooi, K.; Hirotsu, T. Phosphate adsorption on synthetic goethite and akaganeite. *J. Colloid Interface Sci.* **2006**, *298*, 602–608. [CrossRef] [PubMed]
24. Mashile, P.P.; Mpupa, A.; Nomngongo, P.N. Adsorptive removal of microcystin-LR from surface and wastewater using tyre-based powdered activated carbon: Kinetics and isotherms. *Toxicon* **2018**, *145*, 25–31. [CrossRef] [PubMed]
25. Vera Candioti, L.; De Zan, M.M.; Cámara, M.S.; Goicoechea, H.C. Experimental design and multiple response optimization. Using the desirability function in analytical methods development. *Talanta* **2014**, *124*, 123–138. [CrossRef]
26. Murad, E.; Bishop, J.L. The infrared spectrum of synthetic akaganéite, β-FeOOH. *Am. Mineral.* **2000**, *85*, 716–721. [CrossRef]
27. Tufo, A.E.; Larralde, A.L.; Villarroel-Rocha, J.; Sapag, K.; Sileo, E.E. Synthesis and characterization of pure and Al-substituted akaganeites and evaluation of their performance to adsorb As(V). *J. Environ. Chem. Eng.* **2018**, *6*, 7044–7053. [CrossRef]
28. Wilkinson, L. Revising the Pareto chart. *Am. Stat.* **2006**, *60*, 332–334. [CrossRef]
29. Tabachnick, B.G. *Experimental Designs Using ANOVA*; Thomson/Brooks/Cole: Belmont, CA, USA, 2007; ISBN 020532178X.
30. Dunford, R.; Su, Q.; Tamang, E. The pareto principle. *Plymouth Stud. Sci.* **2014**, *7*, 140–148.
31. Nomngongo, P.N.; Ngila, J.C.; Msagati, T.A.M.; Moodley, B. Chemometric optimization of hollow fiber-liquid phase microextraction for preconcentration of trace elements in diesel and gasoline prior to their ICP-OES determination. *Microchem. J.* **2014**, *114*, 141–147. [CrossRef]
32. Mashile, G.P.; Mpupa, A.; Nomngongo, P.N. In-syringe micro solid-phase extraction method for the separation and preconcentration of parabens in environmental water samples. *Molecules* **2018**, *23*, 1450. [CrossRef]
33. Khor, C.P.; bt Jaafar, M.; Ramakrishnan, S. Optimization of Conductive Thin Film Epoxy Composites Properties Using Desirability Optimization Methodology. *J. Optim.* **2016**, *2016*, 1–8. [CrossRef]
34. Nqombolo, A.; Mpupa, A.; Gugushe, A.S.; Moutloali, R.M.; Nomngongo, P.N. Adsorptive removal of lead from acid mine drainage using cobalt-methylimidazolate framework as an adsorbent: Kinetics, isotherm, and regeneration. *Environ. Sci. Pollut. Res.* **2019**, *26*, 3330–3339. [CrossRef]
35. Biata, N.R.; Jakavula, S.; Mashile, G.P.; Nqombolo, A.; Moutloali, R.M.; Nomngongo, P.N. Recovery of gold (III) and iridium (IV) using magnetic layered double hydroxide (Fe3O4/Mg-Al-LDH) nanocomposite: Equilibrium studies and application to real samples. *Hydrometallurgy* **2020**, *197*, 105447. [CrossRef]
36. Gugushe, A.S.; Mpupa, A.; Nomngongo, P.N. Ultrasound-assisted magnetic solid phase extraction of lead and thallium in complex environmental samples using magnetic multi-walled carbon nanotubes/zeolite nanocomposite. *Microchem. J.* **2019**, *149*, 103960. [CrossRef]
37. Ayawei, N.; Ebelegi, A.N.; Wankasi, D. Modelling and Interpretation of Adsorption Isotherms. *J. Chem.* **2017**, *2017*, 1–11. [CrossRef]

38. Gui, W.; Tian, C.; Sun, Q.; Li, S.; Zhang, W.; Tang, J.; Zhu, G. Simultaneous determination of organotin pesticides by HPLC-ICP-MS and their sorption, desorption, and transformation in freshwater sediments. *Water Res.* **2016**, *95*, 185–194. [CrossRef]
39. Riahi, K.; Chaabane, S.; Thayer, B. Ben A kinetic modeling study of phosphate adsorption onto Phoenix dactylifera L. date palm fibers in batch mode. *J. Saudi Chem. Soc.* **2017**, *21*, S143–S152. [CrossRef]
40. Rehman, S.; Adil, A.; Shaikh, A.J.; Shah, J.A.; Arshad, M.; Ali, M.A.; Bilal, M. Role of sorption energy and chemisorption in batch methylene blue and Cu 2+ adsorption by novel thuja cone carbon in binary component system: Linear and nonlinear modeling. *Environ. Sci. Pollut. Res.* **2018**, *25*, 31579–31592. [CrossRef]
41. Sadeghalvad, B.; Khosravi, S.; Azadmehr, A.R. Nonlinear isotherm and kinetics of adsorption of copper from aqueous solutions on bentonite. *Russ. J. Phys. Chem. A* **2016**, *90*, 2285–2291. [CrossRef]
42. Guechi, E.K.; Hamdaoui, O. Sorption of malachite green from aqueous solution by potato peel: Kinetics and equilibrium modeling using non-linear analysis method. *Arab. J. Chem.* **2016**, *9*, S416–S424. [CrossRef]
43. Vieira, R.M.; Vilela, P.B.; Becegato, V.A.; Paulino, A.T. Chitosan-based hydrogel and chitosan/acid-activated montmorillonite composite hydrogel for the adsorption and removal of Pb+ 2 and Ni+ 2 ions accommodated in aqueous solutions. *J. Environ. Chem. Eng.* **2018**, *6*, 2713–2723. [CrossRef]
44. Gugushe, A.S.; Nqombolo, A.; Nomngongo, P.N. Application of Response Surface Methodology and Desirability Function in the Optimization of Adsorptive Remediation of Arsenic from Acid Mine Drainage Using Magnetic Nanocomposite: Equilibrium Studies and Application to Real Samples. *Molecules* **2019**, *24*, 1792. [CrossRef] [PubMed]
45. Vafajoo, L.; Cheraghi, R.; Dabbagh, R.; McKay, G. Removal of cobalt (II) ions from aqueous solutions utilizing the pre-treated 2-Hypnea Valentiae algae: Equilibrium, thermodynamic, and dynamic studies. *Chem. Eng. J.* **2018**, *331*, 39–47. [CrossRef]
46. Nagy, B.; Mânzatu, C.; Măicăneanu, A.; Indolean, C.; Barbu-Tudoran, L.; Majdik, C. Linear and nonlinear regression analysis for heavy metals removal using Agaricus bisporus macrofungus. *Arab. J. Chem.* **2017**, *10*, S3569–S3579. [CrossRef]
47. Jain, S.N.; Shaikh, Z.; Mane, V.S.; Vishnoi, S.; Mawal, V.N.; Patel, O.R.; Bhandari, P.S.; Gaikwad, M.S. Nonlinear regression approach for acid dye remediation using activated adsorbent: Kinetic, isotherm, thermodynamic and reusability studies. *Microchem. J.* **2019**, *148*, 605–615.
48. Aazza, M.; Ahlafi, H.; Moussout, H.; Maghat, H. Adsorption of metha-nitrophenol onto alumina and HDTMA modified alumina: Kinetic, isotherm and mechanism investigations. *J. Mol. Liq.* **2018**, *268*, 587–597. [CrossRef]
49. Hu, X.; Wang, J.; Liu, Y.; Li, X.; Zeng, G.; Bao, Z.; Zeng, X.; Chen, A.; Long, F. Adsorption of chromium (VI) by ethylenediamine-modified cross-linked magnetic chitosan resin: Isotherms, kinetics and thermodynamics. *J. Hazard. Mater.* **2011**, *185*, 306–314.
50. Mashile, G.P.; Mpupa, A.; Nqombolo, A.; Dimpe, K.M.; Nomngongo, P.N. Recyclable magnetic waste tyre activated carbon-chitosan composite as an effective adsorbent rapid and simultaneous removal of methylparaben and propylparaben from aqueous solution and wastewater. *J. Water Process Eng.* **2020**, *33*, 101011.
51. Fu, J.; Chen, Z.; Wang, M.; Liu, S.; Zhang, J.; Zhang, J.; Han, R.; Xu, Q. Adsorption of methylene blue by a high-efficiency adsorbent (polydopamine microspheres): Kinetics, isotherm, thermodynamics and mechanism analysis. *Chem. Eng. J.* **2015**, *259*, 53–61.
52. Liu, Y.; Xiong, Y.; Xu, P.; Pang, Y.; Du, C. Enhancement of Pb (II) adsorption by boron doped ordered mesoporous carbon: Isotherm and kinetics modeling. *Sci. Total Environ.* **2020**, *708*, 134918.
53. Tavlieva, M.P.; Genieva, S.D.; Georgieva, V.G.; Vlaev, L.T. Journal of Colloid and Interface Science Kinetic study of brilliant green adsorption from aqueous solution onto white rice husk ash. *J. Colloid Interface Sci.* **2013**, *409*, 112–122. [PubMed]
54. Minaev, B.F.; Minaeva, V.A. Study of IR spectrum of the 17β-estradiol using quantum-chemical density functional theory. *Biopolym. Cell* **2006**, *22*, 363.
55. Jemutai-Kimosop, S.; Orata, F.; Shikuku, V.O.; Okello, V.A.; Getenga, Z.M. Insights on adsorption of carbamazepine onto iron oxide modified diatomaceous earth: Kinetics, isotherms, thermodynamics, and mechanisms. *Environ. Res.* **2020**, *180*, 108898. [PubMed]
56. Volesky, B. Biosorption and me. *Water Res.* **2007**, *41*, 4017–4029.

57. Tang, P.; Sun, Q.; Zhao, L.; Tang, Y.; Liu, Y.; Pu, H.; Gan, N.; Liu, Y.; Li, H. A simple and green method to construct cyclodextrin polymer for the effective and simultaneous estrogen pollutant and metal removal. *Chem. Eng. J.* **2019**, *366*, 598–607.
58. Sun, W.; Zhang, C.; Xu, N.; Ni, J. Effect of inorganic nanoparticles on 17 b -estradiol and 17 a -ethynylestradiol adsorption by multi-walled carbon nanotubes. *Environ. Pollut.* **2015**, *205*, 111–120.
59. Shi, W.; Li, S.; Chen, B.; Wang, C.; Sun, W. Effects of Fe2O3 and ZnO nanoparticles on 17 b -estradiol adsorption to carbon nanotubes. *Chem. Eng. J.* **2017**, *326*, 1134–1144.

MDPI
St. Alban-Anlage 66
4052 Basel
Switzerland
Tel. +41 61 683 77 34
Fax +41 61 302 89 18
www.mdpi.com

Processes Editorial Office
E-mail: processes@mdpi.com
www.mdpi.com/journal/processes

www.ingramcontent.com/pod-product-compliance
Lightning Source LLC
Chambersburg PA
CBHW040742250926
51850CB00007BB/1381
* 9 7 8 3 0 3 6 5 5 6 2 8 4 *